Retrial Queueing Systems

Jesús R. Artalejo · Antonio Gómez-Corral

Retrial Queueing Systems

A Computational Approach

Jesús R. Artalejo
Department of Statistics
and Operations Research
Faculty of Mathematics
Complutense University of Madrid
28040 Madrid
Spain
jesus_artalejo@mat.ucm.es

Antonio Gómez-Corral
Department of Statistics
and Operations Research
Faculty of Mathematics
Complutense University of Madrid
28040 Madrid
Spain
antonio_gomez@mat.ucm.es

ISBN 978-3-540-78724-2   e-ISBN 978-3-540-78725-9

Library of Congress Control Number: 2008927200

© 2008 Springer-Verlag Berlin Heidelberg

This work is subject to copyright. All rights are reserved, whether the whole or part of the material is concerned, specifically the rights of translation, reprinting, reuse of illustrations, recitation, broadcasting, reproduction on microfilm or in any other way, and storage in data banks. Duplication of this publication or parts thereof is permitted only under the provisions of the German Copyright Law of September 9, 1965, in its current version, and permissions for use must always be obtained from Springer-Verlag. Violations are liable for prosecution under the German Copyright Law.

The use of general descriptive names, registered names, trademarks, etc. in this publication does not imply, even in the absence of a specific statement, that such names are exempt from the relevant protective laws and regulations and therefore free for general use.

Cover design: WMXDesign GmbH, Heidelberg

Printed on acid-free paper

9 8 7 6 5 4 3 2 1

springer.com

To María Jesús, Ana and Rubén, for
their delight in supporting me.

Jesús R. Artalejo

To Gemma, Inés and María, the
nearest persons to my heart.

Antonio Gómez-Corral

Preface

We consider a branch of queueing theory, retrial queueing systems, which is characterized by the following basic assumption: a customer who cannot receive service (due to finite capacity of the system, balking, impatience, etc.) leaves the service area but after some random delay returns to the system again to request service. As a consequence, repeated attempts for service from the pool of unsatisfied customers, called the orbit, are superimposed on the ordinary stream of arrivals of first attempts. Since much of the theory of retrial queues is complex from an analytical viewpoint, we focus our attention in this book to methods for the algorithmic analysis of these models.

Since the pioneering works published in the 1950s, retrial queues have been widely used to provide stochastic modelling of many problems arising in telecommunication, computer networks, and in daily life. Consequently, a vast number of papers have been published in journals oriented to applied probability and stochastic models, statistics and operations research, telecommunication and industrial engineering, and computer science. The growing interest of retrial queues is also reflected in the existence of a series of international workshops on retrial queues which began in Madrid (1998). Subsequent meetings were held in Minsk (1999), Amsterdam (2000), Cochin (2002), Seoul (2004), Miraflores de la Sierra (2006) and Athens (2008). Some recognized international journals have dedicated special issues to this topic. This is the case of the journals *Annals of Operations Research* [97], *European Journal of Operational Research* [113], *Mathematical and Computer Modelling* [66], *Queueing Systems* [659] and *Top* [67].

This book is intended for use by an audience ranging from advanced undergraduates to researchers interested not only in queueing theory, but also in applied probability, stochastic models in operations research, and various engineering disciplines. The required background is a graduate course on stochastic processes, and an interest in algorithmic probability.

The book is divided into three main parts: An Introduction to Retrial Queueing Systems (Part I), Computational Analysis of Performance Descriptors (Part II), and Retrial Queueing Systems Analyzed Through the Matrix-

Analytic Formalism (Part III). Each part is then subdivided to organize the book in nine chapters. We conclude each chapter with some bibliographical notes.

Part I consists of two chapters. Chapter 1 introduces the reader to the broad range of applications of retrial queues. Chapter 2 introduces the basic mathematical formalism, contains descriptions of various major extensions, and compares retrial queues to standard queues with waiting line and queues with losses. We give a survey of main results for both single server and multiserver retrial queues of $M/G/1$ and $M/M/c$ types, and discuss similarities and differences between the retrial queues and their standard counterparts. We demonstrate that, though retrial queues are closely connected with these standard queueing models, they possess unique and distinguishing characteristics.

In Part II, the material is organized in four chapters. Since Part II is devoted to the computational analysis of the main performance measures, we begin in Chapter 3 with an exhaustive analysis of methods for the computation of the limiting distribution of the system state. We specifically focus on the main $M/G/1$ and $M/M/c$ retrial queues. However, we also consider other interesting retrial models: the $M/G/1$ queue with general retrial times, the $Geo/Geo/1$ retrial queue and a multiserver queue with finite population and retrials. Chapter 4 deals with the busy period analysis for the $M/G/1$ retrial queue and the $M/M/c$ retrial queue. For each model, we present theory and computation for the length of the busy period, the number of customers served and the maximal queue length during a busy period. Chapter 5 presents the waiting time analysis for the $M/G/1$ and $M/M/c$ models. It is shown how to compute the waiting time that a customer spends in orbit and the number of repeated attempts made by a tagged customer. Chapter 6 offers the analysis of other descriptors arising from the idiosyncrasies of the retrial feature.

Part III uses the matrix-analytic formalism to analyze a selected number of retrial queues with underlying structured Markov chains. Chapter 7 is a summary of basic results of the matrix-analytic methods. This should be helpful in fixing some notation and the main tools employed in subsequent chapters. Chapter 8 concerns retrial queues with a quasi-birth-and-death structure. We focus on three models: the $MAP/PH/1$ retrial queue, the $MAP/M/c$ retrial queue, and a queue with finite population and service and retrial times of phase-type. For these models, we present the general theory and methods for computing the stationary distribution of the system state and the main characteristics of the busy period and the waiting time. Finally, in Chapter 9 we deal with the computational analysis of the $Geo/Geo/c$ retrial queue and the $BMAP/SM/1$ retrial queue. These models provide, respectively, selected examples of $GI/M/1$ and $M/G/1$ structures.

In the references, we update the existing bibliographical works on retrial queues but, except in a few cases, we restrict this list to papers written in English and published in scientific journals and conference proceedings. We believe the resulting list serves as a useful starting point for readers who are new to the subject.

The authors are grateful to many colleagues and friends who have given assistance and encouragement. We are most grateful to A.N. Dudin, A. Economou, G.I. Falin, J.P. Kharoufeh, Q.L. Li and M.J. López-Herrero, who read parts of the manuscript and made valuable comments and suggestions. In addition, A.N. Dudin and G.V. Tsarenkov were helpful in providing the numerical examples in Subsection 9.2.2. Our special thanks to A. Economou and M.J. López-Herrero who are our coauthors in a number of papers that are a source for this book.

In the course of writing this book, which spans the last four or five years, we have been supported by two research grants from MEC (grant numbers BFM2002-02189 and MTM2005-01248). We thank this funding agency for its sponsorship.

Madrid, *Jesús R. Artalejo*
February 2008 *Antonio Gómez-Corral*

Contents

Part I An Introduction to Retrial Queueing Systems

1 Introduction and Motivating Examples 3
   1.1 Introduction 3
   1.2 Some Examples in Telephone Systems 4
   1.3 Some Examples in Computer Networks 7
   1.4 Bibliographical Notes 10

2 A General Overview 11
   2.1 The Mathematical Formalism 11
   2.2 Comparing Standard and Retrial Queueing Systems 16
       2.2.1 The Main $M/M/c$ Model 16
       2.2.2 The Main $M/G/1$ Model 24
   2.3 Short Description of Some Advanced Retrial Queueing Systems 31
   2.4 Bibliographical Notes 34

Part II Computational Analysis of Performance Descriptors

3 Limiting Distribution of the System State 39
   3.1 The $M/G/1$ Retrial Queue 39
       3.1.1 The Embedded Markov Chain Approach 39
       3.1.2 The Regenerative Approach 41
       3.1.3 The Maximum Entropy Approach 44
   3.2 The $M/G/1$ Queue with General Retrial Times 51
   3.3 The $Geo/G/1$ Retrial Queue 56
   3.4 The $M/M/c$ Retrial Queue 64
       3.4.1 Approximations Based on Truncated Models 64
       3.4.2 Approximations Based on Generalized Truncated
              Models 68
       3.4.3 The RTA Approximation 77

		3.4.4 The Fredericks and Reisner Approximation 80

       3.4.4 The Fredericks and Reisner Approximation 80
       3.4.5 Approximations by Interpolation 84
   3.5 A Multiserver Retrial Queue with Finite Population 87
   3.6 Bibliographical Notes 91

4 **Busy Period** 95
   4.1 The $M/G/1$ Retrial Queue 95
       4.1.1 The Length of the Busy Period...................... 95
       4.1.2 The Number of Customers Served 105
       4.1.3 Maximal Queue Length in a Busy Period 108
   4.2 The $M/M/c$ Retrial Queue 111
       4.2.1 The Length of the Busy Period......................112
       4.2.2 The Number of Customers Served 123
       4.2.3 Maximal Queue Length in a Busy Period 124
   4.3 Bibliographical Notes 129

5 **Waiting Time**...131
   5.1 The $M/G/1$ Retrial Queue 131
       5.1.1 Waiting Time Distribution 131
       5.1.2 The Number of Retrials Made by a Customer 143
   5.2 The $M/M/c$ Retrial Queue 149
       5.2.1 Waiting Time Distribution 149
       5.2.2 The Number of Retrials Made by a Customer 155
   5.3 Bibliographical Notes 158

6 **Other Descriptors** 159
   6.1 Attempts Since the Last Service Completion 159
   6.2 Successful versus Blocked Events 164
   6.3 Server Idle Periods 174
   6.4 Time to Reach a Certain Orbit Level 179
   6.5 Bibliographical Notes 182

Part III Retrial Queueing Systems Analyzed Through the Matrix-Analytic Formalism

---

7 **The Matrix-Analytic Formalism** 187
   7.1 A General Overview 187
       7.1.1 Notation 188
       7.1.2 The Main Structured Markov Chains 188
       7.1.3 The Batch Markovian Arrival Process................190
       7.1.4 The Phase-Type Distribution and the SM Service
             Process 192
   7.2 Some General Tools for QBD Structures 194
       7.2.1 The Finite Case...................................194

		7.2.2	The Matrix-Geometric Distribution 195

        7.2.2 The Matrix-Geometric Distribution 195
        7.2.3 The Computation of the Rate Matrix 197
  7.3 Some General Tools for $GI/M/1$ and $M/G/1$ Structures 198
        7.3.1 The Matrix-Geometric Distribution 198
        7.3.2 The Computation of the Matrix **G** 199
        7.3.3 Asymptotically Quasi-Toeplitz Markov Chains 200
  7.4 Bibliographical Notes 203

8 Selected Retrial Queues with QBD Structure 207
  8.1 The $MAP/PH/1$ Retrial Queue 207
        8.1.1 Stationary Distribution of the System State 207
        8.1.2 Maximal Queue Length in a Busy Period 215
  8.2 The $MAP/M/c$ Retrial Queue 219
        8.2.1 Stationary Distribution of the System State 219
        8.2.2 Busy Period 220
        8.2.3 Waiting Time 228
  8.3 A Queue with Finite Population and PH Service and Retrial Times 233
  8.4 Bibliographical Notes 237

9 Selected Retrial Queues with $GI/M/1$ and $M/G/1$ Structures 241
  9.1 The $Geo/Geo/c$ Retrial Queue 241
        9.1.1 Stationary Distribution of the System State 241
        9.1.2 Busy Period 249
        9.1.3 Waiting Time 254
  9.2 The $BMAP/SM/1$ Retrial Queue 255
        9.2.1 Embedded Markov Chain at Departure Epochs 257
        9.2.2 Limiting Distribution of the System State 261
  9.3 Bibliographical Notes 266

References 269

Author Index 311

Subject Index 315

Part I

An Introduction to Retrial Queueing Systems

1
Introduction and Motivating Examples

1.1 Introduction

In classical queueing theory it is very often assumed that a customer who cannot get service immediately after arrival either joins the waiting line, and then is served according to some queueing discipline, or leaves the system forever. Sometimes impatient customers leave the queue, but it is also assumed that they are leaving the system forever. However, as a matter of fact, the assumption about loss of customers who elected to leave the system is just a first order approximation to a real situation. Usually such a customer after a random time returns to the system and tries to get service again.

We may find queues with returning customers in our daily activities. In retail shopping a customer who finds a long waiting line may wish to do something else and return later on with the hope that the queue dissolves. Similar behavior may be experienced by some impatient customers who entered the waiting line but then discovered that the residual waiting time was too long. In Sections 1.2 and 1.3, we give motivating descriptions of telephone systems and computer networks where retrial queueing systems arise naturally.

Falin and Templeton [288] stressed that the standard queueing models do not take the retrial phenomenon into account and therefore cannot be applied in solving a number of practically important problems. To emphasize this idea, they refer to the book by Kosten [423, page 33] who notes that 'any theoretical result that does not take into consideration this repetition effect should be considered suspect'. Retrial queues (or queues with returning customers, repeated attempts, etc.) have been introduced to solve this deficiency.

The general structure of a retrial queue is shown in Figure 1.1. It is clear from the picture that retrial queues can also be regarded as a special type of queueing networks. In their basic form, these networks contain two nodes: the main node where blocking is possible and a delay node for repeated trials. To describe specific retrial queues with a certain structure and queueing discipline more nodes have to be introduced.

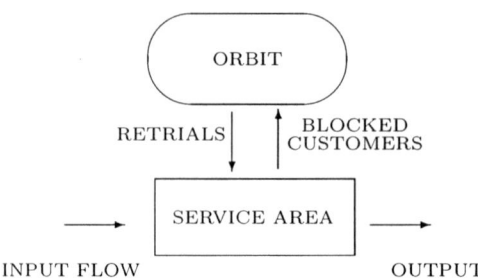

Fig. 1.1. General structure of a retrial queue

Our aim in this book is to discuss algorithmic solutions for queueing systems with retrials. We are interested in obtaining the limiting distribution of the system state because important system descriptors, such as the blocking probability and the mean number of customers making repeated attempts, can be expressed in terms of this limiting law. In a few simple models, it is possible to derive closed-form expressions but very often it is difficult or even impossible to determine the limiting distribution of the system state. Algorithmic methods and approximate techniques are powerful tools for solving the analytically intractable problems. We complete the performance analysis of retrial queues by studying a variety of quality measures including waiting time, busy period and other specific descriptors arising due to the retrial feature. The main focus of this book is on the application of algorithmic methods in studying the $M/G/1$ retrial queue and the $M/M/c$ retrial queue, as well as the use of matrix-analytic techniques [464, 544, 546] to solve some selected retrial queues with QBD, $GI/M/1$ and $M/G/1$ structures. For analytical solutions given in terms of generating functions and Laplace-Stieltjes transforms, we refer interested readers to the textbook by Falin and Templeton [288].

1.2 Some Examples in Telephone Systems

Everybody has experienced that a telephone subscriber who obtains a busy signal repeats the call until the required connection is made. As a result, the flow of calls circulating in a telephone network consists of two parts: the flow of primary calls, which reflects the real wishes of telephone subscribers, and the flow of repeated calls, which is the consequence of the lack of success in previous attempts. These considerations bring into focus the need of the retrial queues as a proper modelling of customer behavior in classical telephone systems. The first specific results appeared in the 1950s, and a very large number of researchers have contributed to the development of this topic. As examples of basic literature on the application of retrial queues to telephone traffic theory, we mention the references [198, 200, 242, 301, 347, 581, 649, 685].

1.2 Some Examples in Telephone Systems

The new developments in telecommunication technology lead to a substantial increase of the retrial phenomenon, which may degrade the performance of the telephone system specially under overload conditions. For example, Kelly [382] explores how the increased use of auto-repeat facilities (repeat-last-number, ring-back-when-free) may affect the traffic. In what follows, we concentrate on the role of the retrials in call centers and in cellular mobile networks.

The basic purpose of a call center is to deliver service by telephone. Since a large volume of requests is received and transmitted, the operation of the call center involves the interaction of both human resources (qualified agents, center managers) and telecommunication equipment (computers, Ethernet, e-mail, fax, voicemail).

A queueing formulation of the call center provides qualitative insight. Indeed, the queueing model can be used for planning and management. For example, the optimal staffing of call centers addresses to the key problem of dimensioning parameters (number of agents, trunks) in order to guarantee maximal profit and a desired grade-of-service measured in terms of acceptable waiting and blocking.

Most major companies use call centers to communicate with their customers. From an operational point of view, a call center can be modelled as a queueing system. The most-widely used models are based on the $M/M/c$ queue but many variants (finite capacity, impatience, tandem queues) must be considered to capture the dynamics of more complicated and modern call centers, which incorporate outbound calls made by the agents to potential customers, flexible and specialized agents, resource-sharing, interconnection with other centers, etc. In Figure 1.2, we illustrate a simple call center including abandonments and retrials.

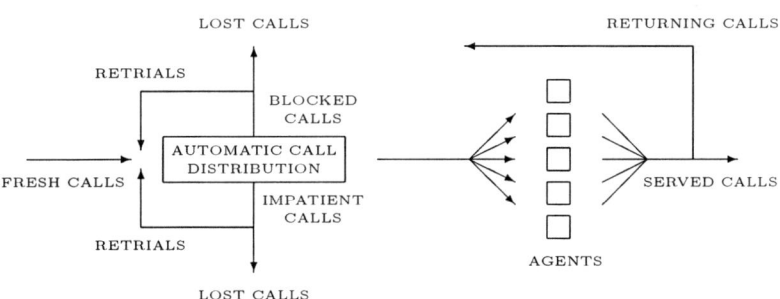

Fig. 1.2. General scheme of a simple call center

In [305], three modes of returning to a call center are distinguished: redials after finding a busy signal, redials after abandoning the queue and revisits after completing service. The latter are mostly delayed returns, which may indicate satisfaction with the received service. They could be considered as a feedback mechanism or indeed as new calls, if the return delays too much. Redials due to blocking and abandonment are most relevant from the point of view of the theory of retrial queues. When a call arrives, it is routed towards the automatic call distribution switch, whose function is to distribute the inbound calls among the free agents. If all qualified agents are busy, the call center may announce an estimated waiting time to customers. Some of them decide to wait for a free agent, while other abandon after some time or immediately upon knowing the foreseen waiting time. A portion of these customers will redial after some random time. In response to this operational scheme, a number of recent publications emphasize the interest in queueing models that address the call center performance, evaluating the interaction of customer balking, impatience and retrials [10, 11, 104, 114, 507, 684].

Figure 1.3 illustrates the transitions among states in the model proposed by Artalejo and Pla [114]. The first coordinate, i, represents the number of calls waiting in the automatic call distribution switch (queue) and the second one, j, denotes the number of calls in the retrial group (orbit). The parameters λ, ν, μ and δ denote, respectively, the rates of fresh calls, call durations, retrials and impatient times. It is assumed that the call center has c agents. The rest of system parameters (i.e., p_i, q_i, r_i, p, q, r, $\bar{q} = 1-q$ and $\bar{r} = 1-r$) correspond to the balking and persistence probabilities.

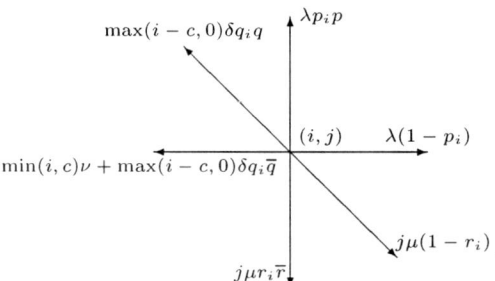

Fig. 1.3. A call center with retrials, balking and impatience

In the rest of this section, we analyze the role of the retrial calls in mobile cellular networks (see Figure 1.4). In a mobile wireless network, the coverage area is partitioned into a certain number of cells. Each cell is served by a base station having a finite number of channels, c. For an efficient use of the radio channel resources, the base station in each cell can handle up to c simultaneous communications. The base station reuses the channels used in other cells, in such a way as the interference of communications taking place in different cells is avoided.

It is known that efficient call handling mechanisms can greatly improve the quality of service and the network performance. Thus, a proper modelling of the mobile cellular network cannot ignore the existence of repeated calls. Single cell load typically consists of new calls initiated by the subscribers inside the bounds of the marked cell, and incoming handover calls which were initiated in the adjacent cell boundary. Since handover calls already use network resources, they should be prioritized with respect to fresh calls. To this end, several approaches can be proposed, such as priority queue of handover calls and guard channels. Guard channels are reserved in the marked cell and can only be used by handover calls. When a new primary call is blocked, the subscriber performs the next attempt with probability H_1. On the other hand, when a repeated call is blocked, the subscriber decides to reattempt with probability H_2. Blocked handover calls are not repeated. These subscribers cross the cell boundary with a call in progress; if they find all the channels busy, then the call is interrupted. The process of call departures from a cell is driven by the minimum between the call duration and the dwell time (i.e., the time that a handover customer needs to cross out the cell). Variants on this operating mode have been developed in several papers [24, 142, 489, 665] that use two-dimensional queueing systems to represent the behavior of such mobile networks.

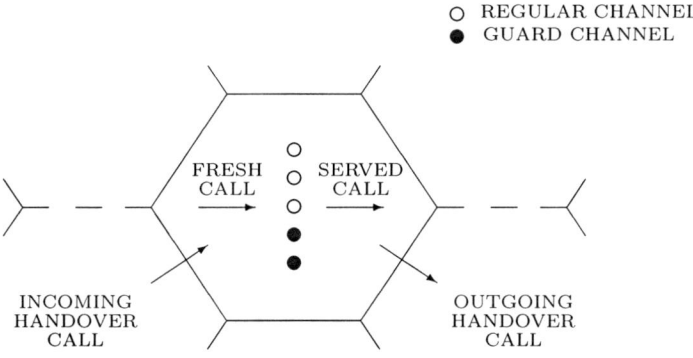

Fig. 1.4. A marked single cell

1.3 Some Examples in Computer Networks

The operational rules of the random access protocols in computer networks provide a major motivation for the design of communication protocols with retransmission control (i.e., with retrials in the queueing terminology). Consider a communication line with slotted time which is shared by several stations. The slot duration equals the transmission time of a single packet of data. If two or more stations are transmitting packets simultaneously, then a collision takes place. As a result, all packets are destroyed and must be retransmitted. The stations involved in the conflict would try to retransmit in the nearest slot, but then a collision occurs with certainty. To avoid this, each station, independently of other stations, may either delay action until the next slot with probability r or retransmit the packet with probability $1 - r$. In other words, each station introduces a random delay before next attempt to transmit the packet. This simple description motivates the interest of the retrial feature in computer networks. In what follows, we briefly summarize the main elements of the random access networks where the retransmission of packets occurs naturally.

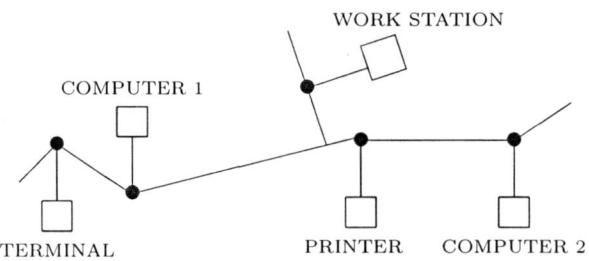

Fig. 1.5. An example of local area network

Computer networks can be classified on the basis of their geographical extent. Here we focus on local area networks (*LANs*) that allow a single organization to connect computers, terminals, peripherals and other devices extended over a moderate geographical area (office, university campus, warehouse) through high-speed, low-noise links. In Figure 1.5, we sketch a generic *LAN*.

The stations (i.e., the nodes of the network) can be connected in different ways. There are four basic structured topologies: bus, ring, star and tree (see Figure 1.6). For the bus topology, all stations attach a transmission medium, or bus. A transmitting station sends a message which propagates the medium in both directions and thus is received at all the stations. In the ring network, the data circulate typically in one direction until the packet returns to the source station, where it is deleted. In the star design, each station attaches to a central node via two-way links, one for transmission and one for reception. We may distinguish two alternative operation modes for the central node. It may retransmit the packet to all stations. In this case, the network is operating as a bus topology although physically is a star. Another alternative is that the central node acts as a switching device. Then, it has capacity to buffer the incoming packets and then retransmit them to the destination station. Finally, the tree topology can be considered as a generalization of the bus topology. The star topology is perhaps the most extended design in the literature for local computer networks with retrials [181, 365, 390, 446].

1.3 Some Examples in Computer Networks   9

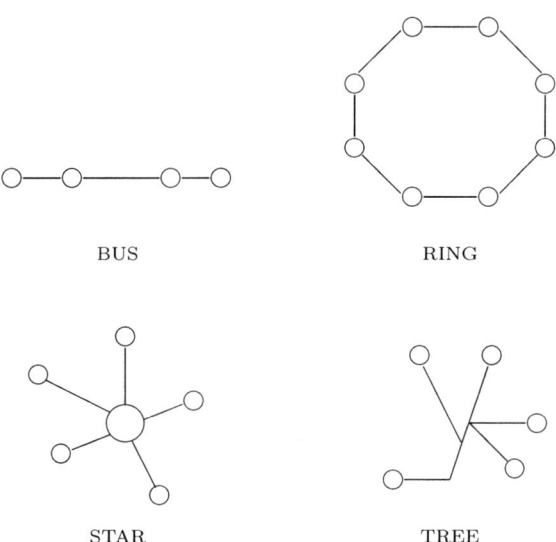

Fig. 1.6. Basic types of *LAN* topology

In the random access networks, each station is free to try the transmission of its packets, independently that another station is also attempting the transmission. In fact, in the pure *ALOHA* protocol a station is free to transmit as soon as it has a packet ready to send. In most *LAN*s, stations are close each other so the propagation delays are short. In this type of networks, it is feasible for a station to listen to the channel. If the channel is sensed to be busy, the station reschedules the transmission of the packet to a later time. This description corresponds to the non-persistent carrier-sense multiple access (*CSMA*) protocol, which arises frequently in the literature as a fundamental source of motivation for the retrial phenomenon in computer networks. It is usual to retry using a backoff algorithm of truncated binary exponential time, such as that employed by Ethernet. One version consists in retrying 15 times, each delayed by a certain integer times the base of the backoff time. The latter is usually taken as twice the end-to-end propagation delay. As a simpler retransmission policy, we mention the geometric scheme described in the illustrating example at the beginning of this section. There are other ways in which the sensed mechanism is used, and this gives rise to persistent *CSMA* protocols. In a persistent protocol, the station keeps sensing the channel until it goes idle. Then, the station transmits with probability p, or the station delays the transmission of the packets with probability $1 - p$. The idea making p lesser than one is to reduce the number of collisions. Clearly, if a collision is detected, the station aborts the packet and tries to repeat the process after the backoff. The most sophisticated protocols include hardware for monitoring while transmitting. If a collision is detected, the packet is promptly aborted and no feedback is needed to inform the transmitting station. However, a signal can be sent to ensure that the other stations are aware of the collision.

Operation of *CSMA* protocols can be slotted or non-slotted. In the former case, it is necessary to synchronize stations to start their transmissions at the beginning of the slot. To conclude, we distinguish between open networks, where stations receive packets from outside the system, and closed networks, where the own station acts as a source of generation of packets. In the former case, it is usual to assume that the external packet flows follow a Poisson process, while the closed networks are typically modelled with the help of the so-called quasi-random input (see Sections 2.3, 3.5 and 8.3).

1.4 Bibliographical Notes

In this introductory chapter the emphasis is put on motivating the fundamental role of the retrials in classical telephone systems, call centers, mobile telephone systems and local computer networks. For descriptions of other practical applications, readers may look into the exhaustive list of references in this book, especially in those contributions published in engineering journals and conference proceedings. Many good books on telephone systems and computer networks are available, as a selected sample we recommend [171, 243, 311, 338, 452, 587, 645, 649, 683, 690] . Specific comments on the retrial literature (books, survey papers) can be found in Section 2.4.

2
A General Overview

In Section 2.1, we present the framework we will use for modelling a retrial queue. To keep the discussion simple, we limit ourselves to the $M/M/c$ and $M/G/1$ retrial queues, even though the material in this section may be readily adapted to advanced retrial queues. In Section 2.2, we establish a comparative analysis of the standard $M/M/c$ and $M/G/1$ models versus their retrial counterparts. Short descriptions of several variants and generalizations of these retrial queues are given in Section 2.3 and, finally, Section 2.4 is devoted to bibliographical notes.

2.1 The Mathematical Formalism

We first consider a multiserver queueing system in which primary customers arrive according to a Poisson flow of rate λ. The service facility consists of c identical servers, and service times are exponentially distributed with parameter ν. If a primary customer finds some server free, he instantly occupies it and leaves the system after service. On the other hand, any customer who finds all servers busy upon arrival is obliged to leave the service area, but he repeats his demand after an exponential time with parameter μ. Thus, we are assuming that the repeated attempts follow the classical retrial policy, where the repetition times of each customer are assumed to be independent and exponentially distributed with intensity μ. We also assume that inter-arrival periods, service times and retrial times are mutually independent.

The system state at time t can be described by means of a bivariate process $\mathcal{X} = \{(C(t), N(t))\,;\, t \geq 0\}$, where $C(t)$ is the number of busy servers and $N(t)$ is the number of customers in orbit. Under the above assumptions the process $\mathcal{X}$ is a regular continuous-time Markov chain with the lattice semi-strip $S = \{0, ..., c\} \times \mathbb{Z}_+$ as the state space. Its infinitesimal transition rates $q_{(i,j)(m,n)}$ are given by

(a) For $0 \leq i \leq c-1$,

$$q_{(i,j)(m,n)} = \begin{cases} \lambda, & \text{if } (m,n) = (i+1,j), \\ i\nu, & \text{if } (m,n) = (i-1,j), \\ j\mu, & \text{if } (m,n) = (i+1,j-1), \\ -(\lambda + i\nu + j\mu), & \text{if } (m,n) = (i,j), \\ 0, & \text{otherwise.} \end{cases}$$

(b) For $i = c$,

$$q_{(c,j)(m,n)} = \begin{cases} \lambda, & \text{if } (m,n) = (c,j+1), \\ c\nu, & \text{if } (m,n) = (c-1,j), \\ -(\lambda + c\nu), & \text{if } (m,n) = (c,j), \\ 0, & \text{otherwise.} \end{cases}$$

By ordering the states as $S = \{(0,0), ..., (c,0), (0,1), ..., (c,1), ...\}$ we can express the infinitesimal generator $\mathbf{Q}$ of the process $\mathcal{X}$ in the following matrix-block form:

$$\mathbf{Q} = \begin{pmatrix} \mathbf{A}_{00} & \mathbf{A}_{01} & & & \\ \mathbf{A}_{10} & \mathbf{A}_{11} & \mathbf{A}_{12} & & \\ & \mathbf{A}_{21} & \mathbf{A}_{22} & \mathbf{A}_{23} & \\ & & \ddots & \ddots & \ddots \end{pmatrix},$$

where $\mathbf{A}_{j,j-1}$, $\mathbf{A}_{jj}$ and $\mathbf{A}_{j,j+1}$ are the following square matrices of order $c+1$:

$$\mathbf{A}_{j,j-1} = \begin{pmatrix} 0 & j\mu & & & \\ & 0 & j\mu & & \\ & & \ddots & \ddots & \\ & & & 0 & j\mu \\ & & & & 0 \end{pmatrix},$$

$$\mathbf{A}_{jj} = \begin{pmatrix} a_{0j} & \lambda & & & \\ \nu & a_{1j} & \lambda & & \\ & 2\nu & a_{2j} & \lambda & \\ & & \ddots & \ddots & \ddots \\ & & & (c-1)\nu & a_{c-1,j} & \lambda \\ & & & & c\nu & a_{cj} \end{pmatrix},$$

$$\mathbf{A}_{j,j+1} = \text{diag}(0, ..., 0, \lambda),$$

with $a_{ij} = -(\lambda + i\nu + (1 - \delta_{ic})j\mu)$, for $0 \leq i \leq c$ and $j \geq 0$, where δ_{ij} is Kronecker's delta defined by

$$\delta_{ij} = \begin{cases} 1, & \text{if } i = j, \\ 0, & \text{otherwise.} \end{cases}$$

Geometrically, the stochastic behavior of the process $\mathcal{X}$ can be represented with the help of the transition diagram shown in Figure 2.1 for the case $c = 3$.

2.1 The Mathematical Formalism 13

It should be noted that random walks on the product of a finite set and the set of non-negative integers (i.e., on a lattice semi-strip) arise in many applications. The best-known family of such walks was introduced by Malyshev [504] and Neuts [543]. The main assumption of their theories is the condition of limited spacial homogeneity

$$\mathbf{A}_{jn} = \mathbf{A}_{n-j}, \quad j \geq j^*,$$

for $n \in \{j-1, j, j+1\}$ and some positive integer j^*. This assumption allows extensive mathematical analysis of both stationary and transient behaviors of the process. In contrast, retrial models operating under the classical retrial policy have transitions from states (i, j) which depend on the second coordinate. The main analytical difficulties and the most interesting properties of retrial queues are connected with this fact. To show the nature of the difficulties in more detail, we now consider the simplest problem; that is, the calculation of the limiting distribution $\mathbf{p} = \{P_{ij}; (i,j) \in S\}$ of the process $\mathcal{X}$. This is usually done with the help of Kolmogorov equations $\mathbf{pQ} = \mathbf{0}'$, where $\mathbf{0}$ denotes the column vector with an infinite number of null entries and $\mathbf{0}'$ is its transpose. Partitioning the limiting probability vector $\mathbf{p}$ as $\mathbf{p} = (\mathbf{p}(0), \mathbf{p}(1), ...)$, where $\mathbf{p}(j) = (P_{0j}, ..., P_{cj})$, we can write Kolmogorov equations in the matrix form

$$\mathbf{p}(j-1)\mathbf{A}_{j-1,j} + \mathbf{p}(j)\mathbf{A}_{jj} + \mathbf{p}(j+1)\mathbf{A}_{j+1,j} = \mathbf{0}'_{c+1}, \quad j \geq 0, \qquad (2.1)$$

where $\mathbf{p}(-1)$ and $\mathbf{A}_{-1,0}$ are defined to be zero, and $\mathbf{0}_{c+1}$ is the column vector of order $c+1$ of 0s.

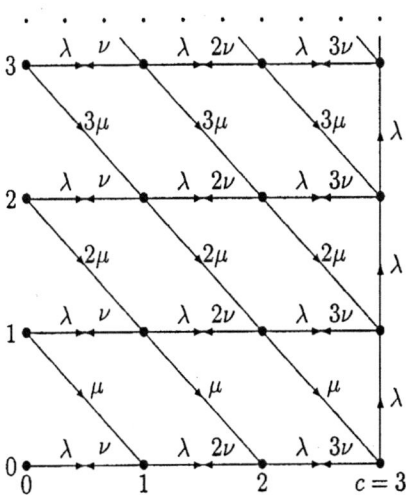

Fig. 2.1. State space and transitions

Alternatively, we may introduce partial generating functions

$$P_i(z) = \sum_{j=0}^{\infty} z^j P_{ij}, \quad 0 \le i \le c,$$

and transform the Kolmogorov equations into the set of differential equations

$$\mu \frac{d\mathbf{p}(z)}{dz} \mathbf{A}(z) = \mathbf{p}(z) \mathbf{B}(z), \tag{2.2}$$

where $\mathbf{p}(z) = (P_0(z), ..., P_c(z))$ and $\mathbf{A}(z)$ and $\mathbf{B}(z)$ are the following square matrices of order $c+1$:

$$\mathbf{A}(z) = \begin{pmatrix} z & -1 & & & \\ & z & -1 & & \\ & & \ddots & \ddots & \\ & & & z & -1 \\ & & & & 0 \end{pmatrix},$$

$$\mathbf{B}(z) = \begin{pmatrix} a_{00} & \lambda & & & \\ \nu & a_{10} & \lambda & & \\ & 2\nu & a_{20} & \lambda & \\ & & \ddots & \ddots & \ddots \\ & & & (c-1)\nu & a_{c-1,0} & \lambda \\ & & & & c\nu & \lambda z + a_{c0} \end{pmatrix}.$$

A point worth mentioning is that, for $c \le 2$, the partial sequences $\{P_{ij}; j \ge 0\}$ satisfy a set of equations of birth-and-death type. We shall return to this remark in Subsection 2.2.1. In the case $c > 2$, we stress that, due to the lack of such a birth-and-death structure, both infinite set of linear equations (2.1) and set of differential equations (2.2) do not have a closed-form solution. Based on equations (2.1) and (2.2) some theoretical approaches provide solutions in terms of contour integrals [200] or as limit of extended continued fractions [560]. However, from a practical point of view, the limiting probabilities P_{ij} cannot be expressed in a tractable form and do not lead to a direct recursive computation.

If $c = 1$, the service distribution can be generalized to follow a general law with probability distribution function $B(x)$ ($B(0) = 0$), Laplace-Stieltjes transform $\beta(s)$ and moments β_k. Now the mathematical model can be viewed as a Markov regenerative process. The main $M/G/1$ retrial queue and many of its variants can be studied by using a variety of methodologies including Markov renewal theory [450], supplementary variable analysis [288], embedded Markov chains [696], etc.

For understanding the physical behavior of the $M/M/c$ and $M/G/1$ retrial queues, it is convenient to analyze the system state at service completion epochs. It should be noted that a server in a standard queueing system is

2.1 The Mathematical Formalism

rendering service in a continuous manner until the queue becomes empty. In contrast, in a retrial queue a server remains unavailable for the system over some interval of time after each service completion. For ease of notation, let us describe this situation for the single server case $c = 1$. At epoch η_{i-1}, the $(i-1)$th customer completes his service and the server becomes free. The next customer enters service after some random time R_i, during which the server is free although there may be customers in the orbit. If the number of customers in orbit at time η_{i-1}, N_{i-1}, is equal to j, then R_i is exponentially distributed with rate $\lambda + j\mu$. Observe that the identity (i.e., primary or orbiting customer) of the customer receiving the ith service time is determined by a competition between two exponential laws of rates λ and $j\mu$. Note also that repeated attempts occurring during the service time S_i that starts at epoch $\xi_i = \eta_{i-1} + R_i$ do not modify the state of the system. At epoch $\eta_i = \xi_i + S_i$ the server becomes idle again. Thus, the evolution of a retrial queue is described in terms of an alternating sequence $\{(R_i, S_i); i \geq 0\}$ of idle and busy periods for the server (see Figure 2.2). This alternating structure is a root of the stochastic decomposition property described in Subsection 2.2.2.

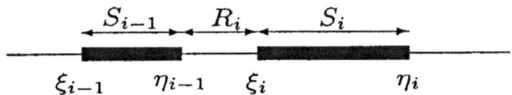

Fig. 2.2. Description of the system behavior

Based on the above figure, the following equation describing the dynamics of the orbit can be derived:

$$N_i = N_{i-1} - B_i + V_i,$$

where V_i is the number of customers arriving during the ith service time, and $B_i = 1$ if the ith customer in service proceeds from the orbit and $B_i = 0$ otherwise. The random vector (R_i, B_i) depends on the history of the system prior to time η_{i-1} only through N_{i-1}, and

$$P(R_i > x, B_i = 1 \mid N_{i-1} = j) = \frac{j\mu}{\lambda + j\mu} e^{-(\lambda + j\mu)x}.$$

On the other hand, the pair (V_i, S_i) does not depend on the history of the system before ξ_i and

$$P(S_i \in (x, x + dx), V_i = n) = e^{-\lambda x} \frac{(\lambda x)^n}{n!} dB(x).$$

The distribution of the channel idle periods R_i is simple for the case $c = 1$. Artalejo and Gómez-Corral [86] discuss the generalization to the multiserver case which is essentially more complicate and interesting. For details, see Section 6.3.

To conclude the section, it may be appropriate to make some comments. The main material concerning other performance characteristics (waiting time, busy period, etc.) is presented in Chapters 4-6. We remark the impossibility of getting analytical solutions for retrial systems in the case of non-exponentially distributed intervals between primary arrivals and inter-retrial periods. It means a significant difference with standard queues which admit closed-form expressions for the model $G/M/c$ and its variants [398]. As an alternative, many efforts have been addressed during the last decade to the numerical investigation of complex retrial queues. In this sense, we especially mention the use of matrix-analytic methods for the investigation of versatile retrial models with inter-arrival and inter-repetition distributions of phase-type, a Markovian arrival process, etc. We shall focus on them in Chapters 7-9.

2.2 Comparing Standard and Retrial Queueing Systems

There exists a rich variety of different single server and multiserver queueing systems with retrials. Although the study of some of them implies a special insight of their peculiarities, in many other cases an extended investigation based on the methods developed for the $M/M/c$ and $M/G/1$ retrial queues may be carried out for structural complex retrial models too. Therefore, we focus on the main models of type $M/M/c$ and $M/G/1$ and give a comparative study of the standard models versus their retrial counterparts.

2.2.1 The Main $M/M/c$ Model

In addition to the process $\mathcal{X} = \{(C(t), N(t)); t \geq 0\}$ which describes the system state for the retrial queue, we now consider a second process $\mathcal{Y} = \{Q(t); t \geq 0\}$ which indicates the number of customers in the system for the standard $M/M/c$ queue (without repeated attempts). Note that the process $\mathcal{Y}$ is a simple birth-and-death process with birth (arrival) rates $\lambda_i = \lambda$, for $i \geq 0$, and death (service) rates $\nu_i = \min(i, c)\nu$, for $i \geq 1$.

As usual, the first question to be investigated is the positive recurrence of $\mathcal{X}$ and $\mathcal{Y}$. It can be shown that both processes are positive recurrent if and only if

$$\rho = \frac{\lambda}{c\nu} < 1. \qquad (2.3)$$

In the case of the process $\mathcal{Y}$, the proof follows from the classical results for the classification of states in birth-and-death processes [398, 450]. The proof for the retrial process $\mathcal{X}$ uses Foster's criterion based on mean drifts. Essentially more interesting is the fact that $\rho = 1$ provides a necessary and sufficient

2.2 Comparing Standard and Retrial Queueing Systems

condition for the null recurrence of $\mathcal{Y}$, whereas the behavior of $\mathcal{X}$ in the case $\rho = 1$ depends on the retrial rate. For instance, if $c = 1$ and $\rho = 1$, then $\mathcal{X}$ is null recurrent if and only if $\mu \geq \nu$ [288].

Let us observe that the positive recurrence condition (2.3) of $\mathcal{X}$ is independent of the retrial rate μ. An intuitive explanation follows by assuming a very congested orbit. Then the idle time R_i converges to zero and the system performs like the standard queue with random order discipline.

If $\rho < 1$, then the steady state of the standard $M/M/c$ queue exists, and the limiting distribution of the number of customers in the system [398] is given by

$$P_j = \begin{cases} P_0 \frac{(c\rho)^j}{j!}, & \text{if } 0 \leq j \leq c, \\ P_0 \frac{c^c \rho^j}{c!}, & \text{if } j > c, \end{cases}$$

where

$$P_0 = \left(\frac{c^c \rho^{c+1}}{c!(1-\rho)} + \sum_{j=0}^{c} \frac{(c\rho)^j}{j!} \right)^{-1}.$$

In particular, in the single server case we have a geometric distribution with parameter ρ

$$P_j = (1-\rho)\rho^j, \quad j \geq 0,$$

whose factorial moments are given by

$$M_k = \sum_{j=k}^{\infty} j(j-1)\ldots(j-k+1)P_j$$

$$= k! \left(\frac{\rho}{1-\rho} \right)^k, \quad k \geq 1.$$

The limiting distribution of the system state for the $M/M/1$ retrial queue [288] is as follows:

$$P_{0j} = \frac{\rho^j}{j!\mu^j}(1-\rho)^{1+\frac{\lambda}{\mu}} \prod_{k=0}^{j-1}(\lambda + k\mu), \quad j \geq 0,$$

$$P_{1j} = \frac{\rho^{j+1}}{j!\mu^j}(1-\rho)^{1+\frac{\lambda}{\mu}} \prod_{k=1}^{j}(\lambda + k\mu), \quad j \geq 0.$$

Thus, we also have the following expression for the limiting distribution of the total number of customers in the system:

$$P_{0j} + (1-\delta_{0j})P_{1,j-1} = \frac{\rho^j}{j!\mu^j}(1-\rho)^{1+\frac{\lambda}{\mu}} \prod_{k=1}^{j}(\lambda+k\mu), \quad j \geq 0,$$

which can be identified as the negative binomial distribution. In this case, the factorial moments are of the form

$$M_k = M_k^0 + M_k^1 + kM_{k-1}^1, \quad k \geq 1,$$

where

$$M_k^i = \sum_{j=k}^{\infty} j(j-1)\ldots(j-k+1)P_{ij}$$

$$= \begin{cases} (1-\rho)\left(\frac{\rho}{\mu(1-\rho)}\right)^k \prod_{j=0}^{k-1}(\lambda+j\mu), & \text{if } i=0, \\ \rho\left(\frac{\rho}{\mu(1-\rho)}\right)^k \prod_{j=1}^{k}(\lambda+j\mu), & \text{if } i=1, \end{cases}$$

for $k \geq 1$, and $M_0^1 = \rho$ represents the probability of blocking.

In the case $c = 2$, the limiting distribution can be expressed in terms of hypergeometric functions [288, 344]. The main feature of such a solution is that each partial sequence $\{P_{ij}; j \geq 0\}$ satisfies a set of relations of birth-and-death type. For example, for $i = 0$ we can write down

$$(\alpha_j + \beta_j)P_{0j} = (1-\delta_{0j})\alpha_{j-1}P_{0,j-1} + \beta_{j+1}P_{0,j+1}, \quad j \geq 0,$$

with birth rates $\alpha_j = \lambda((\lambda+j\mu)^2 + j\nu\mu)$ and death rates $\beta_j = j\nu\mu(3\lambda + 2\nu + 2j\mu)$, for $j \geq 0$. As a result, one readily finds that all the probabilities P_{ij}, for $0 \leq i \leq 2$ and $j \geq 0$, can be expressed through P_{00}, and P_{00} can be written in terms of hypergeometric functions.

The consideration of more than two servers complicates the transitions among states and, as a consequence, the underlying structure of birth-and-death type is not preserved. The particular case $c = 3$ is treated in [316, 396], where the limiting distribution is reduced to finding the probabilities P_{00} and P_{01}, which can be recursively computed.

Now, note that equation (2.2) is the key to get the mean values

$$E[C] = \frac{\lambda}{\nu}, \tag{2.4}$$

$$E[N] = \frac{(\nu+\mu)(\lambda - \nu Var(C))}{\mu(c\nu - \lambda)}, \tag{2.5}$$

where $E[C]$, $E[N]$ and $Var(C)$ are defined as the mean values and variance of $C(t)$ and $N(t)$ as $t \to \infty$ [288]. Formula (2.4) can be thought of as a variant of Little's formula. On the other hand, the use of formula (2.5) reduces the calculation of the mean number of customers in orbit to the variance of the number of busy servers, which is a simpler problem.

2.2 Comparing Standard and Retrial Queueing Systems

Since the recursive computation of P_{ij} cannot be performed, the analysis of numerical methods of calculation of P_{ij} has been the subject matter of many papers. Among them, the so-called generalized truncated models [85, 262, 547] propose to approximate an infinite system, which cannot be solved directly, with the help of another infinite calculable system. The fact that we approximate the original infinite system by another infinite system implies better accuracy than direct methods [643, 685] based on finite truncation of the state space. We give a detailed study of such techniques in Section 3.4.

We now turn our attention to other important performance characteristics such as the busy period and the waiting time. We assume that a busy period is defined as the period starting with the arrival of a customer who finds the system empty and ends at the first service completion epoch at which the system becomes empty again. The busy period analysis is important from the server's point of view and is also helpful in the efficient planning of the system resources. We first analyze the standard $M/M/c$ queue and denote the length of its busy period as L_∞. The existing studies (see [80] and its references) mainly deal with the existence of closed-form solutions for the case $c \leq 2$. The solution is given in terms of the Bessel function of the first kind of order r defined as

$$I_r(x) = \sum_{n=0}^{\infty} \frac{(x/2)^{r+2n}}{(n+r)!\,n!}.$$

In the case $c = 1$, the probability density function of L_∞ is

$$f_{L_\infty}(x) = \frac{e^{-(\lambda+\nu)x}}{x\rho^{1/2}} I_1\left(2x\sqrt{\lambda\nu}\right), \quad x > 0.$$

The first moments of L_∞ can be obtained as a particular case of the corresponding formulas given in Subsection 2.2.2 for the standard $M/G/1$ queue.

If $c = 2$, we have

$$f_{L_\infty}(x) = \frac{e^{-(\lambda+2\nu)x}}{x} \sum_{r=0}^{\infty} (r+1)\left(\frac{\nu}{2\lambda}\right)^{\frac{r+1}{2}} I_{r+1}\left(2x\sqrt{2\lambda\nu}\right), \quad x > 0,$$

and the first moments of L_∞ are

$$E[L_\infty] = \frac{1}{\nu(1-\rho)},$$

$$E[L_\infty^2] = \frac{2-\rho}{\nu^2(1-\rho)^3}.$$

In the case $c > 2$, we do not have explicit expressions. However, from the theory of regenerative processes, we know that the limiting probability of an empty system, P_0, is equal to $(1 + \lambda E[L_\infty])^{-1}$. In [80], the investigation of the Laplace-Stieltjes transform of L_∞ and the computation of its moments is reduced to the recursive solution of some simple systems of linear equations.

For the study of the length of the busy period L_μ in the $M/M/c$ retrial queue, we refer the reader to Subsection 4.2.1 and [105]. Previous results for the density function and transform solutions are unknown. In the case $c = 1$, the moments of L_μ can be recursively computed following the method described by Choo and Conolly [195], whose arguments are mostly based on convolutive equations for first-passage times.

We now consider the virtual waiting time $W(t)$ of a customer who arrives at the system at time t. According to this definition, $W(t)$ means the time that the customer spends waiting at the queue (standard model) or at the orbit (retrial model) excluding the service time. We deal with the system at steady state so we simply denote $W(t)$ by W. Clearly, the distribution of W depends on the service discipline. Retrial queues arising from teletraffic applications operate under a random order access discipline. This means that all calls waiting in the orbit have an identical chance for being allocated to free servers when these become available. The case in which customers are served in order of arrival is of minor interest and its solution is simpler. Thus, in what follows, we assume the random access discipline. Random servicing is complicate due to the overtaking phenomena; that is, it is necessary to consider not only the number of customers present in the system at the time of arrival of a customer whose delay is to be investigated, but also the possibility that customers arriving at later times will compete for free servers. We denote the waiting times for standard and retrial queues respectively as W_∞ and W_μ.

The complementary distribution function of the waiting time in the standard $M/M/c$ queue can be expressed in the form of an integral [649] which can be expanded using Lagrange orthogonal polynomials or McLaurin series methods [202]. This yields the following result:

$$P(W_\infty > x \mid W_\infty > 0) = 1 - c\nu x \frac{1-\rho}{\rho} \ln \frac{1}{1-\rho}$$
$$+ \frac{(c\nu x)^2}{2}(1-\rho)\left(2 - \frac{1-\rho}{\rho}\ln\frac{1}{1-\rho}\right) - \ldots$$

On the other hand, the approximate analysis of W_μ in the $M/M/c$ retrial queue is done in Subsection 5.2.1; see also [93].

To finish this section, we now consider the limit behavior of the $M/M/c$ retrial queue under high and low rate of retrials. This is an important feature due to lack of analytical formulas for the main performance characteristics, since limit theorems allow us to understand the influence of the repeated attempts in some domains of the system parameters. Besides, limit results provide insight into correspondence between retrial queues and the classical queueing models with waiting line and losses.

First, we consider the case of high rate of retrials. As $\mu \to \infty$ (i.e., the intervals between successive retrials tend to zero) the $M/M/c$ retrial queue can be viewed as the standard one with waiting line. This general heuristic remark can be transformed into a rigorous mathematical result in several ways. For

2.2 Comparing Standard and Retrial Queueing Systems

example, denoting by $P_{ij}(\infty)$ the limiting probability that the number of busy servers equals i and the queue length equals j in the standard $M/M/c$ queue, and by $P_{ij}(\mu)$ the corresponding probability in the $M/M/c$ retrial queue, we have that

$$\lim_{\mu \to \infty} P_{ij}(\mu) = P_{ij}(\infty), \quad 0 \leq i \leq c, \ j \geq 0,$$

where

$$P_{00}(\infty) = \left(\frac{c^c \rho^{c+1}}{c!(1-\rho)} + \sum_{j=0}^{c} \frac{(c\rho)^j}{j!} \right)^{-1},$$

$$P_{ij}(\infty) = \begin{cases} P_{00}(\infty) \frac{(c\rho)^i}{i!}, & \text{if } 0 \leq i \leq c-1, \ j = 0, \\ P_{00}(\infty) \frac{\rho^{c+j} c^c}{c!}, & \text{if } i = c, \ j \geq 0, \\ 0, & \text{otherwise.} \end{cases}$$

In this sense, it is interesting to obtain asymptotic expansions for limiting performance characteristics in a power series in the mean time between successive retrials $1/\mu$ [270]. The first term of such an expansion is the corresponding performance characteristic for the standard $M/M/c$ queue, and the second one describes the influence of retrials and hence is of special interest. To illustrate this, we first consider the limiting blocking probability B_μ of the retrial queue defined as the probability that all servers are busy. Then, we have

$$B_\mu = B_\infty + \frac{(c-1)\nu - \lambda + \nu B_\infty}{\mu} \ln(1-\rho) B_\infty + o\left(\frac{1}{\mu}\right),$$

where B_∞ is the blocking probability in the standard $M/M/c$ queue given by

$$B_\infty = \frac{\frac{(c\rho)^c}{(c-1)!}}{\frac{(c\rho)^c}{(c-1)!} + c(1-\rho) \sum_{i=0}^{c-1} \frac{(c\rho)^i}{i!}}.$$

If we focus on the limiting number of customers in orbit N_μ of the $M/M/c$ retrial queue, then such an asymptotic property is of the form

$$E[N_\mu] = E[N_\infty] + Z E[N_\infty] + o\left(\frac{1}{\mu}\right),$$

where $E[N_\infty] = \rho(1-\rho)^{-1} B_\infty$ corresponds to the standard counterpart and

$$Z = \frac{\lambda - (c(c-1)\nu + \lambda(\lambda \nu^{-1} - B_\infty - 2(c-1))) \ln(1-\rho)}{\lambda \mu \nu^{-1}}.$$

We present some numerical examples to illustrate how the multiserver retrial queue approaches its standard counterpart. In Figures 2.3 to 2.5, we show the influence of the retrial rate μ and the traffic intensity ρ on the mean values of W_μ and L_μ, and the blocking probability B_μ. We consider an $M/M/c$ queue with $c = 5$ and service rate $\nu = 1.0$. In Figures 2.3 and 2.5, the curves, which in decreasing order correspond to $\rho = 0.75$, 0.5 and 0.25, show that, as is to be expected, $E[W_\mu]$ and $E[L_\mu]$ decrease with increasing values of μ and approach the values $E[W_\infty]$ and $E[L_\infty]$, respectively. On the other hand, in Figure 2.4 the curves show that B_μ increases as a function of μ and converges to the blocking probability B_∞ of the standard queue. Here, our numerical computations are performed by using the generalized truncated approach proposed by Artalejo and Pozo [85].

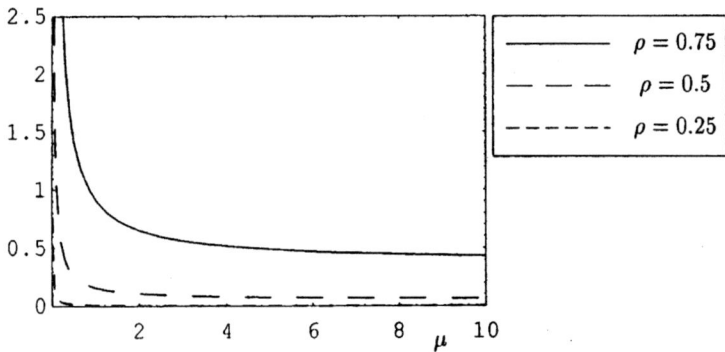

Fig. 2.3. $E[W_\mu]$ versus μ and ρ

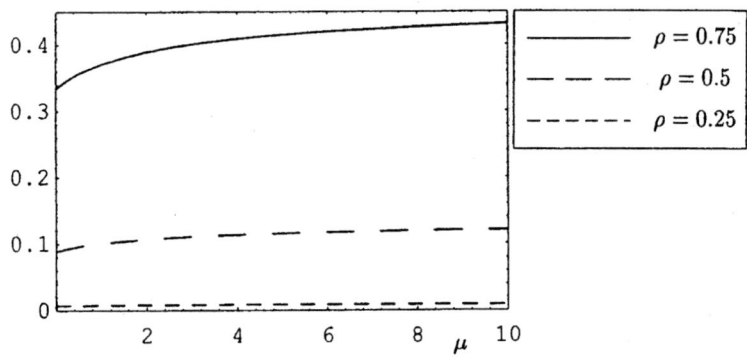

Fig. 2.4. B_μ versus μ and ρ

On the other hand, the limit behavior of retrial queues as $\mu \to 0$ is of interest on account of the weak dependence of the limiting distribution

2.2 Comparing Standard and Retrial Queueing Systems

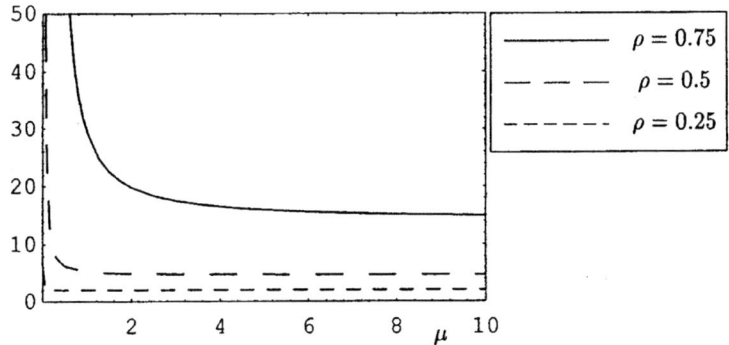

Fig. 2.5. $E[L_\mu]$ versus μ and ρ

$\{P_{i.}(\mu); 0 \leq i \leq c\}$ of the number of busy servers upon μ, where the marginal probability $P_{i.}(\mu)$ is defined as

$$P_{i.}(\mu) = \sum_{j=0}^{\infty} P_{ij}(\mu).$$

This fact is numerically illustrated in [288]. Because for complex retrial queues $\lim_{\mu \to 0} P_{i.}(\mu)$ can be found more simply than $\lim_{\mu \to \infty} P_{i.}(\mu)$, it is reasonable to use this limit as an approximation of $P_{i.}(\mu)$ for all $\mu > 0$. For the $M/M/c$ retrial queue in steady state, as $\mu \to 0$, the distribution of the number of busy servers converges to the corresponding distribution for the standard Erlang loss system $M/M/c/0$ (which is a truncated Poisson distribution), but with increased arrival rate $\Lambda = \lambda + r$, where r is the unique positive root of the polynomial equation

$$r \sum_{i=0}^{c-1} \left(\frac{\lambda+r}{\nu}\right)^i \frac{1}{i!} = \lambda \left(\frac{\lambda+r}{\nu}\right)^c \frac{1}{c!}.$$

The additional arrival rate r equals $\lim_{\mu \to 0} \mu E[N]$ and can be thought of as a load formed by sources of repeated calls. This result shows that distinguishing between the cases $\mu = 0$ and $\mu \to 0$ is important. If $\mu = 0$, then the blocked customers do not send repeated attempts at all. Thus, the retrial queue becomes the standard Erlang loss system with the same arrival rate λ with limiting distribution

$$P_{i.}(0) = \frac{\frac{(c\rho)^i}{i!}}{\sum_{i=0}^{c} \frac{(c\rho)^i}{i!}}, \quad 0 \leq i \leq c.$$

We also note that $N(t)$ converges to ∞, as $t \to \infty$. In contrast, if $\mu \to 0$, then the retrial queue in steady state can be viewed as the standard Erlang loss system but with the increased arrival rate $\Lambda = \lambda + r$.

2.2.2 The Main $M/G/1$ Model

We next consider the single server case so the service time distribution is allowed to follow a general law. The subsequent necessary notation was introduced in Section 2.1. We assume that $\rho = \lambda\beta_1 < 1$ so our queueing models are stable and the limiting probabilities

$$P_j = \lim_{t\to\infty} P(Q(t)=j), \quad j \geq 0,$$
$$P_{ij} = \lim_{t\to\infty} P(C(t)=i, N(t)=j), \quad (i,j) \in \{0,1\} \times \mathbb{Z}_+,$$

exist and are positive.

Both sequences $\{P_j; j \geq 0\}$ and $\{P_{ij}; i \in \{0,1\}, j \geq 0\}$ can be computed recursively with the help of the following equations [285, 450]:

$$P_0 = 1 - \rho,$$
$$P_1 = \frac{1-a_0}{a_0} P_0,$$
$$P_2 = \frac{1-a_0-a_1}{a_0}(P_0 + P_1),$$
$$P_{j+1} = \frac{1-\sum_{i=0}^{j} a_i}{a_0}\sum_{i=0}^{j} P_i + \sum_{i=2}^{j} P_i \sum_{k=j-i+2}^{j} \frac{a_k}{a_0}, \quad j \geq 2,$$
$$P_{0j} = \frac{\lambda}{\lambda + j\mu}\pi_j, \quad j \geq 0,$$
$$P_{1j} = \frac{(j+1)\mu}{\lambda}P_{0,j+1}, \quad j \geq 0,$$
$$\pi_0 = (1-\rho)\exp\left\{-\frac{\lambda}{\mu}\int_0^1 \frac{1-\beta(\lambda-\lambda u)}{\beta(\lambda-\lambda u)-u}du\right\},$$
$$\pi_j = \sum_{i=0}^{j}\pi_i\frac{\lambda}{\lambda+i\mu}a_{j-i} + \sum_{i=1}^{j+1}\pi_i\frac{i\mu}{\lambda+i\mu}a_{j-i+1}, \quad j \geq 0,$$

where

$$a_j = \int_0^\infty e^{-\lambda x}\frac{(\lambda x)^j}{j!}dB(x), \quad j \geq 0,$$

is the probability that exactly j customers arrive during a service time. Indeed, the sequence $\{\pi_j; j \geq 0\}$ corresponds to the distribution of the embedded Markov chain at service completion epochs. The above recursions provide an effective method for computing P_j, P_{ij} and π_j. The reader is referred to Subsection 3.1.1 for a rigorous proof of them.

An alternative solution in terms of the generating functions

$$P(z) = \sum_{j=0}^{\infty} z^j P_j,$$

2.2 Comparing Standard and Retrial Queueing Systems

$$P_i(z) = \sum_{j=0}^{\infty} z^j P_{ij}, \quad i \in \{0, 1\},$$

is given by

$$P(z) = \frac{(1-\rho)(1-z)\beta(\lambda - \lambda z)}{\beta(\lambda - \lambda z) - z}, \tag{2.6}$$

$$P_0(z) = (1-\rho) \exp\left\{ -\frac{\lambda}{\mu} \int_z^1 \frac{1 - \beta(\lambda - \lambda u)}{\beta(\lambda - \lambda u) - u} du \right\}, \tag{2.7}$$

$$P_1(z) = \frac{1 - \beta(\lambda - \lambda z)}{\beta(\lambda - \lambda z) - z} P_0(z). \tag{2.8}$$

Note that the solution for both standard and retrial queues are given in terms of the Laplace-Stieltjes transform of the service time but the retrial model exhibits a more complex expression mainly due to the integral arising in the right-hand side of (2.7).

In particular, the corresponding expectations are given by

$$E[Q] = \rho + \frac{\lambda^2 \beta_2}{2(1-\rho)},$$

$$E[C] = \rho,$$

$$E[N] = \frac{\lambda^2}{1-\rho}\left(\frac{\beta_1}{\mu} + \frac{\beta_2}{2}\right).$$

Here, $E[Q]$ denotes the mean value of $Q(t)$ as $t \to \infty$.

In Section 2.1, we remarked the existence of a sequence of idle periods in which the server is unavailable for the system. Due to this, a retrial queue can be considered as a special type of vacation model in which the vacation begins at the end of each service time [49]. To exploit this fact, we assume the stationary regime and denote the process $(C(t), N(t))$ as (C_μ, N_μ) to remark the dependence on the retrial rate μ. Let (C_∞, N_∞) be the corresponding formulation for the standard $M/G/1$ queue. This vector represents the server state and the number of customers in queue at steady state, so that $Q(t)$ equals $C_\infty + N_\infty$.

From equations (2.6)-(2.8) we observe that the vector (C_μ, N_μ) can be represented as a sum of two independent random vectors as follows

$$(C_\mu, N_\mu) = (C_\infty, N_\infty) + (0, N_\mu^0),$$

where N_μ^0 represents the number of customers in orbit given that the server is free; that is, $N_\mu^0 \equiv N_\mu | C_\mu = 0$. Hence, its generating function is $P_0(z)/(1-\rho)$.

An application of the stochastic decomposition property yields explicit relationships among the factorial moments of the number of customers in orbit in the $M/G/1$ retrial queue and the factorial moments in the standard $M/G/1$ queue [49]. In particular, we have that

$$M_k(\mu) = M_k(\infty) + \sum_{n=1}^{k} \binom{k}{n} n! M_{k-n}(\infty) \sum_{j=1}^{n} \frac{1}{j!} \left(\frac{\lambda}{\mu(1-\rho)} \right)^j$$

$$\times \sum_{\mathbf{x} \in X_j(n)} \frac{1}{\mathbf{x}!} \prod_{i=1}^{j} (M_{x_i-1}(\infty) - (1-\rho)\delta_{1,x_i}),$$

where $X_j(n)$ is the set of all vectors $\mathbf{x} = (x_1, ..., x_j) \in \mathbb{Z}_+^j$ such that $x_i > 0$ and $x_1 + ... + x_j = n$, and $\mathbf{x}! = x_1!...x_j!$.

Moreover, we can estimate the following measure of proximity between the limiting distributions of the $M/G/1$ queues with and without retrials

$$D = \sum_{i=0}^{1} \sum_{j=0}^{\infty} |P_{ij}(\mu) - P_{ij}(\infty)|,$$

as follows:

$$2(1 - \rho - \pi_0) < D < 2 \left(1 - \frac{\pi_0}{1-\rho} \right).$$

Figure 2.6 illustrates the effect on D of varying the retrial rate μ and the traffic intensity ρ. We consider service times governed by an Erlang (E_k) law

$$B(x) = 1 - \sum_{j=0}^{k-1} e^{-\nu x} \frac{(\nu x)^j}{j!}, \quad x \geq 0,$$

with parameters $k = 3$ and $\nu = 0.75$. The numerical examples show how D decreases to zero with increasing values of μ. The convergence of D to zero is faster for systems with lower traffic intensity ρ.

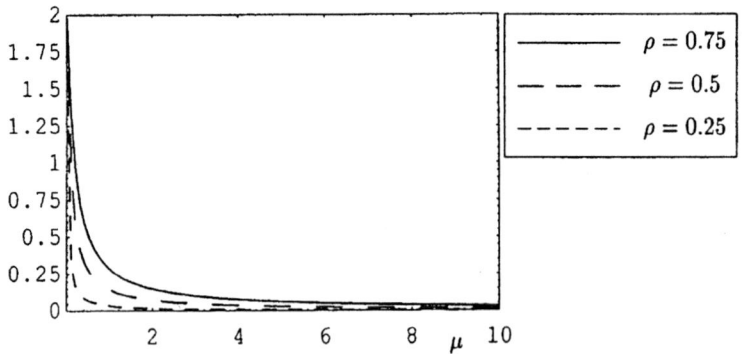

Fig. 2.6. D versus μ and ρ for an $M/E_3/1$ retrial queue

2.2 Comparing Standard and Retrial Queueing Systems

Several methodologies can be used to analyze the busy period of the standard $M/G/1$ queue. In particular, the length of a busy period L_∞ is independent of the queueing discipline so we can change the order in which customers are served and create a last-come-first-served discipline. In this way, the analysis of L_∞ is connected to a branching process in which each customer arriving during the first service time generates a sub-busy period distributed as the initial busy period under study [398]. This yields the equation for the Laplace-Stieltjes transform of L_∞

$$L_\infty^*(s) = \beta(s + \lambda - \lambda L_\infty^*(s)). \tag{2.9}$$

The functional equation (2.9) can be easily differentiated to find the first moments of L_∞. In particular, we have

$$E[L_\infty] = \frac{\beta_1}{1-\rho},$$

$$E[L_\infty^2] = \frac{\beta_2}{(1-\rho)^3}.$$

On the other hand, the structure of the busy period L_μ of the $M/G/1$ retrial queue and its analysis in terms of Laplace-Stieltjes transforms have been investigated by several methods [252, 288]. Thus, we have

$$L_\mu^*(s) = \frac{\int_0^{L_\infty^*(s)} \frac{\beta(s+\lambda-\lambda u)}{e(s,u)(\beta(s+\lambda-\lambda u)-u)} du}{\int_0^{L_\infty^*(s)} \frac{du}{e(s,u)(\beta(s+\lambda-\lambda u)-u)}}, \quad s > 0, \tag{2.10}$$

where $L_\infty^*(s)$ is the Laplace-Stieltjes transform for the busy period in the standard queue without retrials given in (2.9) and $e(s, u)$ is defined as

$$e(s,u) = \exp\left\{\frac{1}{\mu}\int_0^u \frac{s+\lambda-\lambda\beta(s+\lambda-\lambda v)}{\beta(s+\lambda-\lambda v)-v} dv\right\}, \quad 0 \leq u < L_\infty^*(s).$$

The above expression provides a theoretical solution but it has serious limitations in practice. For instance, we cannot compute the first moments of L_μ by direct differentiation. The expectation follows easily from the theory of regenerative processes and is given by

$$E[L_\mu] = \frac{1}{\lambda}\left(\frac{1}{P_{00}} - 1\right). \tag{2.11}$$

A direct method of calculation for the second moment [75] yields

$$E[L_\mu^2] = \frac{1}{P_{00}}\left(\frac{1}{(1-\rho)^2}\left(\frac{2\rho\beta_1}{\mu}+\beta_2\right) - \int_0^1 \frac{2}{\lambda\mu(\beta(\lambda-\lambda u)-u)}\right.$$
$$\left.\times\left(1 - \frac{\lambda(1-u)\beta'(\lambda-\lambda u)}{\beta(\lambda-\lambda u)-u} - \frac{1}{1-\rho}\exp\left\{\frac{\lambda}{\mu}\int_u^1 \frac{1-\beta(\lambda-\lambda v)}{\beta(\lambda-\lambda v)-v}dv\right\}\right)du\right). \tag{2.12}$$

Furthermore, it does not seem possible to numerically invert $L_\mu^*(s)$ by applying well-known algorithms [1, 2, 3] because formula (2.10) is derived when s is a real value, and such algorithmic methods require to evaluate the Laplace-Stieltjes transform at any desired complex s.

By assuming exponential service times, $L_\mu^*(s)$ reduces to a simpler form, which can be numerically evaluated. The busy period of the single server retrial queue can also be connected to branching processes but with more complex structure [336].

The analysis for the waiting time of the standard $M/G/1$ queue with random order of service [653] leads to the following expression for the Laplace-Stieltjes transform of W_∞:

$$W_\infty^*(s) = 1 - \rho + \frac{\lambda(1-\rho)}{s} \int_{L_\infty^*(s)}^1 \frac{(1-u)(\beta(\lambda - \lambda u) - \beta(s + \lambda - \lambda u))}{(\beta(\lambda - \lambda u) - u)(u - \beta(s + \lambda - \lambda u))}$$

$$\times \exp\left\{-\int_u^1 \frac{dv}{v - \beta(s + \lambda - \lambda v)}\right\} du.$$

The first two moments are given by

$$E[W_\infty] = \frac{\lambda \beta_2}{2(1-\rho)},$$

$$E[W_\infty^2] = \frac{2\lambda \beta_3}{3(1-\rho)(2-\rho)} + \frac{\lambda^2 \beta_2^2}{(1-\rho)^2(2-\rho)}.$$

There are algorithms for numerically computing $P(W_\infty > t \mid W_\infty > 0)$. Nevertheless, except for a few special service time distributions, they are complex procedures. For more information, see [202].

The formula for the Laplace-Stieltjes transform of W_μ in the $M/G/1$ retrial queue [283] is still more formidable:

$$W_\mu^*(s) = 1 - \rho + \frac{\lambda(1-\rho)}{s} \int_{L_\infty^*(s)}^1 \frac{(1-u)(\beta(\lambda - \lambda u) - \beta(s + \lambda - \lambda u))}{(\beta(\lambda - \lambda u) - u)(u - \beta(s + \lambda - \lambda u))}$$

$$\times \exp\left\{\frac{1}{\mu}\int_u^1 \left(\frac{s + \mu + \lambda - \lambda v}{\beta(s + \lambda - \lambda v) - v} - \frac{\lambda - \lambda v}{\beta(\lambda - \lambda v) - v}\right) dv\right\} du, \tag{2.13}$$

for $s > 0$. Since s is real, we remark again that most typical techniques for its numerical inversion do not apply.

Nevertheless, the mean waiting time can be easily obtained with the help of Little's formula

$$E[W_\mu] = E[W_\infty] + \frac{\rho}{\mu(1-\rho)}. \tag{2.14}$$

Recently, Artalejo et al. [83] have obtained the following formula for the second moment of W_μ:

$$E[W_\mu^2] = E[W_\infty^2] + \frac{\lambda \beta_2}{\mu} \left(\frac{2}{(1-\rho)^2(2-\rho)} + \frac{\rho}{(1-\rho)^2} \right) + \frac{2\rho}{\mu^2(1-\rho)^2}.$$
(2.15)

Of course, as $\mu \to \infty$, the above formulas for the retrial queue agree with the corresponding results for the standard $M/G/1$ queue. Before dealing in more detail with limit theorems for the $M/G/1$ retrial queue, we mention the possibility of studying discrete processes closely related to L_μ and W_μ. In the case of L_μ, the study can be extended to the number of customers arriving to the system during the length of a busy period [288, 491, 492]. It is also natural to measure the waiting time by the number of retrials made by a primary customer entering the system at time t, before he starts service [110, 267, 283]. In Sections 4.1 and 5.1, we will provide a detailed numerical analysis of L_μ and W_μ and their related discrete measures [108, 109].

Even for the standard $M/G/1$ queue the type of distribution of the queue length in steady state is unknown. The Pollaczek-Khintchine equation (2.6) gives only the generating function of the distribution in terms of $\beta(s)$. However, under the heavy traffic analysis the queue length distribution can be approximated by an exponential law. To be more exact, as λ varies in such a way that $\rho \to 1-$, the distribution of the scaled queue length $(1-\rho)Q(t)$ weakly converges to an exponential distribution with mean $\beta_2/(2\beta_1^2)$. A similar result holds for the $M/G/1$ retrial queue [252], but now the limiting distribution of the scaled number of customers in orbit $(1-\rho)N(t)$ is gamma with mean

$$\frac{\beta_2}{2\beta_1^2} + \frac{1}{\mu\beta_1}$$

and variance

$$\frac{\beta_2^2}{4\beta_1^4} + \frac{\beta_2}{2\mu\beta_1^3}.$$

Since a retrial queueing model involves one additional parameter μ, another interesting limit situations are $\mu \to \infty$ (short intervals between retrials) and $\mu \to 0$ (long intervals between retrials). Of course, the corresponding counterparts for the standard queue do not exist.

In the case $\mu \to \infty$, it follows easily from (2.6)-(2.8) that the steady state distribution of the number of customers in orbit converges to the corresponding distribution in the standard $M/G/1$ queue.

As $\mu \to 0$, the number of customers in orbit is asymptotically Gaussian [288] with mean

$$\frac{\lambda\rho}{\mu(1-\rho)}$$

and variance

$$\frac{2\lambda\rho(1-\rho) + \lambda^3\beta_2}{2\mu(1-\rho)^2}.$$

The effect of the retrial rate μ and the service time distribution on the mean values of N_μ and L_μ is illustrated in Figures 2.7 and 2.8. In each figure we fix $\rho = 0.75$ and present three curves which correspond to several service time distributions. Specifically, the numerical examples concern with exponentially distributed service times of rate $\nu = 1.0$, hyperexponential (H_2) service times with parameters $p = 0.25$, $\nu_1 = 2.0$ and $\nu_2 = 6.0$, and E_k service times with $k = 3$ and $\nu = 1.5$. The hyperexponential law follows the probability distribution function

$$B(x) = p(1 - e^{-\nu_1 x}) + (1-p)(1 - e^{-\nu_2 x}), \quad x \geq 0.$$

In Figure 2.7, it is shown that $E[N_\mu]$ is a decreasing function of μ. Indeed, it is an increasing function of ρ. The H_2 case is represented by the highest curve, and the lowest curve concerns with the E_k case. In Figure 2.8, we focus on the mean length of a busy period which, as a function of μ, exhibits a similar monotone behavior.

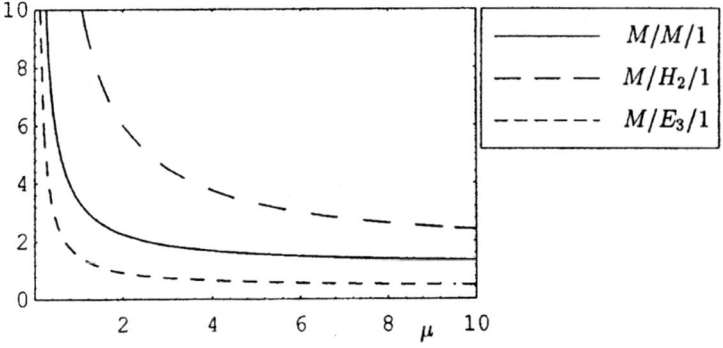

Fig. 2.7. $E[N_\mu]$ versus μ

Finally, we consider the departure process which is defined as the sequence of times $\{\eta_i; i \geq 0\}$ at which customers complete service and leave the system. Equivalently, we consider the sequence of inter-departure times $T_i = \eta_i - \eta_{i-1}$. Taking into account that the departure process from a queueing system can be the input process for another queueing system, it is important to find conditions under which the departure process is Poisson or, at least, it is a renewal process. For the standard $M/G/1$ queue in steady state the departure process is a renewal one if and only if the service time distribution is exponential, in which case the process is Poisson. The departure process for the $M/G/1$ retrial queue is never a renewal process, except in the trivial case of instantaneous service times.

The first two moments of the inter-departure times in the standard queue [653] are

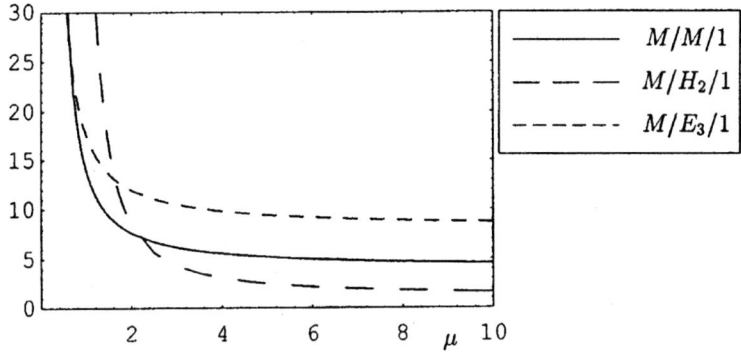

Fig. 2.8. $E[L_\mu]$ versus μ

$$E[T_i] = \frac{1}{\lambda},$$
$$Var(T_i) = \frac{1}{\lambda^2} + \beta_2 - 2\beta_1^2,$$

and, in the case of the $M/G/1$ retrial queue [253], we have

$$E[T_i] = \frac{1}{\lambda},$$
$$Var(T_i) = \frac{1}{\lambda^2} + \beta_2 - 2\beta_1^2 - \frac{2(1-\rho)}{\lambda^2}$$
$$+ \frac{2(1-\rho)}{\lambda\mu} \int_0^1 u^{\frac{\lambda}{\mu}-1} \exp\left\{\frac{\lambda}{\mu} \int_1^u \frac{1-\beta(\lambda-\lambda v)}{\beta(\lambda-\lambda v) - v} dv\right\} du.$$

Note that $E[T_i]$ does not depend on μ. It is consequence of the fact that, in steady state, the rate of the departure flow must be equal to the rate of the input flow.

2.3 Short Description of Some Advanced Retrial Queueing Systems

In complex models of computer and communication systems the repeated attempts are combined with a variety of queueing phenomena leading to a large number of variants and generalizations of the main retrial queues described in Section 2.2. In investigating variants, we should distinguish between basic structural properties which are extended analogues of the derivations for the main models of type $M/M/c$ and $M/G/1$, and in particular significant properties of each specific model. For purposes of illustration, we now briefly describe a few of such variants.

Batch arrivals. It is very common the consideration of communication systems at which customers arrive in batches [248, 460]. In batch arrival retrial queues it is assumed that at every arrival epoch a batch of k primary customers arrives with probability c_k. If $c = 1$ and the channel is busy at the arrival epoch, then the whole group joins the orbit, whereas if the channel is free, then one of the arriving customers starts his service and the rest form sources of repeated calls. The literature dealing with matrix-analytic methods frequently employs a batch Markovian arrival process as input stream [167, 229, 394, 479].

Finite population. It is frequently assumed that the flow of primary arrivals is Poisson. Usually this means that primary arrivals are generated by a very large number of sources and each of them generates primary calls very seldom. From this point of view a model with Poisson input flow is a model with an infinite number of sources. However, in many practical situations it is important to take into account the fact that the rate of generation of new primary calls decreases as the number of customers in the system increases. Examples of this behavior arise from the performance analysis of hybrid fiber-coax, cellular networks and star-like local area networks with collision avoidance circuits [357, 365, 390, 665]. This can be done with the help of finite source models where each individual source generates its own flow of primary demands [111, 289, 290].

Generalized retrial policies. In many applications to telephony a call receiving a busy signal is not allowed to await for the termination of the busy condition. In this context, each blocked call generates a source of repeated requests for service independently of the rest of calls in the orbit. Thus, the classical retrial policy assumes that the probability of a repeated attempt during the interval $(t, t + dt)$, given that j calls are in orbit at time t, is $j\mu dt + o(dt)$. In contrast to this, some applications to computer and communication networks are based on the feature that the time between successive repeated attempts is controlled by an electronic device and, consequently, is independent of the number of customers applying for service, so the probability of a repeated attempt during $(t, t + dt)$, given that the orbit is no empty, is $\alpha dt + o(dt)$. This second type of discipline is called constant retrial policy [79, 181, 293, 298]. Both models can be treated in a unified way by defining the linear retrial policy [58]. We also mention the consideration of a quadratic retrial policy [81], and a retrial policy dealing with an infinitely increasing retrial rate [167].

A variant of the constant retrial policy consists in assuming a full access rule [73] defined as follows: when there are $j \geq 1$ customers in orbit, a signal is sent out in accordance with an exponential law of rate α and the number k of idle servers is reported back; then $\min(j,k)$ customers in orbit are taken for service. Such a full access rule contributes to a better use of the system resources. A more complicate policy of full access type can be found in [174].

2.3 Short Description of Some Advanced Retrial Queueing Systems   33

Multiclass queues. In the main models it is assumed homogeneity of characteristics such as the arrival process, the service time and inter-retrial time distributions. However, in practice, these characteristics may differ widely for different subscriber groups. This leads to multiclass retrial queues [120, 261, 336, 363, 444, 524]. Multiclass models are far more difficult for mathematical analysis than single class models because now the joint queue length process is a random walk on the multidimensional integer lattice $\mathbb{Z}_+^n$ rather than on $\mathbb{Z}_+$.

Negative arrivals and disasters. Whereas an ordinary customer in the case of blocking increases by one the orbit size, a negative arrival has the effect of a signal that induces an ordinary customer in orbit, if any is present, to leave immediately the system [69, 140, 426]. An extension of this first description results in retrial queues with disasters [63, 224, 610], where a negative arrival removes all the work from the system.

Nonpersistent customers. Suppose that a calling subscriber after some unsuccessful retrials decides to abandon the system. This practical variant can be modelled with the help of the so-called persistence function $\{H_j; j \geq 0\}$, where H_j represents the probability that after the jth attempt fails, a subscriber will make the $(j+1)$th one. For further details, see [288, Sections 3.3 and 4.2], [697] and references therein.

Priorities. The basic retrial queue with priority classes is characterized by the existence of two different types of primary customers who arrive according to independent streams. Demands from the first stream are identified as high priority customers, and they are queued and served according to some discipline. In the case of blocking, any low priority customer (from the second stream) immediately leaves the service area and subsequently retries until he finds the server free. As a consequence, high priority customers have non-preemptive priority over low priority customers [179, 190, 284]. A major motivation for a preemptive rule, where the repeated demands have priority over the waiting line [78], comes from daily life situations where a single agent must simultaneously attend to a waiting line and a telephone.

Retrials due to balking, impatience and breakdowns. Most queueing systems with retrials are motivated by computer and telecommunication applications where a repeated attempt appears due to blocking in a system with limited service capacity. However, the existence of retrials can be due to other reasons. For instance, [164, 299] study a single server system with retrials where the repeated attempts are due to impatience of the queueing customers. A second possibility is provided by the consideration of mixed models with waiting line and orbit [287], where a customer finding a long queue upon arrival may decide to attend another secondary job and come back later hoping to find a shorter queue. As related work, we mention the existence of models where retrials are due to breakdowns of unreliable servers [599].

Server breakdowns. As in standard queueing models, the servers of a retrial queue might be subject to failures and repairs. In standard queues it is generally assumed that the customer whose service is interrupted will be able to complete his service when the interruption is cleared. Three different effects of interruption policies can be handled: preemptive resume, preemptive repeat identical and preemptive repeat different. The behavior of the interrupted customers in a retrial queue is essentially different. They might either leave the system or join the retrial group. But no order is assigned to customers in the retrial group so the new independent service of the displaced customer will begin as soon as he gains service again.

Retrial queues that take into account servers failures and repairs were introduced by Aissani [13] and Kulkarni and Choi [449]. As a selection of the related literature we mention [18, 38, 428, 479, 599, 699].

The retrial literature is vast and rich so it is possible to find a great number of variants and generalizations including discrete-time queues [30, 112, 128, 135, 187, 472, 474, 553, 656, 701], overloading systems [39, 40], polling systems [459, 461, 462], retrial queueing networks [91, 320, 525, 570, 655], vacations [57, 96, 193, 458, 471], etc. The interested reader may find useful material about the above variants and many other retrial models in the monograph [288], the papers [64, 65, 322] and references therein.

2.4 Bibliographical Notes

The monograph of Falin and Templeton provides a comprehensive reference and gives a good account of analytical results for the main retrial models of type $M/G/1$ [288, Chapter 1] and $M/M/c$ [288, Chapter 2]. The emphasis is put on those performance measures and mathematical methods yielding concise expressions (i.e., explicit formulas or transform solutions) and on the asymptotic analysis and extensions to more complicate retrial models [288, Chapters 3 and 4]. The examination of algorithmic aspects is taken into account, mainly in the context of the $M/M/c$ retrial queue, but it is less involved than corresponding considerations in this book. Some other textbooks include chapters or sections providing a general overview of the main retrial features [158, 160, 450, 512, 581, 649, 689]. For a review of the main results and literature the reader is also referred to the survey papers [64, 65, 82, 268, 279, 322, 451, 696].

The survey papers [268, 279, 451, 696] summarize many of the developments in the analysis of retrial queues appeared until 1997. The papers by Falin [268, 279] give a general overview not only of the main models but also of many variants and generalizations. Yang and Templeton [696] focus on the embedded Markov chain approach and the stochastic decomposition property. In [451] some open problems are presented making emphasis on stochastic monotonicity and optimal control problems. On the other hand, Artalejo and Falin [82] concentrate on the comparative analysis between standard and retrial queues. In fact, the treatment given in this chapter owes most to reference [82]. The papers [64, 65, 322] provide bibliographical information. In [64] the criterion is an easy accessibility so the list only includes books and papers published in journals written in English. The readers interested in translated papers, languages other than English, theses, and proceedings of conferences are referred to the monograph [288]. The paper [65] presents a classified bibliography of the work performed in the decade 1990-1999. In a similar manner, the paper [322] claims to be a bibliographical guide to the use of matrix-analytic techniques in retrial queues. Finally, two more surveys summarize most of the developments in the analysis of retrial queues with a finite population [59] and retrial queues with priorities [190].

The influence of the retrial rate in the classification of states of the $M/M/1$ retrial queue is an interesting fact which has been discussed in detail in [288, Subsection 1.3.1]. On the other hand, the key to investigate the stability of the $M/G/1$ retrial queue is the study of the positive recurrence condition for the embedded Markov chain at departure epochs. Since the arrival stream is a Poisson process, it follows from Burke's theorem (see [202, Section 5.3] or [450, Theorem 7.1]) that the limiting probabilities P_{ij} exist and are positive if and only if the embedded chain is positive recurrent.

Most vacation models deal with the case of exhaustive policies [221, 308, 653, 658]; that is, the system must be empty when the server starts a vacation. In contrast to this, the retrial queues provide an interesting example of vacation models where after each service completion the server remains idle despite there are customers in the system [49, 86, 302]. The combination of the retrial phenomenon and classical vacation rules like the N and T-policies yields a double stochastic decomposition property [57]. Its probabilistic interpretation leads to three contributions which are the result of the classical vacation, the repeated attempts and the system size of the standard queue without vacations.

There are various elementary algorithms for the numerical integration of functions. For instance, in Figure 2.6, we use a trapezoidal rule (subroutine TRAPZD in [573, Chapter 4]) to numerically evaluate the integral in the formula for π_0.

Part II

Computational Analysis of Performance Descriptors

3
Limiting Distribution of the System State

This chapter deals with the computational analysis of the limiting distribution of the system state in some basic retrial queues. In Section 3.1, we present recursive schemes (embedded Markov chain approach and regenerative approach) and an alternative approach based on maximum entropy techniques for the computation of the limiting probabilities in the main $M/G/1$ retrial queue. Sections 3.2 and 3.3 extend the study to models with general retrial times and the discrete-time case, respectively. Section 3.4 discusses approximations for the main $M/M/c$ retrial queue based on finite and infinite models (see Subsections 3.4.1 and 3.4.2). In Subsections 3.4.3-3.4.5, other approximating assumptions are introduced. We then discuss numerically the goodness of these approximations. In Section 3.5, we show how to deal with a multiserver model with finite population. The bibliographical comments are given in Section 3.6.

We stress that the mathematical descriptions and notations regarding to the $M/G/1$ and $M/M/c$ retrial queues were given in Section 2.1.

3.1 The $M/G/1$ Retrial Queue

3.1.1 The Embedded Markov Chain Approach

In Subsection 2.2.2, we indicated how the computation of the limiting probabilities $\{P_{ij}; i \in \{0,1\}, j \geq 0\}$ can be reduced to find the limiting distribution $\{\pi_j; j \geq 0\}$ of the number of customers in orbit at the service completion epochs. The next result gives a proof.

Theorem 3.1. *If $\rho = \lambda \beta_1 < 1$, then*

$$P_{0j} = \frac{\lambda}{\lambda + j\mu} \pi_j, \quad j \geq 0, \tag{3.1}$$

$$P_{1j} = \frac{(j+1)\mu}{\lambda} P_{0,j+1}, \quad j \geq 0, \tag{3.2}$$

where $\{\pi_j; j \geq 0\}$ is calculated recursively from the equalities

$$\pi_0 = (1-\rho)\exp\left\{-\frac{\lambda}{\mu}\int_0^1 \frac{1-\beta(\lambda-\lambda u)}{\beta(\lambda-\lambda u)-u}du\right\}, \tag{3.3}$$

$$\pi_j = \sum_{i=0}^{j} \pi_i \frac{\lambda}{\lambda+i\mu} a_{j-i} + \sum_{i=1}^{j+1} \pi_i \frac{i\mu}{\lambda+i\mu} a_{j-i+1}, \quad j \geq 0, \tag{3.4}$$

$$a_j = \int_0^\infty e^{-\lambda x}\frac{(\lambda x)^j}{j!}dB(x), \quad j \geq 0. \tag{3.5}$$

Proof. Let η_n be the time at which the nth service completion occurs. The sequence $N_n = N(\eta_n+)$ constitutes an embedded Markov chain. Its one-step transition probability matrix has elements

$$r_{ij} = \frac{\lambda}{\lambda+i\mu} a_{j-i} + \frac{i\mu}{\lambda+i\mu} a_{j-i+1}, \quad i \geq 0, \; j \geq i-1,$$

where a_j corresponds to the probability that j primary customers arrive during a service time and is given by formula (3.5).

Following [288, Section 1.3] we recall that $\{N_n; n \geq 1\}$ is positive recurrent if and only if $\rho < 1$. Since the arrival input is a Poisson process, it follows from the *PASTA* property (Poisson arrivals see time averages, see [688] or [689, Section 5.16]) and Burke's theorem that the limiting probabilities $\{P_{ij}; i \in \{0,1\}, j \geq 0\}$ exist and are positive under the same condition $\rho < 1$.

The Kolmogorov equations $\pi_j = \sum_{i=0}^{\infty} \pi_i r_{ij}$, for $j \geq 0$, lead to equation (3.4). We now observe that the generating function of $\{\pi_j; j \geq 0\}$ is

$$\Pi(z) = \frac{(1-\rho)(1-z)\beta(\lambda-\lambda z)}{\beta(\lambda-\lambda z)-z}\exp\left\{-\frac{\lambda}{\mu}\int_z^1 \frac{1-\beta(\lambda-\lambda u)}{\beta(\lambda-\lambda u)-u}du\right\}.$$

Thus, putting $z = 0$ in $\Pi(z)$ we get formula (3.3) for π_0.

Equation (3.2) follows by equating the flow rate into and the flow rate out of the subset $\{(i,k); i \in \{0,1\}, k \leq j\}$ and after an appeal to *PASTA* property. Finally, we use the fact that $\pi_j = P_{0j} + (1-\delta_{0j})P_{1,j-1}$ and (3.2) to obtain the expression (3.1). $\square$

The computation of the integral in (3.5) is reduced to finite sums for the case of the most usual service time distributions; see also Section 3.6. We remark the following relationships which will be useful in the sequel:

$$a_0 = 1 - \lambda \hat{a}_0,$$
$$a_j = \lambda(\hat{a}_{j-1} - \hat{a}_j), \quad j \geq 1,$$

where $\hat{a}_j$ is defined as

$$\hat{a}_j = \int_0^\infty e^{-\lambda x}\frac{(\lambda x)^j}{j!}(1-B(x))dx, \quad j \geq 0.$$

We next list some cases where tractable expressions for $\hat{a}_j$ can be given.

(i) *Deterministic service times.* Assume that the service time is a constant $D > 0$. Then, we have

$$\hat{a}_j = \frac{1}{\lambda}\left(1 - \sum_{i=0}^{j} e^{-\lambda D}\frac{(\lambda D)^i}{i!}\right), \quad j \geq 0.$$

(ii) *Hyperexponential service times.* Suppose that $B(x)$ is defined as follows:

$$B(x) = \sum_{i=1}^{m} p_i(1 - e^{-\nu_i x}), \quad x \geq 0,$$

where $p_i > 0$ and $\sum_{i=1}^{m} p_i = 1$. In this case, $\hat{a}_j$ is given by

$$\hat{a}_j = \sum_{i=1}^{m} \frac{p_i}{\lambda + \nu_i}\left(\frac{\lambda}{\lambda + \nu_i}\right)^j, \quad j \geq 0.$$

(iii) *Erlang service times.* Suppose that the service time distribution is E_m with $m \in \{1, 2, ...\}$. Then, $\hat{a}_j$ can be expressed as

$$\hat{a}_j = \frac{\lambda^j}{j!(\lambda+\nu)^{j+1}}\sum_{i=0}^{m-1}\frac{(i+j)!}{i!}\left(\frac{\nu}{\lambda+\nu}\right)^i, \quad j \geq 0.$$

3.1.2 The Regenerative Approach

We next use the so-called regenerative approach [662, Subsection 9.2.1] to derive a recursive algorithm for the computation of the limiting probabilities. We will follow the formulation given by De Kok [204], but the approach can be generalized to more complex retrial queues.

The regenerative approach involves a basic result from the theory of regenerative processes, an appeal to *PASTA* property and a simple crossing argument. Then, a stable recursive scheme is derived. Following the system description in [204], we assume that the stream of primary arrivals is modelled as a state dependent Markovian process; that is, the arrival process is Poisson with rate λ_{ij}, when the system is in state (i, j). The orbit capacity K can be finite or infinite. The existence of the limiting probabilities is assumed.

Theorem 3.2. *The limiting probabilities $\{P_{ij}; i \in \{0,1\}, j \in \{0, ..., K\}\}$ can be computed recursively from the equalities*

$$\lambda_{1,j-1}P_{1,j-1} = j\mu P_{0j}, \quad 1 \leq j \leq K, \quad (3.6)$$

$$(1 - (1-\delta_{jK})\lambda_{1j}B_{jj})P_{1j} = (1-\delta_{0j})\sum_{k=1}^{j}\lambda_{1,k-1}\frac{\lambda_{0k}+k\mu}{k\mu}A_{kj}P_{1,k-1}$$
$$+ \lambda_{00}A_{0j}P_{00}, \quad 0 \leq j \leq K, \quad (3.7)$$

$$P_{00} = 1 - \sum_{j=1}^{K} P_{0j} - \sum_{j=0}^{K} P_{1j}, \quad (3.8)$$

where

$$A_{kj} = \frac{\lambda_{0k}}{\lambda_{0k} + k\mu} B_{kj} + \frac{k\mu}{\lambda_{0k} + k\mu} B_{k-1,j}, \quad 0 \leq k \leq j \leq K. \quad (3.9)$$

The auxiliary quantities B_{kj} are defined as the expected amount of time that during a service time j customers are in orbit, given that at the beginning of the service k customers were in orbit.

Proof. Let a regeneration cycle be the time elapsed between two consecutive primary arrivals finding the system empty. The process $\{(C(t), N(t)); t \geq 0\}$ regenerates itself at these epochs. We define some random variables

T : the length of a cycle,
T_{ij} : the amount of time in a cycle during which the system state is (i,j),
N_{0j} : the number of service completions in a cycle at which j customers in orbit are left behind.

By the theory of regenerative processes, the probabilities P_{ij} can be interpreted as the long-run fraction of time that the process $(C(t), N(t))$ is in the state (i,j). In fact, we have

$$P_{ij} = \frac{E[T_{ij}]}{E[T]}, \quad i \in \{0,1\}, \ 0 \leq j \leq K. \quad (3.10)$$

We observe that the number of transitions from state $(0,j)$ is equal to the number of transitions into $(0,j)$ in a cycle $(0,T]$. Since the primary arrival process is Poisson, we use *PASTA* property to get the following relation between $E[T_{0j}]$ and $E[N_{0j}]$:

$$(\lambda_{0j} + j\mu)E[T_{0j}] = E[N_{0j}], \quad 0 \leq j \leq K. \quad (3.11)$$

From the definition of T_{1j} and N_{0j}, and using Wald's identity, we obtain

$$E[T_{1j}] = \sum_{k=0}^{\min(j+1,K)} E[N_{0k}] A_{kj}, \quad 0 \leq j \leq K, \quad (3.12)$$

where A_{kj} is defined as the expected amount of time that during a service time j customers are in orbit, given that the previous service completion left k customers in orbit.

The derivation of (3.6) is parallel to formula (3.2) for the case $\lambda_{ij} = \lambda$ and $K = \infty$. It follows from a simple level crossing argument over the jth level of the orbit.

Observe that $E[T_{00}] = \lambda_{00}^{-1}$, so we find the relation $E[T] = (\lambda_{00} P_{00})^{-1}$. We now combine (3.6) and (3.10)-(3.12) to obtain the recursive relation (3.7). It should be pointed out that $A_{j+1,j}$ does not exit when $j = K < \infty$. Otherwise, we obtain $A_{j+1,j}$ by noting that the service in progress is initiated by the reattempt of a customer in orbit. Further, the probability that this service exceeds time x and that no primary arrivals occur during $(0, x)$ is $e^{-\lambda_{1j}x}(1 - B(x))$ so we have

$$A_{j+1,j} = \frac{(j+1)\mu}{\lambda_{0,j+1} + (j+1)\mu} B_{jj},$$

where $B_{jj} = \int_0^\infty e^{-\lambda_{1j}x}(1 - B(x))dx$.

It remains to specify the calculations of the rest of coefficients A_{kj}, for $0 \leq k \leq j \leq K$. Clearly, the quantities A_{kj} and B_{kj} are connected by the relation (3.9). Observe that an infinitesimal interval $(x, x + dx)$ contributes to B_{kj}, for $j \leq K - 1$, if the service time is still in progress at time x and $j - k$ primary customers arrive at the system during $(0, x)$. The derivation of B_{kK} is similar, but we allow the number of primary arrivals in $(0, x)$ to be greater than or equal to $K - k$.

Finally, we get formula (3.8) for P_{00} by using the normalization condition $\sum_{i=0}^{1} \sum_{j=0}^{K} P_{ij} = 1$. □

We next give explicit expressions for the quantities B_{kj} in a few particular cases.

(i) *The M/G/1 retrial queue.* In this case, we have that

$$B_{kj} = \hat{a}_{j-k}, \quad 0 \leq k \leq j.$$

In Subsection 3.1.1 we considered some common choices for the service time distribution and gave explicit expressions for $\hat{a}_j$, for $j \geq 0$.

(ii) *The M/G/1 retrial queue with finite capacity.* In this case, $\lambda_{1K} = 0$. Now the quantities B_{kj}, for $0 \leq k \leq j \leq K < \infty$, are given by

$$B_{kj} = \begin{cases} \hat{a}_{j-k}, & \text{if } j < K, \\ \sum_{i=K-k}^{\infty} \hat{a}_i, & \text{if } j = K. \end{cases}$$

For the same service time distributions of Subsection 3.1.1, we readily derive the following tractable expressions for B_{kK}:

(ii.1) *Deterministic service times*

$$B_{kK} = D - \frac{1 - \delta_{kK}}{\lambda} \sum_{i=0}^{K-k-1} \left(1 - \sum_{n=0}^{i} e^{-\lambda D} \frac{(\lambda D)^n}{n!}\right), \quad 0 \leq k \leq K.$$

(ii.2) *Hyperexponential service times*

$$B_{kK} = \sum_{i=1}^{m} \frac{p_i}{\nu_i} \left(\frac{\lambda}{\lambda + \nu_i}\right)^{K-k}, \quad 0 \leq k \leq K.$$

(ii.3) *Erlang service times*

$$B_{kK} = \frac{m}{\nu} - \frac{1 - \delta_{kK}}{\lambda + \nu} \sum_{n=0}^{m-1} \frac{1}{n!} \left(\frac{\nu}{\lambda + \nu}\right)^n \sum_{i=0}^{K-k-1} \frac{(i+n)!}{i!} \left(\frac{\lambda}{\lambda + \nu}\right)^i,$$

for $0 \leq k \leq K$.

44   3 Limiting Distribution of the System State

(iii) *Single server system with finite population.* We consider a queueing model in which there are M sources of primary customers. When a source is free it may generate a primary arrival in the interval $(t, t+dt)$ with probability αdt, as $dt \to 0$. The description of the retrial discipline and the service mechanism agrees with the assumptions given for the $M/G/1$ retrial queue. We observe that the arrival rates are then given by $\lambda_{ij} = (M-i-j)\alpha$, for $i \in \{0,1\}$ and $0 \le i+j \le M$, and the orbit capacity is then $K = M-1$. Using the arguments given in Theorem 3.2, we now find

$$B_{kj} = \int_0^\infty \binom{M-k-1}{j-k} \left(1 - e^{-\alpha x}\right)^{j-k} e^{-(M-j-1)\alpha x}(1 - B(x))dx,$$

for $0 \le k \le j \le M-1$.

3.1.3 The Maximum Entropy Approach

Most textbooks on queueing theory provide some 'classical' techniques including the general framework of birth-and-death processes and methods of solution for non-Markovian stochastic processes, such as embedded Markov chains, supplementary variables, matrix-analytic techniques, etc. An elegant alternative is given by information theoretic methods that use the principles of maximum entropy (*PME*) and minimum cross-entropy, if a prior distribution is available, to estimate probability distributions given information in the form of known mean values. We refer the reader to the survey paper by Kouvatsos [424] and references therein.

A novel reader having a first approach to the literature could feed the idea that maximum entropy solutions only provide a reasonable approximation to the true (but complex) queueing system modelled by classical techniques. Such an interpretation of the information theoretic techniques is poor and trivial. The aim of the *PME* is to provide a self-contained method of inference for estimating uniquely an unknown probability distribution. The maximum entropy distribution gives the most random solution; that is, it introduces the minimum additional information beyond what is implied in the original available mean constraints. It should be pointed out that information theoretic analysis neither pretends to replace the classical queueing solutions nor to be an approximation to those classical results. The idea is just to apply the maximum entropy formalism in order to get the widest probability distribution subject to the known constraints. Hence, when in what follows we present classical queueing results (given in terms of exact distributions or numerical inversions) versus maximum entropy solutions, we only claim to display two alternative tools for analyzing an unique real underlying queueing phenomenon. It is so far of our intention to suggest a possible (philosophical or numerical) superiority of the classical methodology over the maximum entropy approach or vice versa.

In this section, we focus on the maximum entropy analysis for the distribution of the system state in the $M/G/1$ retrial queue. However, the maximum entropy approach will be also helpful to analyze discrete-time retrial queues (see Section 3.3) and other queueing descriptors (see Subsections 4.1.1 and 5.1.1).

We next summarize the maximum entropy formalism [424, 616]. The general theory is common for both the discrete and continuous cases. Thus, we simply denote by $f(x)$ the corresponding probability mass function or density function associated with the queueing performance measure under study. We assume that $f(x)$ takes values in a state space $\mathcal{S}$, so we have the normalization condition

$$\int_{\mathcal{S}} f(x)dx = 1. \tag{3.13}$$

The known information about $f(x)$ can be expressed in terms of mean value constraints of the form

$$\int_{\mathcal{S}} F_k(x)f(x)dx = F_k, \quad 1 \le k \le m, \tag{3.14}$$

for known functions $F_k(x)$ and known values F_k. We note that the structural form of the constraints (3.14) covers important special cases such as

(i) $F_k(x) = x^k$ (central moment of order k).
(ii) $F_k(x) = I_{(-\infty, x_k]}(x)$ (value of the distribution function at the point x_k).
(iii) $F_k(x) = e^{-s_k x}$ (value of the Laplace-Stieltjes transform or the moment generating function at the point s_k).

The *PME* states that, of all the distributions satisfying the mean value constraints (3.13) and (3.14), the minimal prejudiced is the one maximizing the Shannon's entropy functional

$$H(f) = -\int_{\mathcal{S}} f(x) \ln f(x) dx. \tag{3.15}$$

Suppose that a prior distribution $g(x)$ is given as current estimate. Then, the principle of minimum cross-entropy generalizes the *PME* by stating that, of all the distributions satisfying the mean constraints, the minimum cross-entropy solution is chosen by minimizing the functional

$$H(f, g) = \int_{\mathcal{S}} f(x) \ln \frac{f(x)}{g(x)} dx. \tag{3.16}$$

In fact, the *PME* corresponds to the particular case when the prior distribution $g(x)$ in (3.16) is uniformly distributed on the state space $\mathcal{S}$.

The maximization of $H(f)$ can be carried out with the help of the method of Lagrange's multipliers. If there exists a distribution that maximizes the entropy (3.15) and satisfies the constraints (3.13) and (3.14), then it has the following form [424, 616]:

3 Limiting Distribution of the System State

$$\hat{f}(x) = \exp\left\{-\hat{\alpha}_0 - \sum_{k=1}^{m} \hat{\alpha}_k F_k(x)\right\}, \quad x \in \mathcal{S}, \quad (3.17)$$

where $\hat{\alpha}_k$, for $0 \leq k \leq m$, are the Lagrangian multipliers. In particular, $\hat{\alpha}_0$ is determined from the normalization condition (3.13), so we obtain

$$\exp\{\hat{\alpha}_0\} = \int_{\mathcal{S}} \exp\left\{-\sum_{k=1}^{m} \hat{\alpha}_k F_k(x)\right\} dx. \quad (3.18)$$

The rest of Lagrangian multipliers satisfy the relations

$$-\frac{\partial \hat{\alpha}_0}{\partial \hat{\alpha}_k} = F_k, \quad 1 \leq k \leq m. \quad (3.19)$$

In general, it is not possible to solve (3.19) for $\hat{\alpha}_k$ explicitly. As an exception, we mention the special case where $m = 1$ and $F_1(x) = x$, which yields the explicit distribution

$$\hat{f}_1(x) = \frac{1}{F_1} e^{-x/F_1}, \quad x \in \mathcal{S}. \quad (3.20)$$

Suppose that we add the second moment as an additional constraint, then the pair $(\hat{\alpha}_1, \hat{\alpha}_2)$ must be computed numerically. To this end, by combining (3.14) and (3.17), we observe that a standard method for finding the optimal $\hat{\alpha}_k$ is to solve the system

$$\int_{\mathcal{S}} (F_i(x) - F_i) \exp\left\{-\sum_{k=1}^{m} \alpha_k (F_k(x) - F_k)\right\} dx = 0, \quad 1 \leq i \leq m. \quad (3.21)$$

The above equations (3.21) for the Lagrangian multipliers are implicit and non-linear. It can be proved then that the problem of solving (3.21) is equivalent to minimizing the potential function

$$F(\alpha_1, ..., \alpha_m) = \ln \int_{\mathcal{S}} \exp\left\{-\sum_{k=1}^{m} \alpha_k (F_k(x) - F_k)\right\} dx, \quad (3.22)$$

or, alternatively, the balanced function

$$G(\alpha_1, ..., \alpha_m) = \sum_{i=1}^{m} p_i \left(\int_{\mathcal{S}} (F_i(x) - F_i) \exp\left\{-\sum_{k=1}^{m} \alpha_k (F_k(x) - F_k)\right\} dx\right)^2, \quad (3.23)$$

where $p_i > 0$ and $\sum_{i=1}^{m} p_i = 1$.

The balanced function $G(\alpha_1, ...\alpha_m)$ in (3.23) takes the value 0 at the optimal solution $(\hat{\alpha}_1, ..., \hat{\alpha}_m)$ which provides a computational advantage over the potential function F in (3.22). For computing the minimum in (3.23) we will employ Nelder and Mead's algorithm [540] which is a method of direct search and does not involve derivatives, thus avoiding problems arising when the Hessian of G is algorithmically almost singular. A complete discussion of this technical problem can be found in [9].

After the preceding preliminaries we are ready to apply the maximum entropy methodology to the distribution of the system state in the $M/G/1$ retrial queue [285]. Firstly, we assume that the available information consists of the marginal distribution of the server state and the partial expectations of the number of customers in orbit, so we know that

$$M_0^0 = 1 - \rho, \quad M_0^1 = \rho, \tag{3.24}$$

$$M_1^0 = \frac{\lambda \rho}{\mu}, \quad M_1^1 = \frac{\lambda^2 \beta_2}{2(1-\rho)} + \frac{\lambda \rho^2}{\mu(1-\rho)}. \tag{3.25}$$

Distinguishing the server state is important in order to provide a more detailed information. Then, the constraints (3.24) play the role of the normalization condition (3.13). According to (3.20), by using (3.24) and (3.25) we expect to find a first order maximum entropy solution $\{\hat{P}_{ij}^1; i \in \{0,1\}, j \geq 0\}$ of geometric type. This is formalized in the following result.

Proposition 3.3. *If the available information is given by M_k^i, for $i \in \{0,1\}$ and $k \in \{0,1\}$, then according to the PME the estimation of the probability distribution of the system state is*

$$\hat{P}_{0j}^1 = \frac{(M_0^0)^2}{M_0^0 + M_1^0} \left(\frac{M_1^0}{M_0^0 + M_1^0} \right)^j, \quad j \geq 0, \tag{3.26}$$

$$\hat{P}_{1j}^1 = \frac{(M_0^1)^2}{M_0^1 + M_1^1} \left(\frac{M_1^1}{M_0^1 + M_1^1} \right)^j, \quad j \geq 0. \tag{3.27}$$

Proof. It is sufficient to consider the case $i = 0$. Applying the method of Lagrangian multipliers we get a solution $\hat{P}_{0j}^1$ of the form

$$\hat{P}_{0j}^1 = uv^j, \quad j \geq 0.$$

Since $\{\hat{P}_{0j}^1; j \geq 0\}$ satisfies the constraints M_0^0 and M_1^0, we find that

$$u = \frac{(M_0^0)^2}{M_0^0 + M_1^0}, \quad v = \frac{M_1^0}{M_0^0 + M_1^0}.$$

This proves the desired expression (3.26). □

According to the geometric structural form, the first order estimations (3.26) and (3.27) are decreasing sequences. Nevertheless, the limiting probabilities $\{P_{ij}; j \geq 0\}$ may have one mode at any arbitrary level of the orbit, we say j_i^*, for $i \in \{0,1\}$. In particular, in the case of the $M/M/1$ retrial queue, the distribution is unimodal [50] and the modes are given by

$$\tilde{j}_0^* = \begin{cases} 0, & \text{if } \lambda\rho < \mu, \\ \left[\frac{(\lambda-\mu)\rho}{\mu(1-\rho)}\right], & \text{if } \lambda\rho \geq \mu, \end{cases}$$

$$\tilde{j}_1^* = \left[\frac{\lambda\rho}{\mu(1-\rho)}\right],$$

where $[x]$ is the integer part of x. We observe that if $(\lambda - \mu)\rho/\mu(1-\rho)$ (respectively, $\lambda\rho/\mu(1-\rho)$) is integer, then $\tilde{j}_0^* - 1$ (respectively, $\tilde{j}_1^* - 1$) is also a mode.

In the light of the information about the modes, it is clear that the first order estimation will be accurate only when $\tilde{j}_0^* = \tilde{j}_1^* = 0$, which amounts to the inequality $\lambda\rho < \mu(1-\rho)$.

To illustrate the above comments, in Table 3.1 we consider an $M/M/1$ retrial queue with a small retrial rate $\mu = 0.05$, so that the distribution is sparse and $\tilde{j}_0^* = \tilde{j}_1^* = 6$. The maximum entropy solution $\hat{P}_{ij}^1$ gives a bad estimation of the probabilities P_{ij}. Hence, the necessity of deriving new estimations of the system state is clear.

Two initial reasons justify the use of two moment estimations. Firstly, in Subsection 2.2.2 it was mentioned that the number of customers in orbit is asymptotically Gaussian, as $\mu \to 0$. This fact agrees with the structural form of the maximum entropy distribution; see (3.17) and (3.18). Furthermore, by treating j as a continuous variable, we easily see that the k-moment estimation has at most $k - 1$ relative extremes.

The second order estimation is based on the partial moments $m_k^i = \sum_{j=0}^\infty j^k P_{ij}$, for $k = 0, 1$ and 2, given by

$$m_k^i = M_k^i, \quad i \in \{0,1\}, \ k \in \{0,1\},$$
$$m_2^0 = M_1^0 + \frac{\lambda}{\mu} M_1^1, \tag{3.28}$$
$$m_2^1 = \frac{\lambda^3 \beta_3}{3(1-\rho)} + \frac{\lambda^4 \beta_2^2}{4(1-\rho)^2} + \frac{\lambda^2 \beta_2}{2(1-\rho)} + \frac{\lambda^3 \beta_2}{2\mu(1-\rho)^2} + \frac{\lambda\rho}{\mu(1-\rho)}$$
$$+ \frac{\lambda^4}{(1-\rho)^2}\left(\frac{\beta_1}{\mu} + \frac{\beta_2}{2}\right)^2 - m_2^0. \tag{3.29}$$

By adapting the maximum entropy formalism to the case under consideration, we see that the Lagrangian multipliers can be obtained by minimizing the potential functions

$$F_i(\alpha_1^i, \alpha_2^i) = \ln \sum_{j=0}^\infty \exp\left\{-\sum_{k=1}^2 \alpha_k^i \left(j^k - \frac{m_k^i}{m_0^i}\right)\right\}, \quad i \in \{0,1\}. \tag{3.30}$$

The computation of the infinite series on the right-hand side of (3.30) implies the consideration of a truncation threshold K which can be determined with the help of Tchebychev's inequality.

3.1 The M/G/1 Retrial Queue

Table 3.1. $M/M/1$ retrial queue with $(\lambda, \mu) = (1.0, 0.05)$ and $\rho = 0.25$

j	P_{0j}	P_{1j}	$\hat{P}^1_{0j}$	$\hat{P}^1_{1j}$	$\hat{P}^2_{0j}$	$\hat{P}^2_{1j}$	$\hat{P}^{2,1}_{0j}$	$\hat{P}^{2,1}_{1j}$
0	0.00237	0.00059	0.09782	0.03125	0.01010	0.00285	0.00186	0.00043
1	0.01189	0.00312	0.08506	0.02734	0.01912	0.00545	0.01141	0.00294
2	0.03121	0.00858	0.07397	0.02392	0.03260	0.00941	0.03265	0.00899
3	0.05723	0.01645	0.06432	0.02093	0.05002	0.01471	0.05984	0.01734
4	0.08226	0.02468	0.05593	0.01831	0.06911	0.02077	0.08341	0.02520
5	0.09872	0.03085	0.04863	0.01602	0.08597	0.02652	0.09719	0.03048
6	0.10283	0.03342	0.04229	0.01402	0.09628	0.03061	0.09975	0.03240
7	0.09549	0.03222	0.03677	0.01227	0.09707	0.03194	0.09277	0.03119
8	0.08057	0.02819	0.03197	0.01073	0.08812	0.03013	0.07944	0.02763
9	0.06266	0.02271	0.02780	0.00939	0.07203	0.02570	0.06317	0.02274
10	0.04543	0.01703	0.02418	0.00822	0.05300	0.01982	0.04687	0.01747
11	0.03097	0.01200	0.02102	0.00719	0.03511	0.01382	0.03253	0.01257
12	0.02000	0.00800	0.01828	0.00629	0.02094	0.00871	0.02116	0.00847
13	0.01231	0.00507	0.01589	0.00550	0.01124	0.00496	0.01290	0.00536
14	0.00725	0.00308	0.01382	0.00481	0.00543	0.00255	0.00738	0.00318
15	0.00411	0.00179	0.01202	0.00421	0.00236	0.00119	0.00396	0.00177
SE	3.05175		3.51236		3.07099		3.05262	

The second order estimations $\hat{P}^2_{ij}$ in Table 3.1 have modes at the seventh level of the orbit. The last row of the table gives the value of Shannon's entropy (3.15) for the limiting distribution and the maximum entropy estimations. As expected, we observe that the entropy decreases when we increase the number of known moments.

It should be noted that the moments m^i_k (see (3.28) and (3.29)) are obtained by taking derivatives of the partial generating function $P_i(z)$, for $i \in \{0, 1\}$, at the point $z = 1$. Hence, it should be interesting to improve the estimation by considering any other constraint providing information related to another different point z_0. To this end, we consider $P_i(z_0)$ which satisfies the structural form described in (3.14). Now the maximum entropy solution has the form

$$\hat{P}^{2,1}_{ij} = \exp\left\{-\left(\hat{\alpha}^i_0 + j\hat{\alpha}^i_1 + j^2\hat{\alpha}^i_2 + z_0^j \hat{\alpha}^i_{2,1}\right)\right\}, \quad i \in \{0,1\}, \quad j \geq 0.$$

The Lagrangian coefficients $(\hat{\alpha}^i_0, \hat{\alpha}^i_1, \hat{\alpha}^i_2, \hat{\alpha}^i_{2,1})$, for $i = 0$ and 1, can be computed after a new appeal to the use of Nelder and Mead's algorithm.

The entries in Table 3.1 show that the estimation improves when we employ $\hat{P}^{2,1}_{ij}$, with $z_0 = 0.55$, instead of $\hat{P}^2_{ij}$. In particular, the estimations $\hat{P}^{2,1}_{ij}$ fit the modes of the limiting probabilities P_{ij}.

Another different possibility is to employ as constraints the relationships (3.2), which express the conservation of flow across the level j of the orbit. The details and some numerical examples can be found in [285].

50   3 Limiting Distribution of the System State

Table 3.2. $M/M/1$ retrial queue with $(\lambda, \mu) = (0.9, 5.0)$ and $\rho = 0.9$

j	P_{0j}	P_{1j}	$\hat{P}^1_{0j}$	$\hat{P}^1_{1j}$	$\hat{P}^2_{0j}$	$\hat{P}^2_{1j}$	$\hat{P}^{2,1}_{0j}$	$\hat{P}^{2,1}_{1j}$
0	0.06606	0.05946	0.03816	0.07745	0.04492	0.07163	0.06453	0.05889
1	0.01070	0.06314	0.02360	0.07078	0.02383	0.06647	0.01386	0.06307
2	0.00568	0.06194	0.01459	0.06469	0.01294	0.06162	0.00556	0.06289
3	0.00371	0.05909	0.00902	0.05912	0.00719	0.05709	0.00315	0.06030
4	0.00265	0.05558	0.00557	0.05403	0.00408	0.05285	0.00215	0.05657
5	0.00200	0.05182	0.00344	0.04938	0.00237	0.04889	0.00163	0.05243
6	0.00155	0.04804	0.00213	0.04513	0.00141	0.04519	0.00131	0.04827
7	0.00123	0.04435	0.00131	0.04125	0.00086	0.04174	0.00109	0.04427
8	0.00099	0.04081	0.00081	0.03770	0.00053	0.03852	0.00092	0.04051
9	0.00081	0.03746	0.00050	0.03445	0.00034	0.03553	0.00078	0.03702
10	0.00067	0.03432	0.00031	0.03149	0.00022	0.03274	0.00067	0.03381
11	0.00056	0.03139	0.00019	0.02878	0.00014	0.03014	0.00058	0.03085
12	0.00047	0.02868	0.00011	0.02630	0.00010	0.02774	0.00050	0.02814
13	0.00039	0.02617	0.00007	0.02404	0.00007	0.02550	0.00043	0.02566
14	0.00033	0.02385	0.00004	0.02197	0.00005	0.02343	0.00037	0.02339
15	0.00028	0.02173	0.00002	0.02008	0.00003	0.02151	0.00032	0.02132
SE	3.52914		3.56020		3.55273		3.53329	

In Table 3.2 we consider a second numerical example in which the system parameters are chosen to fix the traffic intensity $\rho = 0.9$. As a consequence of increasing the value of ρ, the limiting probabilities P_{ij} become sparse. Thus, a good estimation typically demands the use of higher truncation thresholds. For example, to calculate $\{\hat{P}^{2,1}_{1j}; j \geq 0\}$ we take $K = 80$. The maximum entropy solution $\hat{P}^{2,1}_{ij}$ based on the values $P_i(0.55)$, for $i \in \{0, 1\}$, seems to be an accurate estimation. In particular, it fits the modes $j^*_0 = 0$ and $j^*_1 = 1$ of the limiting distribution P_{ij}.

The preceding numerical examples deal with a model with exponential service times. However, this assumption is not restrictive. In [285] similar conclusions are obtained for other service time distributions. In fact, once the mean value constraints are fixed, the formalism is independent of the service time distribution.

We conclude this section with some practical tips for the computation of Lagrangian multipliers. Some numerical results for sensitivity analysis are also presented in Table 3.3. Although the potential function F is a strictly convex function and therefore a Newton-Raphson method should converge for any initial guess $(\alpha_1, ..., \alpha_m)$, there are some practical problems. Due to the exponential kernel of the maximum entropy solution (3.17), F becomes asymptotically linear along some directions. Thus, its Hessian eventually becomes algorithmically singular when the initial guess for the multiplier is chosen close to the asymptotic region. It is also typical that F has a long valley in

Table 3.3. Sensitivity analysis of the function G on the Lagrangian multipliers

	$(\lambda,\mu,\rho) = (1.0, 0.05, 0.25)$		$(\lambda,\mu,\rho) = (0.9, 5.0, 0.9)$	
	$i = 0$	$i = 1$	$i = 0$	$i = 1$
$(\hat{\alpha}_1^i + \varepsilon/2, \hat{\alpha}_2^i + \varepsilon/2)$	68.93654	94.68714	911.19975	1884403.59571
$(\hat{\alpha}_1^i - \varepsilon/2, \hat{\alpha}_2^i + \varepsilon/2)$	52.75803	73.28129	857.83477	1745129.89862
$(\hat{\alpha}_1^i - \varepsilon/2, \hat{\alpha}_2^i - \varepsilon/2)$	72.95420	100.44831	6224.68875	14647723.46844
$(\hat{\alpha}_1^i + \varepsilon/2, \hat{\alpha}_2^i - \varepsilon/2)$	55.66468	77.50923	5413.78114	12915677.37585
$(\hat{\alpha}_1^i + \varepsilon, \hat{\alpha}_2^i, \hat{\alpha}_{2,1}^i)$	0.75063	0.94927	0.57966	4596.29723
$(\hat{\alpha}_1^i - \varepsilon, \hat{\alpha}_2^i, \hat{\alpha}_{2,1}^i)$	0.75457	0.95421	0.60634	4843.18357
$(\hat{\alpha}_1^i, \hat{\alpha}_2^i + \varepsilon, \hat{\alpha}_{2,1}^i)$	177.71642	242.76019	184.56806	2869660.85947
$(\hat{\alpha}_1^i, \hat{\alpha}_2^i - \varepsilon, \hat{\alpha}_{2,1}^i)$	207.74396	285.33873	1189.34103	65279520.68163
$(\hat{\alpha}_1^i, \hat{\alpha}_2^i, \hat{\alpha}_{2,1}^i + \varepsilon)$	0.00025	0.00024	0.00105	0.27528
$(\hat{\alpha}_1^i, \hat{\alpha}_2^i, \hat{\alpha}_{2,1}^i - \varepsilon)$	0.00025	0.00024	0.00105	0.27548

some direction, then the gradient of F is in the direction of the valley, but not necessarily in the direction of the optimal solution.

The discussion for the balanced function G is analogous but we recall that $G(\hat{\alpha}_1, ..., \hat{\alpha}_m) = 0$. Thus, in Table 3.3 we deal with G and discuss the sensitivity of the Lagrangian multipliers based on small changes in some determined directions. For the scenarios in Tables 3.1 and 3.2, we allow the optimal solutions $(\hat{\alpha}_1^i, \hat{\alpha}_2^i)$ and $(\hat{\alpha}_1^i, \hat{\alpha}_2^i, \hat{\alpha}_{2,1}^i)$ to change along ten directions described in the first column of the table. Then, the entries give the value of G for the choice $\varepsilon = 10^{-3}$. This sensitivity analysis shows the very strong incidence of the initial guess. The effect is stronger when we perturb the Lagrangian multiplier $\hat{\alpha}_2^i$ associated with the second order moment, and also when the distribution is very sparse (see the case $i = 1$ for the example with $\rho = 0.9$).

In Chapters 4 and 5 devoted to the busy period and waiting time analysis, we will use again the maximum entropy approach and consequently the numerical discussion will be extended.

3.2 The $M/G/1$ Queue with General Retrial Times

Investigation of retrial queues with non-exponential retrial times is motivated by real computer and telecommunication networks, where retrial times can hardly be exponentially distributed.

For the $M/G/1$ retrial queue, we generalize the classical retrial policy by allowing inter-retrial times to follow a general distribution. The first and foremost reference here is the work of Yang et al. [700], who assume that the time between two consecutive attempts by the same customer in orbit is drawn from a common distribution function $T(x)$, with first moment $1/\mu$, and is independent of other previous attempts and of other random variables involved in the system. The inherent difficulty with non-exponential retrial

52   3 Limiting Distribution of the System State

times is caused by the fact that the queueing model must keep track of the elapsed retrial time for each of possibly a very large number of customers in orbit.

We start by introducing some notation and establishing basic relations at pre-arrival, post-departure and arbitrary epochs. Let η_n^* be the time at which the nth primary customer arrives. Assume that the conditions for the queue to be stable as $t \to \infty$ are satisfied, and define the limiting probabilities

$$\pi_{ij}^* = \lim_{n \to \infty} P(C(\eta_n^* -) = i, N(\eta_n^* -) = j), \quad i \in \{0,1\}, \ j \geq 0.$$

If we apply Burke's theorem and *PASTA* property, then we readily see that

$$\pi_j = \pi_{0j}^* + (1 - \delta_{0j})\pi_{1,j-1}^*, \quad j \geq 0, \tag{3.31}$$
$$P_{ij} = \pi_{ij}^*, \quad i \in \{0,1\}, \ j \geq 0. \tag{3.32}$$

Therefore, finding the limiting probabilities $\{P_{0j} + (1 - \delta_{0j})P_{1,j-1}; j \geq 0\}$ is equivalent to obtain the limiting probabilities $\{\pi_j; j \geq 0\}$.

The basis of much of the analysis that follows is a stochastic decomposition property of the $M/G/1$ retrial queue with general retrial times which further allows us to develop a method for approximating the limiting probabilities $\{P_{ij}; i \in \{0,1\}, j \geq 0\}$ in steady state.

Theorem 3.4. *The generating function of the number of customers in the system in steady state satisfies*

$$P_0(z) + zP_1(z) = P(z)\frac{P_0(z)}{1 - \rho}, \tag{3.33}$$

where $\rho = \lambda\beta_1$ is the probability in steady state that the server is busy, and $P(z)$ corresponds to the generating function of the number of customers in the standard $M/G/1$ queue, which was given in (2.6).

Proof. For the number N_{n+1} of customers in orbit at the time of the $(n+1)$th departure, we have the fundamental equation

$$N_{n+1} = N_n - \delta(N_n, \mathbf{X}^n) + V_{n+1}, \tag{3.34}$$

where $\delta(N_n, \mathbf{X}^n) = 1$ if the $(n+1)$th served customer proceeds from the orbit and $\delta(N_n, \mathbf{X}^n) = 0$ otherwise. Here, $\mathbf{X}^n$ is the vector of elapsed retrial times of the $N_n > 0$ customers in orbit at time η_n+. If $N_n = 0$, then we let $\delta(N_n, \mathbf{X}^n)$ be 0 with probability one.

Since $N_n - \delta(N_n, \mathbf{X}^n)$ and V_{n+1} are independent, (3.34) leads to

$$E\left[z^{N_{n+1}}\right] = E\left[z^{N_n - \delta(N_n, \mathbf{X}^n)}\right] E\left[z^{V_{n+1}}\right],$$

from which it follows, by (3.31) and (3.32), that

3.2 The M/G/1 Queue with General Retrial Times

$$P_0(z) + zP_1(z) = \beta(\lambda - \lambda z) \lim_{n \to \infty} E\left[z^{N_n - \delta(N_n, \mathbf{X}^n)}\right]. \quad (3.35)$$

To determine the limit in (3.35), we define the joint distribution of $(N_n, \mathbf{X}^n)$ by $f_j^n(x_1, ..., x_j)$ on $\{N_n = j\}$ and, for simplicity, we write $f_j^n(\mathbf{x})$ instead of $f_j^n(x_1, ..., x_j)$. Then, we get

$$E\left[z^{N_n - \delta(N_n, \mathbf{X}^n)}\right] = \sum_{j=0}^{\infty} \int_{[0,\infty)^j} E\left[z^{N_n - \delta(N_n, \mathbf{X}^n)} \middle| (N_n, \mathbf{X}^n) = (j, \mathbf{x})\right] f_j^n(\mathbf{x}) d\mathbf{x}$$

$$= \sum_{j=0}^{\infty} z^{j-1} \int_{[0,\infty)^j} (zP(\delta(N_n, \mathbf{X}^n) = 0|(N_n, \mathbf{X}^n) = (j, \mathbf{x}))$$

$$+ 1 - P(\delta(N_n, \mathbf{X}^n) = 0|(N_n, \mathbf{X}^n) = (j, \mathbf{x}))) f_j^n(\mathbf{x}) d\mathbf{x}$$

$$= \sum_{j=0}^{\infty} z^{j-1} \int_{[0,\infty)^j} f_j^n(\mathbf{x}) d\mathbf{x} + \left(1 - \frac{1}{z}\right) \sum_{j=0}^{\infty} z^j$$

$$\times \int_{[0,\infty)^j} P(\delta(N_n, \mathbf{X}^n) = 0|(N_n, \mathbf{X}^n) = (j, \mathbf{x})) f_j^n(\mathbf{x}) d\mathbf{x},$$

where the integral $\int_{[0,\infty)^j} f_j^n(\mathbf{x}) d\mathbf{x}$ equals $P(N_n = j)$. Similarly, we may notice that

$$\int_{[0,\infty)^j} P(\delta(N_n, \mathbf{X}^n) = 0|(N_n, \mathbf{X}^n) = (j, \mathbf{x})) f_j^n(\mathbf{x}) d\mathbf{x}$$

corresponds to the joint probability that $\{N_n = j\}$ and the $(n+1)$th served customer proceeds from the outside.

Observe that an arriving primary customer finds the state $(0, j)$ if and only if the last departure left behind j customers in orbit and a primary arrival occurred before any of the j customers in orbit retried for service. Thus, by letting $n \to \infty$, we have

$$\pi_{0j}^* = \lim_{n \to \infty} \int_{[0,\infty)^j} P(\delta(N_n, \mathbf{X}^n) = 0|(N_n, \mathbf{X}^n) = (j, \mathbf{x})) f_j^n(\mathbf{x}) d\mathbf{x}, \quad j \geq 0.$$

Thus, using (3.32) we obtain that the limit claimed in (3.35) is precisely

$$\frac{1}{z}(P_0(z) + zP_1(z)) + \left(1 - \frac{1}{z}\right) P_0(z).$$

Substituting in (3.35) and solving for $P_0(z) + zP_1(z)$ yields (3.33), which completes the proof. □

Theorem 3.4 shows that the well-known stochastic decomposition property of the $M/G/1$ retrial queue still holds when retrial times are no longer exponentially distributed, but does not compute the limiting probabilities $\{P_{ij}; i \in \{0,1\}, j \geq 0\}$. To this end, Yang et al. [700] develop an effective approximation based on the forward recurrence time from the renewal theory. More precisely, they propose to approximate the distribution function of the elapsed retrial time by the limiting distribution of the forward recurrence time given by

$$m(x) = \mu \int_0^x (1 - T(u))du, \quad x \geq 0.$$

Therefore, $f_j^n(\mathbf{x})$ is expected to be well approximated by

$$\hat{f}_j^n(\mathbf{x}) = P(N_n = j)\mu^j \prod_{i=1}^{j}(1 - T(x_i)). \tag{3.36}$$

Theorem 3.5. *By assuming the approximating assumption (3.36), we obtain the following approximations $\{\hat{P}_{ij}; i \in \{0,1\}, j \geq 0\}$ of the exact limiting probabilities:*

$$\hat{P}_{0j} = w_j b_j v_0, \quad j \geq 0, \tag{3.37}$$

$$\hat{P}_{1j} = v_{j+1} - \hat{P}_{0,j+1}, \quad j \geq 0, \tag{3.38}$$

where $w_0 = 1$ and

$$w_j = \frac{1}{(1-\rho)(1-b_j)} \sum_{k=1}^{j} P_k b_{j-k} w_{j-k}, \quad j \geq 1,$$

$$b_j = \int_0^\infty \lambda e^{-\lambda x}(1 - m(x))^j dx, \quad j \geq 0,$$

$$v_0 = \left(\sum_{j=0}^{\infty} w_j\right)^{-1},$$

$$v_j = w_j v_0, \quad j \geq 1.$$

Proof. Since we may write down

$$\pi_{0j}^* = \lim_{n \to \infty} \int_{[0,\infty)^j} f_j^n(\mathbf{x}) \int_0^\infty \lambda e^{-\lambda t} \prod_{i=1}^{j} \frac{1 - T(x_i + t)}{1 - T(x_i)} dt d\mathbf{x},$$

the approximation (3.36) and equalities (3.31) and (3.32) lead after some algebra to

$$\hat{\pi}_{0j}^* = \hat{P}_{0j} = (P_{0j} + (1 - \delta_{0j})P_{1,j-1})b_j, \quad j \geq 0. \tag{3.39}$$

Equation (3.39) gives an approximate relationship for $j \geq 1$, but it becomes exact for $j = 0$. A second exact relation between the sequences $\{\pi_{0j}^*; j \geq 0\}$ and $\{P_{0j} + (1 - \delta_{0j})P_{1,j-1}; j \geq 0\}$ can be derived from Theorem 3.4. By virtue of (3.33), we have

3.2 The $M/G/1$ Queue with General Retrial Times

$$P_{0j} + (1-\delta_{0j})P_{1,j-1} = \frac{1}{1-\rho}\sum_{k=0}^{j} P_k \pi_{0,j-k}^*, \quad j \geq 0.$$

Therefore, the problem reduces to solve the system of equations

$$\hat{P}_{0j} = v_j b_j, \quad j \geq 0, \tag{3.40}$$

$$v_j = \frac{1}{1-\rho}\sum_{k=0}^{j} P_k \hat{P}_{0,j-k}, \quad j \geq 0, \tag{3.41}$$

$$\sum_{j=0}^{\infty} v_j = 1, \tag{3.42}$$

where the unknown values of v_j are approximations of $P_{0j} + (1-\delta_{0j})P_{1,j-1}$, for $j \geq 0$.

By using induction it is easy to check that the above system of equations (3.40)-(3.42) has a solution of the form

$$\hat{P}_{0j} = w_j b_j v_0, \quad j \geq 0, \tag{3.43}$$

$$v_0 = \left(\sum_{j=0}^{\infty} w_j\right)^{-1}, \tag{3.44}$$

$$v_j = w_j v_0, \quad j \geq 1. \tag{3.45}$$

This proves that the approximations $\hat{P}_{0j}$ are as claimed in (3.37). Finally, formula (3.38) for $\hat{P}_{1j}$ follows trivially. □

The uniqueness of the solution to the system (3.43)-(3.45), when the retrial time is non zero, can be reduced to show that the series $\sum_{j=0}^{\infty} w_j$ converges (see [700, Section 5]).

Corollary 3.6. *In the case of exponentially distributed retrial times the approximation becomes exact.*

Proof. If $T(x) = 1 - e^{-\mu x}$, for $x \geq 0$, then $m(x) = T(x)$ and $b_j = \lambda/(\lambda+j\mu)$, for $j \geq 0$. Thus, (3.39) turns into

$$\hat{P}_{0j} = (P_{0j} + (1-\delta_{0j})P_{1,j-1})\frac{\lambda}{\lambda+j\mu}, \quad j \geq 0.$$

By (3.31) and (3.32), it immediately results that

$$\hat{P}_{0j} = \pi_j \frac{\lambda}{\lambda+j\mu}, \quad j \geq 0,$$

which implies that $\hat{P}_{0j} = P_{0j}$ by (3.1). Then, formula (3.41) for v_j also becomes exact and $\hat{P}_{1j}$ in (3.38) turns into the exact limiting probability P_{1j}. □

Yang et al. [700] point out that the approximate solution performs well in approximating the mean and the variance of the number of customers in the system.

We report here a set of examples used in [700, Section 7] for $M/E_2/1$ retrial queues with

$$B(x) = 1 - e^{-2x} - 2xe^{-2x}, \quad x \geq 0.$$

We consider that

$$T(x) = 1 - e^{-2\mu x} - 2\mu x e^{-2\mu x}, \quad x \geq 0,$$

for Erlang retrial times, and

$$T(x) = 1 - p_1 e^{-\theta_1 x} - (1-p_1) e^{-\theta_2 x}, \quad x \geq 0,$$

for hyperexponential retrial times. The parameters p_1, θ_1 and θ_2 are determined by

$$p_1 = \frac{1}{K+1}, \quad \theta_1 = \frac{2\mu}{K+1}, \quad \theta_2 = K\theta_1,$$

where $K = c_v^2 + \sqrt{c_v^4 - 1}$ and $c_v > 1$ is the coefficient of variation of the retrial times.

Table 3.4 lists both exact and approximate values of the mean number of customers in the system and of its variance for queues with $c_v = 1.5$ (in the hyperexponential case), $\rho = 0.1, 0.3$ and 0.6, and $\mu = 1.0, 3.3$ and 10.0. Exact results are obtained by using a Gauss-Seidel iterative method; see e.g. [160, Section 3.4] and [700].

Yang et al. [700] attribute the accuracy of their approximation to the fact that retrial and service time distributions are relatively close to the exponential distribution. Numerical results in Table 3.4 seem to agree with the intuition that the approximation is expected to improve as μ approaches to ∞. See [700, Section 7] for more detailed comments revealing basic qualitative properties of the approximate solution.

3.3 The $Geo/G/1$ Retrial Queue

In this section, we study the limiting probabilities of the $Geo/G/1$ retrial queue. The motivation to investigate discrete-time queues is that they are more appropriate than their continuous-time counterparts for modelling many communication and computer systems where the basic units are digital; that is, bits, packets, machine cycle time, etc. These systems where time is slotted include asynchronous transfer mode (ATM), $ALOHA$ protocol, time-division multiple access ($TDMA$) and others. Interested readers are referred to the bibliography [171, 177, 358, 654, 690] for a comprehensive review.

Table 3.4. Mean and variance of the number of customers in the system for $M/E_2/1$ retrial queues

ρ	μ		E_2 retrial times		H_2 retrial times	
			Mean	Variance	Mean	Variance
0.1	1.0	Exact	0.1170	0.1233	0.1216	0.1284
		Approx.	0.1166	0.1230	0.1250	0.1318
	3.3	Exact	0.1106	0.1164	0.1129	0.1192
		Approx.	0.1106	0.1164	0.1136	0.1199
	10.0	Exact	0.1090	0.1147	0.1098	0.1155
		Approx.	0.1090	0.1147	0.1099	0.1156
0.3	1.0	Exact	0.5030	0.6138	0.5463	0.6793
		Approx.	0.4977	0.6260	0.5695	0.7068
	3.3	Exact	0.4263	0.5272	0.4467	0.5502
		Approx.	0.4259	0.5269	0.4522	0.5563
	10.0	Exact	0.4062	0.5014	0.4151	0.5124
		Approx.	0.4062	0.5014	0.4158	0.5129
0.6	1.0	Exact	2.0582	4.1652	2.2736	4.4377
		Approx.	2.0297	4.1048	2.3307	4.5429
	3.3	Exact	1.4999	2.9560	1.6049	3.0954
		Approx.	1.4964	2.9428	1.6239	3.1220
	10.0	Exact	1.3504	2.6270	1.3925	2.6682
		Approx.	1.3466	2.6015	1.3962	2.6716

The $Geo/G/1$ retrial queue was initially investigated by Yang and Li [701]. In what follows, we extend their methodology making emphasis on the computation of the limiting probabilities.

We assume that time is slotted and that queueing events (primary arrivals, retrials and departures) occur at the slot boundaries. For convenience, we suppose that departures occur just before the slot boundaries. In contrast, primary arrivals and retrials occur at the moment immediately after slot boundaries. In what follows, this policy is called generalized early arrival scheme (*G-EAS*). Primary customers arrive according to a geometric arrival process with probability p. If the server is busy, the arriving customer joins the orbit. The time between successive retrials follows a geometric distribution with probability $1 - r$, where $r \in [0, 1)$ corresponds to the probability that a customer in orbit will not make a retrial in a slot. If the server is idle, then in case of simultaneous coalescence of primary customers and retrials at the same slot boundary, one of them occupies the server and the rest of customers go to the orbit. Obviously, the identity of the selected customer does not modify the system state distribution. Service times follow a general distribution $\{s_k; k \geq 1\}$, where s_k is the probability that the service time of a customer consists of k slots. The corresponding generating function is denoted by $\beta(x)$ and the nth factorial moment by $\tilde{\beta}_n$. We also assume that the flow of primary arrivals, the retrials and the service times are mutually independent.

3 Limiting Distribution of the System State

Let $t+$ be the instant immediately after the tth slot. At time $t+$ the system state can be described by the process $X_t = (C_t, \xi_t, N_t)$, where C_t represents the server state (0 when no customer is in service and 1 if the server is busy), and N_t is the number of customers in orbit. If $C_t = 1$, then ξ_t denotes the remaining time of the service in progress. The limiting distributions at times t and $t-$ can be readily derived from that at time $t+$. Thus, we focus on the limiting distribution of $\{X_t; t \geq 0\}$.

It can be shown that $\{X_t; t \geq 0\}$ is an irreducible Markov chain which is positive recurrent if the load of the system defined by $\rho = p\tilde{\beta}_1$ is less than one [90]. Our objective is to investigate the limiting probabilities

$$\pi_s = \lim_{t \to \infty} P(X_t = s), \quad s \in \mathcal{S},$$

where the state space is defined to be $\mathcal{S} = \mathcal{S}_0 \cup \mathcal{S}_1$ with $\mathcal{S}_0 = \{(0, j); j \geq 0\}$ and $\mathcal{S}_1 = \{(1, k, j); k \geq 1, j \geq 0\}$. To this end, we define the generating functions

$$\Pi_0(z) = \sum_{j=0}^{\infty} z^j \pi_{0j},$$

$$\Pi_1(x, z) = \sum_{k=1}^{\infty} \sum_{j=0}^{\infty} x^k z^j \pi_{1kj}.$$

Following Yang and Li [701] we establish the following result.

Theorem 3.7. *If $\rho < 1$, then*

$$\Pi_0(z) = (1 - \rho) \frac{\prod_{k=1}^{\infty} G(r^k z)}{\prod_{k=1}^{\infty} G(r^k)}, \tag{3.46}$$

$$\Pi_1(x, z) = \frac{px(1 - z)(\beta(x) - \beta(\bar{p} + pz))}{(x - \bar{p} - pz)(\beta(\bar{p} + pz) - z)} \Pi_0(z), \tag{3.47}$$

where $\bar{p} = 1 - p$ and

$$G(z) = \bar{p} \frac{\beta(\bar{p} + pz) - (\bar{p} + pz)z}{(\bar{p} + pz)(\beta(\bar{p} + pz) - z)}.$$

Proof. First of all, we consider the Kolmogorov equations

$$\pi_{0j} = \bar{p}r^j \pi_{0j} + \bar{p}r^j \pi_{11j}, \quad j \geq 0,$$

$$\pi_{1kj} = ps_k \pi_{0j} + \bar{p}\left(1 - r^{j+1}\right) s_k \pi_{0,j+1} + ps_k \pi_{11j} + \bar{p}\left(1 - r^{j+1}\right) s_k \pi_{1,1,j+1}$$
$$+ (1 - \delta_{0j})p\pi_{1,k+1,j-1} + \bar{p}\pi_{1,k+1,j}, \quad k \geq 1, \; j \geq 0,$$

and the normalization condition

$$\sum_{j=0}^{\infty} \pi_{0j} + \sum_{k=1}^{\infty} \sum_{j=0}^{\infty} \pi_{1kj} = 1. \tag{3.48}$$

By introducing generating functions, we obtain

$$\Pi_0(z) = \bar{p}\Pi_0(rz) + \bar{p}\Pi_{11}(rz), \tag{3.49}$$

$$\Pi_{1k}(z) = \frac{z-1}{z}ps_k\Pi_0(z) + \frac{\bar{p}+pz}{z}s_k\Pi_{11}(z) + (\bar{p}+pz)\Pi_{1,k+1}(z), \tag{3.50}$$

for $k \geq 1$, where $\Pi_{1k}(z) = \sum_{j=0}^{\infty} z^j \pi_{1kj}$. Hence, using (3.50) we derive

$$\frac{x-\bar{p}-pz}{x}\Pi_1(x,z) = p\frac{z-1}{z}\beta(x)\Pi_0(z) + \frac{\bar{p}+pz}{z}(\beta(x)-z)\Pi_{11}(z). \tag{3.51}$$

Substituting $x = \bar{p} + pz$ in (3.51), we have that

$$\Pi_{11}(z) = \frac{p(1-z)\beta(\bar{p}+pz)}{(\bar{p}+pz)(\beta(\bar{p}+pz)-z)}\Pi_0(z). \tag{3.52}$$

We note that if $\rho < 1$, then the inequality $\beta(\bar{p}+pz) > z$ holds for $0 \leq z < 1$. Furthermore, we have

$$\lim_{z \to 1} \frac{1-z}{\beta(\bar{p}+pz)-z} = \frac{1}{1-\rho}.$$

Thus, the generating function $\Pi_{11}(z)$ is well defined in $z \in [0,1)$ and it can be extended by continuity in $z = 1$. By (3.51) and (3.52), we obtain that $\Pi_1(x, z)$ is as we claimed in (3.47).

We now turn the attention to $\Pi_0(z)$. Substituting $\Pi_{11}(z)$ into (3.49), we get

$$\Pi_0(z) = G(rz)\Pi_0(rz). \tag{3.53}$$

A recursive application of (3.53) leads to

$$\Pi_0(z) = \Pi_0(0) \prod_{k=1}^{\infty} G(r^k z). \tag{3.54}$$

From (3.54) and the normalizing condition (3.48), we obtain $\Pi_0(1) = 1 - \rho$ and the expression (3.46) for $\Pi_0(z)$.

To conclude the proof, we need to show that the infinite product in (3.54) converges to a finite value. Since $G(z)$ can be expressed as $G(z) = 1 + F(z)$, where $F(z) \geq 0$, for $0 \leq z \leq 1$, if $\rho < 1$, we observe that the infinite product converges if and only if the series $\sum_{k=1}^{\infty} F(r^k z)$ converges. The latter is clear since $\lim_{k \to \infty} F(r^{k+1}z)/F(r^k z) = r < 1$. □

Corollary 3.8. *(i) The partial generating function of the number of customers in orbit when the server is busy is given by*

$$\Pi_1(1, z) = \frac{1 - \beta(\bar{p}+pz)}{\beta(\bar{p}+pz)-z}\Pi_0(z). \tag{3.55}$$

(ii) The probability generating function of the total number of customers in the system is given by

$$\Pi_0(z) + z\Pi_1(1,z) = \frac{(1-z)\beta(\bar{p}+pz)}{\beta(\bar{p}+pz)-z}\Pi_0(z).$$

(iii) The partial factorial moments of the limiting distribution up the order two are given by

$$M_0^0 = 1 - \rho, \qquad (3.56)$$

$$M_0^1 = \rho, \qquad (3.57)$$

$$M_1^0 = (1-\rho)\sum_{k=1}^{\infty}\frac{G'(r^k)}{G(r^k)}r^k, \qquad (3.58)$$

$$M_1^1 = \frac{p^2\tilde{\beta}_2}{2(1-\rho)} + \rho\sum_{k=1}^{\infty}\frac{G'(r^k)}{G(r^k)}r^k, \qquad (3.59)$$

$$M_2^0 = (1-\rho)\left(\left(\sum_{k=1}^{\infty}\frac{G'(r^k)}{G(r^k)}r^k\right)^2 + \sum_{k=1}^{\infty}\frac{G''(r^k)G(r^k)-(G'(r^k))^2}{(G(r^k))^2}r^{2k}\right), \qquad (3.60)$$

$$M_2^1 = \frac{\rho}{1-\rho}M_2^0 + \frac{p^4\tilde{\beta}_2^2}{2(1-\rho)^2} + \frac{p^3\tilde{\beta}_3}{3(1-\rho)} + \frac{p^2\tilde{\beta}_2}{1-\rho}\sum_{k=1}^{\infty}\frac{G'(r^k)}{G(r^k)}r^k. \qquad (3.61)$$

The proof of the corollary is standard and thus overlooked. However, at this point we remark that the computation of the preceding formulas for the partial generating functions and the moments M_k^i will become a keystone for the numerical inversion and the maximum entropy approach developed in the sequel.

The next result provides recursive formulas for the limiting probabilities of the Markov chain $\{X_t; t \geq 0\}$.

Proposition 3.9. *If $\rho < 1$, then the limiting probabilities of the system state are given by*

$$\pi_{00} = \frac{1-\rho}{\prod_{k=1}^{\infty}G(r^k)}, \qquad (3.62)$$

$$\pi_{0j} = \frac{\sum_{n=0}^{j-1}(d_{j-n} - r^n b_{j-n})r^{j-n}\pi_{0n}}{(1-r^j)\beta(\bar{p})}, \quad j \geq 1, \qquad (3.63)$$

$$\pi_{1j} = \frac{1-\beta(\bar{p})}{\beta(\bar{p})}\pi_{0j} + \frac{1-\delta_{0j}}{\beta(\bar{p})}\sum_{n=0}^{j-1}d_{j-n}(\pi_{0n}+\pi_{1n}), \quad j \geq 0, \qquad (3.64)$$

where $\pi_{1j} = \sum_{k=1}^{\infty}\pi_{1kj}$ *and*

$$b_m = \sum_{j=m}^{\infty} \binom{j}{m} B_{j+1} \bar{p}^{j+1-m} p^{m+1}, \quad m \geq 1,$$

$$d_m = \sum_{j=m}^{\infty} \binom{j}{m} B_j \bar{p}^{j-m} p^{m+1}, \quad m \geq 1,$$

$$B_j = \sum_{k=j+1}^{\infty} s_k, \quad j \geq 0.$$

Proof. A sketch of the proof is as follows. We consider two auxiliary generating functions defined by

$$\Gamma(z) = \sum_{m=0}^{\infty} z^m \gamma_m = \frac{\bar{p}}{1-z}\left(\frac{\beta(\bar{p}+pz)}{\bar{p}+pz} - z\right),$$

$$\Omega(z) = \sum_{m=0}^{\infty} z^m \omega_m = \frac{\beta(\bar{p}+pz) - z}{1-z}.$$

After an appropriate series expansion of the generating function $\beta(\bar{p}+pz)$ we can express $\Gamma(z)$ and $\Omega(z)$ in the form $\Gamma(z) = b_0 - \sum_{m=1}^{\infty} z^m b_m$ and $\Omega(z) = d_0 - \sum_{m=1}^{\infty} z^m d_m$. Thus, we have $\gamma_0 = b_0$, $\omega_0 = d_0$, $\gamma_m = -b_m$ and $\omega_m = -d_m$, for $m \geq 1$. It can be shown that $\gamma_0 = \omega_0 = \beta(\bar{p})$.

Since $G(z) = \Gamma(z)/\Omega(z)$, we have from (3.53) that

$$\Pi_0(z)\Omega(rz) = \Pi_0(rz)\Gamma(rz).$$

By equating the coefficients of z^j in the series expansion of the above formula, we find that

$$\sum_{n=0}^{j} \pi_{0n} \omega_{j-n} r^{j-n} = \sum_{n=0}^{j} \pi_{0n} r^n \gamma_{j-n} r^{j-n}, \quad j \geq 0,$$

which yields the recursive formula (3.63) for $\{\pi_{0j}; j \geq 1\}$. Formula (3.62) for π_{00} is trivial from (3.46).

Now we rewrite (3.55) as

$$\Pi_1(1,z)\Omega(z) = \Pi_0(z)(1 - \Omega(z)),$$

which leads to

$$\sum_{n=0}^{j} \pi_{1n} \omega_{j-n} = \pi_{0j}(1 - \omega_0) - (1 - \delta_{0j}) \sum_{n=0}^{j-1} \pi_{0n} \omega_{j-n}, \quad j \geq 0.$$

Solving this expression for π_{1j} we derive (3.64). □

The coefficients b_m and d_m of the recursive formulas in Proposition 3.9 are expressed in terms of double infinite series. Hence, substantial simplifications

Table 3.5. $E[N]$ versus r for three discrete-time retrial queues

r	Geo/Geo/1 queue $\rho = 0.25$	$\rho = 0.5$	$\rho = 0.75$	Geo/D/1 queue $\rho = 0.5$	Geo/Pois/1 queue $\rho = 0.5$
0.0	0.04166	0.25000	1.12500	0.16666	0.09836
0.05	0.04261	0.25451	1.13741	0.17137	0.10218
0.1	0.04370	0.25977	1.15239	0.17676	0.10650
0.15	0.04493	0.26590	1.17036	0.18292	0.11140
0.2	0.04633	0.27304	1.19190	0.18999	0.11698
0.25	0.04795	0.28139	1.21772	0.19813	0.12337
0.3	0.04982	0.29119	1.24876	0.20757	0.13074
0.35	0.05199	0.30276	1.28623	0.21859	0.13931
0.4	0.05455	0.31653	1.33180	0.23157	0.14937
0.45	0.05759	0.33310	1.38768	0.24703	0.16132
0.5	0.06126	0.35328	1.45704	0.26572	0.17573
0.55	0.06577	0.37826	1.54438	0.28869	0.19340
0.6	0.07143	0.40983	1.65652	0.31754	0.21555
0.65	0.07873	0.45079	1.80411	0.35477	0.24411
0.7	0.08849	0.50582	2.00490	0.40456	0.28226
0.75	0.10218	0.58333	2.29084	0.47445	0.33576
0.8	0.12274	0.70014	2.72572	0.57947	0.41609
0.85	0.15705	0.89552	3.45838	0.75475	0.55011
0.9	0.22573	1.28725	4.93517	1.10565	0.81831
0.95	0.43188	2.46423	9.38760	2.15897	1.62321
0.99	2.08138	11.88623	45.08252	10.58759	8.06341

Table 3.6. $Geo/D/1$ retrial queue with $(p, D, r) = (0.3, 3, 0.8)$ and $\rho = 0.9$

j	$\tilde{\pi}_{0j}$	$\tilde{\pi}_{1j}$	$\hat{\pi}^1_{0j}$	$\hat{\pi}^1_{1j}$	$\hat{\pi}^2_{0j}$	$\hat{\pi}^2_{1j}$	$\hat{\pi}^{2,1}_{0j}$	$\hat{\pi}^{2,1}_{1j}$
0	0.00842	0.01613	0.02193	0.11906	0.01223	0.04800	0.00855	0.01560
1	0.01502	0.04424	0.01712	0.10331	0.01379	0.05714	0.01470	0.04304
2	0.01730	0.07239	0.01336	0.08964	0.01441	0.06555	0.01734	0.07383
3	0.01607	0.09195	0.01043	0.07778	0.01396	0.07250	0.01626	0.09493
4	0.01314	0.10027	0.00814	0.06749	0.01254	0.07730	0.01324	0.10233
5	0.00987	0.09885	0.00635	0.05856	0.01043	0.07944	0.00986	0.09890
6	0.00699	0.09080	0.00496	0.05081	0.00805	0.07870	0.00693	0.08923
7	0.00475	0.07924	0.00387	0.04409	0.00575	0.07516	0.00469	0.07699
8	0.00312	0.06657	0.00302	0.03826	0.00381	0.06919	0.00309	0.06447
9	0.00200	0.05434	0.00236	0.03319	0.00234	0.06140	0.00199	0.05284
10	0.00126	0.04339	0.00184	0.02880	0.00133	0.05252	0.00126	0.04261
SE	3.02084		3.22128		3.06271		3.02136	

leading to a finite recursive scheme can be expected only for a few simple service times particularizations. We next solve this drawback by employing

Table 3.7. $Geo/Pois/1$ retrial queue with $(p, \lambda, r) = (1 - e^{-0.5}, 0.5, 0.975)$ and $\rho = 0.5$

j	$\tilde{\pi}_{0j}$	$\tilde{\pi}_{1j}$	$\hat{\pi}_{0j}^1$	$\hat{\pi}_{1j}^1$	$\hat{\pi}_{0j}^2$	$\hat{\pi}_{1j}^2$	$\hat{\pi}_{0j}^{2,1}$	$\hat{\pi}_{1j}^{2,1}$
0	0.02588	0.02151	0.12092	0.11543	0.03800	0.03319	0.02632	0.02190
1	0.07244	0.06396	0.09167	0.08878	0.06617	0.05935	0.07118	0.06283
2	0.10530	0.09854	0.06950	0.06828	0.09153	0.08509	0.10581	0.09899
3	0.10580	0.10476	0.05269	0.05252	0.10057	0.09778	0.10660	0.10553
4	0.08255	0.08633	0.03995	0.04039	0.08778	0.09007	0.08251	0.08631
5	0.05329	0.05877	0.03028	0.03107	0.06086	0.06651	0.05290	0.05838
6	0.02960	0.03438	0.02296	0.02389	0.03352	0.03937	0.02940	0.03416
7	0.01454	0.01776	0.01741	0.01838	0.01466	0.01868	0.01454	0.01777
8	0.00644	0.00827	0.01319	0.01413	0.00509	0.00710	0.00651	0.00835
9	0.00261	0.00351	0.01000	0.01087	0.00140	0.00216	0.00266	0.00358
10	0.00098	0.00138	0.00758	0.00836	0.00030	0.00052	0.00100	0.00140
SE	2.71016		3.00493		2.72291		2.71021	

numerical inversion of the partial generating functions and maximum entropy methodologies.

Previously, in Table 3.5, we show the effect of the probability r on the mean number of customers in orbit. To evaluate $E[N] = M_1^0 + M_1^1$ we need to approximate the infinite series in (3.58) and (3.59). By using an integral approximation [656], we obtain the estimation

$$\sum_{k=1}^{\infty} \frac{G'(r^k)}{G(r^k)} r^k \approx \sum_{k=1}^{n_0(\varepsilon)} \frac{G'(r^k)}{G(r^k)} r^k + \frac{1}{1-r} \int_0^{r^{n_0(\varepsilon)+1}} \frac{G'(u)}{G(u)} du$$

$$= \sum_{k=1}^{n_0(\varepsilon)} \frac{G'(r^k)}{G(r^k)} r^k + \frac{\ln G(r^{n_0(\varepsilon)+1})}{1-r},$$

where, for some $\varepsilon > 0$, $n_0(\varepsilon)$ is the first positive integer such that $r^{n_0(\varepsilon)+1} < \varepsilon$.

We first consider a $Geo/Geo/1$ retrial queue with $q = 0.5$; that is, q is the probability of completing the service time in progress in a given slot. Then, we choose p to set $\rho = 0.25$, 0.5 and 0.75. As was expected, $E[N]$ is an increasing function of r and ρ. To illustrate the influence of the service time distribution we consider the systems $Geo/D/1$ and $Geo/Pois/1$ with retrials. For the former we assume that each service time requires $D = 3$ slots. The Poissonian service time distribution follows the law $s_k = e^{-\lambda}\lambda^k/(1-e^{-\lambda})k!$, for $k \geq 1$, with $\lambda = 0.5$.

In Table 3.6 we give a numerical inversion and maximum entropy estimations for the limiting probabilities of a $Geo/D/1$ retrial queue. The system parameters were chosen to fix the load at the value $\rho = 0.9$. We carry out the numerical recovery of probabilities π_{ij} by using the discrete-time fast Fourier transform method [2, 573, 662]. In order to evaluate the generating functions

$\Pi_0(z)$ and $\Pi_1(1,z)$, we use (3.55) and the following integral approximation:

$$\Pi_0(z) \approx \pi_{00} \exp\left\{\sum_{k=1}^{n_0(\varepsilon)} \ln G(r^k z) + \frac{1}{1-r}\int_0^{r^{n_0(\varepsilon)+1}} \frac{\ln G(uz)}{u} du\right\},$$

where

$$\pi_{00} \approx (1-\rho)\exp\left\{-\left(\sum_{k=1}^{n_0(\varepsilon)} \ln G(r^k) + \frac{1}{1-r}\int_0^{r^{n_0(\varepsilon)+1}} \frac{\ln G(u)}{u} du\right)\right\}.$$

The inverse probabilities $\tilde{\pi}_{ij}$ have modes at the points $j_0^* = 2$ (when the server is idle) and $j_1^* = 4$ (server busy). We also give an alternative solution based on the maximum entropy techniques discussed in Subsection 3.1.3.

The mean value constraints under consideration are the moments given in (3.56)-(3.61) and the value of the generating functions $\Pi_0(z)$ and $\Pi_1(1,z)$ at the point $z_0 = 0.6$. The numerical evaluation of the infinite series arising in (3.60) for M_2^0 is based on the following approximation:

$$\sum_{k=1}^{\infty} \frac{G''(r^k)G(r^k) - (G'(r^k))^2}{(G(r^k))^2} r^{2k} \approx \sum_{k=1}^{n_0(\varepsilon)} \frac{G''(r^k)G(r^k) - (G'(r^k))^2}{(G(r^k))^2} r^{2k}$$
$$+ \frac{1}{1-r}\int_0^{r^{n_0(\varepsilon)+1}} \frac{G''(u)G(u) - (G'(u))^2}{(G(u))^2} u\,du.$$

As far as we introduce more constraints the maximum entropy solutions $\hat{\pi}_{ij}$ approach the inverse probabilities. The modes are fitted and Shannon's entropy converges to the entropy of the underlying theoretical distribution.

In Table 3.6 the load ρ is moderately high. Another way to have a sparse distribution consists in choosing a low retrial probability. To this end, in Table 3.7 we consider a $Geo/Pois/1$ queue with $r = 0.975$. We note that the modes are $j_0^* = j_1^* = 3$. The numerical results in the table show once more that the maximum entropy approach provides good estimations of the limiting probabilities.

3.4 The $M/M/c$ Retrial Queue

3.4.1 Approximations Based on Truncated Models

In Subsection 2.2.1 we emphasized the absence of explicit formulas and recursive schemes for the computation of the limiting probabilities of the $M/M/c$ retrial queue when $c > 2$. This difficulty motivates the implementation of approximations numerically tractable and accurate. Here, we first focus on the most natural and traditional procedure consisting in placing a fictitious

3.4 The M/M/c Retrial Queue

limit L in the orbit capacity. In this way, the initial infinite state space $\mathcal{S} = \{0, ..., c\} \times \mathbb{Z}_+$ is replaced by the finite state space $\mathcal{S}^W = \{0, ..., c\} \times \{0, ..., L\}$. Starting from the pioneering paper by Wilkinson [685], this direct truncation scheme has been widely used in the numerical analysis of retrial queues.

Let $\mathcal{X}^W = \{(C^W(t), N^W(t)); t \geq 0\}$ be the Markovian process obtained under the truncation assumption. Since the state space $\mathcal{S}^W$ is finite, the process $\mathcal{X}^W$ is always positive recurrent. The limiting probabilities

$$P_{ij}^W = \lim_{t \to \infty} P\left(C^W(t) = i, N^W(t) = j\right), \quad (i,j) \in \mathcal{S}^W,$$

can be found by solving the following Kolmogorov equations:

$$(\lambda + i\nu + j\mu)P_{ij}^W = \lambda P_{i-1,j}^W + (j+1)\mu P_{i-1,j+1}^W$$
$$+ (i+1)\nu P_{i+1,j}^W, \quad 0 \leq i \leq c-1, \ 0 \leq j \leq L-1, \quad (3.65)$$
$$(\lambda + c\nu)P_{cj}^W = \lambda P_{c-1,j}^W + (j+1)\mu P_{c-1,j+1}^W$$
$$+ \lambda P_{c,j-1}^W, \quad 0 \leq j \leq L-1, \quad (3.66)$$
$$(\lambda + i\nu + L\mu)P_{iL}^W = \lambda P_{i-1,L}^W + (i+1)\nu P_{i+1,L}^W, \quad 0 \leq i \leq c-1, \quad (3.67)$$
$$c\nu P_{cL}^W = \lambda P_{c-1,L}^W + \lambda P_{c,L-1}^W, \quad (3.68)$$

which verify the normalization condition

$$\sum_{i=0}^{c} \sum_{j=0}^{L} P_{ij}^W = 1. \quad (3.69)$$

For notational convenience, we take $P_{ij}^W = 0$, for $(i,j) \notin \mathcal{S}^W$.

Equations (3.65)-(3.68) can be expressed under a block tridiagonal matrix form where each block consists of the unknowns P_{ij}^W, for $0 \leq i \leq c$, corresponding to the level j of the orbit. Formulas (3.65) and (3.67) also lead to tridiagonal linear systems with scalar elements. The algorithmic solution in the following theorem [288, Subsection 2.4.5] exploits this fact.

Theorem 3.10. *The probabilities* $\{P_{ij}^W; (i,j) \in \mathcal{S}^W\}$ *can be computed from the following steps:*

Step 1. Introduce new variables $\{r_{ij}^W; (i,j) \in \mathcal{S}^W\}$ defined by $r_{ij}^W = P_{ij}^W / P_{0L}^W$.
Then $r_{0L}^W = 1$.
Step 2. Put $j = L$. Compute recursively r_{iL}^W, for $i = 1, ..., c$, by

$$r_{iL}^W = \frac{(\lambda + (i-1)\nu + L\mu)r_{i-1,L}^W - \lambda r_{i-2,L}^W}{i\nu}. \quad (3.70)$$

Step 3. Put $j = j - 1$. Calculate the value of r_{cj}^W from

$$r_{cj}^W = \frac{(j+1)\mu}{\lambda} \sum_{i=0}^{c-1} r_{i,j+1}^W, \quad 0 \leq j \leq L-1. \quad (3.71)$$

Compute recursively r_{ij}^W, for $i = c-1, ..., 0$, by

$$r_{ij}^W = \frac{D_{ij} - \gamma_{ij} r_{i+1,j}^W}{b_{ij} + \beta_{0j}}, \tag{3.72}$$

where

$$\alpha_{ij} = -\lambda, \quad \beta_{ij} = \lambda + i\nu + j\mu,$$
$$\gamma_{ij} = -(i+1)\nu, \quad \omega_{ij} = (j+1)\mu r_{i-1,j+1}^W,$$
$$b_{0j} = 0, \quad b_{ij} = \frac{i\nu(b_{i-1,j} + j\mu)}{b_{i-1,j} + \beta_{0j}}, \quad 1 \le i \le c-1,$$
$$D_{0j} = 0, \quad D_{ij} = \omega_{ij} - \frac{\alpha_{ij} D_{i-1,j}}{b_{i-1,j} + \beta_{0j}}, \quad 1 \le i \le c-1.$$

Step 4. Repeat Step 3 until $j = 0$.
Step 5. Calculate the initial probabilities $\{P_{ij}^W; (i,j) \in \mathcal{S}^W\}$ as

$$P_{ij}^W = \frac{r_{ij}^W}{\sum_{(m,n) \in \mathcal{S}^W} r_{mn}^W}. \tag{3.73}$$

Proof. By definition of r_{ij}^W, we have $r_{0L}^W = 1$. Equation (3.70) follows trivially from (3.67). Then, we find the relationship (3.71) by equating the flow rate into and out of the level j of the orbit. Now we observe that equation (3.65) has the tridiagonal form

$$\alpha_{ij} r_{i-1,j}^W + \beta_{ij} r_{ij}^W + \gamma_{ij} r_{i+1,j}^W = \omega_{ij}, \quad 0 \le i \le c-1.$$

The above linear system can be written in the form (3.72), so that it is convenient to calculate r_{ij}^W, for $0 \le i \le c-1$, in reverse order starting from r_{cj}^W. The procedure is completed by repeating Step 3 while $j \ge 1$. Finally, we recover the original unknowns P_{ij}^W from (3.73). □

It is interesting to remark that the main recursion described in Step 3 only deals with algebraic operations involving positive terms, which guarantees that we work with a stable recursive scheme.

In general, the recursive scheme proposed in Theorem 3.10 provides a simple and efficient procedure for the computation of the limiting probabilities. However, at high levels of congestion, a direct truncation approach is computationally very demanding. This occurs when the process $\mathcal{X}^W$ is used to approximate the initial system under heavy traffic ($\lambda \to c\nu$) and/or low retrial rate ($\mu \to 0$). For these special, but important particular cases the truncation level L becomes very large. Several approximate infinite models [85, 262, 547] have been developed to overcome this drawback. We next briefly present an alternative approach due to Stepanov [643] which is usually performed over a more sophisticated finite state space. A discussion on another type of approximate models based on infinite state spaces is postponed to the next subsection.

3.4 The $M/M/c$ Retrial Queue

As a result of dealing with a congested system, the limiting distribution of the system state becomes sparse and emerges the necessity of calculating probabilities of many states of extremely small values. Stepanov [643] suggests the use of more refined state spaces by taking away those states with negligible probability. For example, given a small value ε, it is natural to consider the special case of truncated state spaces comprising states $(i,j) \in \mathcal{S}$ such that $P_{ij} \geq \varepsilon$. Figure 3.1 illustrates some possibilities for the choice of the state space. Obviously, the borders are strongly dependent on the set of system parameters under consideration.

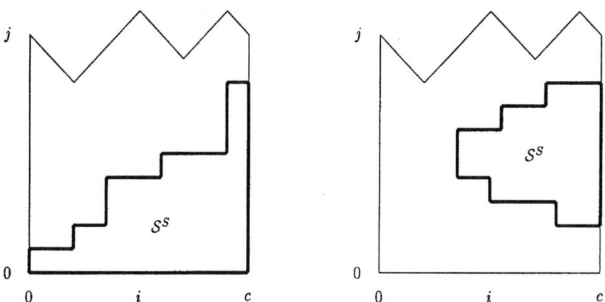

Fig. 3.1. Stepanov's truncated models

We may define the truncated state space $\mathcal{S}^S$ as any arbitrary proper subset of $\mathcal{S}$, whose border $\mathcal{B}$ is defined as the subset of $\mathcal{S}^S$ from which the process $\mathcal{X}$ may leave $\mathcal{S}^S$ in one transition. We also denote as $\mathcal{X}^S = \{(C^S(t), N^S(t)); t \geq 0\}$ the resulting truncated model based on each concrete choice of truncated state space. In what follows, we focus on a concrete specification called 'mixed scheme for the basic case' by Stepanov [643]. First, a traditional truncation of type $\mathcal{X}^W$ is taken into account, so that the orbit is truncated at the level J. It means that those primary customers finding the state (c, J) upon arrival must leave the system. The second step refines the state space by taking special actions at selected epochs. In particular, in the case of the mixed scheme, we add a new extra customer who occupies a free server each time that after a service completion the process $\mathcal{X}$ should move out of the state space $\mathcal{S}^S$. In this way, we arrive to a set of Kolmogorov equations for the limiting probabilities

$$P_{ij}^S = \lim_{t \to \infty} P\left(C^S(t) = i, N^S(t) = j\right), \quad (i,j) \in \mathcal{S}^S,$$

where $\mathcal{S}^S = \{(i,j) \in \mathcal{S}; 0 \leq j \leq J, L(j) \leq i \leq c\}$. Here, $L(j)$ is a non-decreasing integer function defined on $\{0, ..., c\}$. We need to assume that $L(0) = 0$ and $L(j) < c$.

If $(i,j) \in \mathcal{B}$, then we have

$$(\lambda + j\mu)P_{ij}^S = (i+1)\nu P_{i+1,j}^S, \quad L(j) \leq i < c, \tag{3.74}$$

$$c\nu P_{cJ}^S = \lambda P_{c-1,J}^S + \lambda P_{c,J-1}^S. \tag{3.75}$$

On the other hand, if $(i,j) \in \mathcal{S}^S - \mathcal{B}$, then the Kolmogorov equation for (i,j) coincides with the corresponding equation in the original process $\mathcal{X}$.

Equation (3.74) is explained taking into account that state (i,j) belongs to the border $\mathcal{B}$, so we introduce an extra arrival if a service completion occurs. Furthermore, (i,j) is accessible only from $(i+1,j)$ since $(i,j) \in \mathcal{B}$. Similar arguments explain the flow balance in equation (3.75).

For a more exhaustive study, we refer the reader to the paper [643], and references therein. The reader can find there information about how to determine the borders of the truncated state space or how to construct upper bounds for the error caused by the truncation.

In the next subsection, after introducing generalized truncated models, we discuss numerically the accuracy of the different approximate systems through their speed of convergence to the main performance measures of the $M/M/c$ retrial queue.

3.4.2 Approximations Based on Generalized Truncated Models

In this subsection, we use generalized truncated models to improve the numerical computation of the limiting probabilities $\{P_{ij}; 0 \leq i \leq c, j \geq 0\}$. The main idea of a generalized truncation is to approximate an infinite intractable system with the help of another infinite, but calculable system. The fact that we approximate the original infinite system by another infinite one will provide better accuracy than the use of truncated models (see Subsection 3.4.1) based on a finite truncation of the state space. We next investigate three proposals of such generalized truncated models.

Falin [262] introduces a simple generalized model by assuming that the retrial rate becomes infinite when the number of customers in orbit exceeds a certain level M. It means that, from the level M up, the system performs as an standard $M/M/1$ queue with arrival rate λ and service rate $c\nu$. Let $\mathcal{X}^F = \{(C^F(t), N^F(t)); t \geq 0\}$ be the process describing the system state in such an approximate model. We notice that the retrial rate, given that the number of customers in orbit is j, is given by

$$\mu_j = \begin{cases} j\mu, & \text{if } 0 \leq j \leq M, \\ \infty, & \text{if } j \geq M+1. \end{cases}$$

It is easy to see that $\mathcal{X}^F$ is a bivariate continuous-time Markov chain with state space $\mathcal{S}^F = \{0, ..., c-1\} \times \{0, ..., M\} \cup \{c\} \times \mathbb{Z}_+$. The condition $\lambda < c\nu$ is again necessary and sufficient for the positive recurrence of $\mathcal{X}^F$. The state space and transitions among states are shown in Figure 3.2 for the case $c = 3$ and $M = 2$.

For $0 \leq i \leq c$ and $0 \leq j \leq M$, the limiting probabilities

$$P_{ij}^F = \lim_{t \to \infty} P\left(C^F(t) = i, N^F(t) = j\right)$$

coincide up to a normalization constant with the corresponding limiting probabilities of the model $\mathcal{X}^W$

$$P_{ij}^F = \frac{P_{ij}^W}{1 + \frac{\lambda}{c\nu - \lambda} P_{cM}^W}.$$

For the case $i = c$ and $j > M$, we have the relationship [288, Subsection 2.5.2]

$$P_{cj}^F = \left(\frac{\lambda}{c\nu}\right)^{j-M} P_{cM}^F.$$

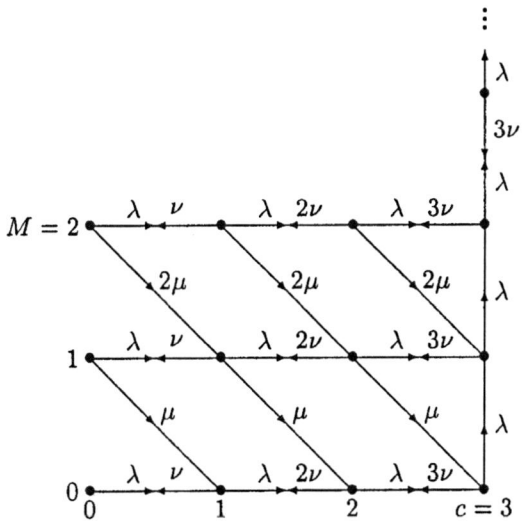

Fig. 3.2. State space and transitions of process $\mathcal{X}^F$

Neuts and Rao [547] give a second possibility to get a numerically tractable approximation. To this end, they assume that the number of customers in orbit who are allowed to conduct retrials is restricted to an appropriate number N, so the retrial rate is

$$\mu_j = \min(j, N)\mu, \quad j \geq 0.$$

This assumption yields an approximate process $\mathcal{X}^{NR} = \{(C^{NR}(t), N^{NR}(t));\ t \geq 0\}$ which can be seen as a quasi-birth-and-death (QBD) process with

70   3 Limiting Distribution of the System State

a large number of boundary states. Since QBD processes have been dealt widely in the literature [464, 544], the methods for determining the positive recurrence condition and for computing the limiting probabilities

$$P_{ij}^{NR} = \lim_{t \to \infty} P(C^{NR}(t) = i, N^{NR}(t) = j), \quad (i,j) \in \mathcal{S}^{NR},$$

with $\mathcal{S}^{NR} = \{0,...,c\} \times \mathbb{Z}_+$, are well investigated.

We notice that $\mathcal{S}^{NR}$ is identical to the initial state space $\mathcal{S}$ of the $M/M/c$ retrial queue. However, the positive recurrence condition $\lambda < c\nu$ does not hold for $\mathcal{X}^{NR}$ because the retrial rate is homogeneous from the level N up. Following the general theory for QBD processes, it can be proved that the process $\mathcal{X}^{NR}$ is positive recurrent if and only if

$$\frac{\lambda + N\mu}{c!} \left(\frac{\lambda + N\mu}{\nu}\right)^c < N\mu \sum_{k=0}^{c} \frac{1}{k!} \left(\frac{\lambda + N\mu}{\nu}\right)^k.$$

We shall return to the underlying QBD structure in Section 7.2.

We next introduce one more generalized method of approximation by Artalejo and Pozo [85]. Let μ_{ij} be the retrial rate when the system state is (i,j). Then, we let μ_{ij} be

$$\mu_{ij} = \begin{cases} \infty, & \text{if } 0 \leq i \leq c-2, \ j \geq K+1, \\ j\mu, & \text{otherwise.} \end{cases}$$

Let $\mathcal{X}^{AP} = \{(C^{AP}(t), N^{AP}(t)); t \geq 0\}$ be the Markov process associated with this truncation assumption. We notice that $\mathcal{X}^{AP}$ takes values on $\mathcal{S}^{AP} = \{0,...,c-2\} \times \{0,...,K\} \cup \{c-1,c\} \times \mathbb{Z}_+$. The infinitesimal generator $\mathbf{Q}^{AP}$ is a variant of the generator $\mathbf{Q}$ given in Section 2.1 for the main $M/M/c$ retrial queue. Specifically, for $i = c-1$ and $j \geq K+1$, its elements are given by

$$q_{(c-1,j)(m,n)}^{AP} = \begin{cases} \lambda, & \text{if } (m,n) = (c,j), \\ (c-1)\nu, & \text{if } (m,n) = (c-1,j-1), \\ j\mu, & \text{if } (m,n) = (c,j-1), \\ -(\lambda + (c-1)\nu + j\mu), & \text{if } (m,n) = (c-1,j), \\ 0, & \text{otherwise.} \end{cases}$$

The rest of infinitesimal transition rates are equal to the corresponding values in the $M/M/c$ retrial queue.

Note that the process $\mathcal{X}^{AP}$ is a natural generalization of the truncation model $\mathcal{X}^F$. Comparing state spaces $\mathcal{S}^F$ and $\mathcal{S}^{AP}$ and transitions given in Figures 3.2 and 3.3, we may conclude that the process $\mathcal{X}^{AP}$ is more similar, in a graphical sense, to the original process $\mathcal{X}$ modelling the $M/M/c$ retrial queue. At this point, we observe that the retrial rate μ_{ij} in process $\mathcal{X}^{AP}$ is non-homogeneous in j. This fact agrees with the non-homogeneous retrial behavior of $\mathcal{X}$, and it is a distinguished feature with respect to the truncated models $\mathcal{X}^F$ and $\mathcal{X}^{NR}$ which are homogeneous from the truncation level and

3.4 The $M/M/c$ Retrial Queue

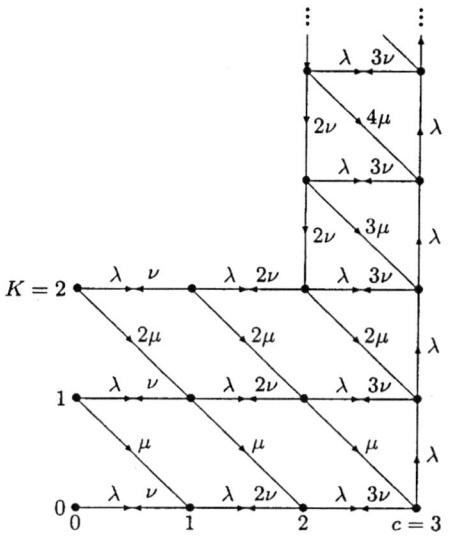

Fig. 3.3. State space and transitions of process $\mathcal{X}^{AP}$

up. These heuristic comments give support to the approximate process $\mathcal{X}^{AP}$. We also notice that $\mathcal{X}^{AP}$ cannot be enlarged to allow transitions to the states of $\{(c-2,j); j > K\}$ because this leads to an analytically intractable model.

The necessary and sufficient condition for the positive recurrence of process $\mathcal{X}^{AP}$ is $\lambda < c\nu$. Under such a condition, the limiting probabilities

$$P_{ij}^{AP} = \lim_{t \to \infty} P\left(C^{AP}(t) = i, N^{AP}(t) = j\right), \quad (i,j) \in \mathcal{S}^{AP},$$

exist and are positive. From the Kolmogorov equations for states $(c-1,j)$ and (c,j), for $j \geq K+1$, we get after some algebra

$$P_{c-1,j}^{AP} = P_{cK}^{AP} \left(\frac{\lambda}{c\nu}\right)^{j-K} \frac{c\nu}{(c-1)\nu + (K+1)\mu} \prod_{n=K+1}^{j-1} \frac{\alpha+n}{\beta+n}, \quad (3.76)$$

$$P_{cj}^{AP} = P_{cK}^{AP} \left(\frac{\lambda}{c\nu}\right)^{j-K} \frac{(c-1)\nu + (j+1)\mu}{(c-1)\nu + (K+1)\mu} \prod_{n=K+1}^{j} \frac{\alpha+n}{\beta+n}, \quad (3.77)$$

where $\alpha = (\lambda + (c-1)\nu)/\mu$ and $\beta = 1 + (c-1)(\lambda + c\nu)/c\mu$. The computation of the limiting probabilities P_{ij}^{AP}, for $j \leq K$, is a minor variant of the recursive scheme described in Theorem 3.10, where L and r_{ij}^W are replaced by K and r_{ij}^{AP}, respectively. Step 2 holds for $1 \leq i \leq c-1$, and r_{cK}^{AP} is computed as

$$r_{cK}^{AP} = \frac{(\lambda + (c-1)\nu + K\mu)r_{c-1,K}^{AP} - \lambda r_{c-2,K}^{AP}}{c\nu + (\lambda(c-1)\nu)/((c-1)\nu + (K+1)\mu)}.$$

72   3 Limiting Distribution of the System State

In the next result, we show how the main performance measures of $\mathcal{X}^{AP}$ can be expressed in terms of hypergeometric functions

$$F(a,b,c;z) = \sum_{j=0}^{\infty} \frac{z^j}{j!} \prod_{k=0}^{j-1} \frac{(a+k)(b+k)}{c+k}.$$

Theorem 3.11. *If the process $\mathcal{X}^{AP}$ is positive recurrent, then:*
  (i) *The partial generating functions $P_i^{AP}(z)$, for $0 \leq i \leq c$, are given by*

$$P_i^{AP}(z) = \sum_{j=0}^{K} z^j P_{ij}^{AP}, \quad 0 \leq i \leq c-2, \tag{3.78}$$

$$P_{c-1}^{AP}(z) = \sum_{j=0}^{K} z^j P_{c-1,j}^{AP} + P_{cK}^{AP} \frac{\lambda z^{K+1}}{(c-1)\nu + (K+1)\mu}$$

$$\times F\left(1, \alpha + K + 1, \beta + K + 1; \frac{\lambda z}{c\nu}\right), \tag{3.79}$$

$$P_c^{AP}(z) = \sum_{j=0}^{K} z^j P_{cj}^{AP} + P_{cK}^{AP} \frac{\lambda z^{K+1}(\alpha + K + 1)}{c\nu(\beta + K + 1)}$$

$$\times \left(F\left(1, \alpha + K + 2, \beta + K + 2; \frac{\lambda z}{c\nu}\right) \right.$$

$$\left. + \frac{\mu}{(c-1)\nu + (K+1)\mu} F\left(2, \alpha + K + 2, \beta + K + 2; \frac{\lambda z}{c\nu}\right) \right). \tag{3.80}$$

  (ii) *The blocking probability $B^{AP} = \sum_{j=0}^{\infty} P_{cj}^{AP}$ and the mean number of customers in orbit $E\left[N^{AP}\right] = \sum_{(i,j) \in \mathcal{S}^{AP}} j P_{ij}^{AP}$ are given by*

$$B^{AP} = P_c^{AP}(1), \tag{3.81}$$

$$E\left[N^{AP}\right] = \sum_{i=0}^{c-2} \sum_{j=0}^{K} j P_{ij}^{AP} + P_{c-1}^{\prime AP}(1) + P_c^{\prime AP}(1). \tag{3.82}$$

Proof. Clearly expression (3.78) is true by definition. Formulas (3.79) and (3.80) follow readily from (3.76) and (3.77) by splitting the orbit coordinate into the subsets $\{0,...,K\}$ and $\{K+1, K+2,...\}$. Formulas (3.81) and (3.82) for B^{AP} and $E[N^{AP}]$ are trivial from the expressions for the partial generating functions. □

Finally, we note that the probability P_{cK}^{AP} can be determined from the normalization condition

$$\sum_{i=0}^{c-2} \sum_{j=0}^{K} P_{ij}^{AP} + P_{c-1}^{AP}(1) + P_c^{AP}(1) = 1,$$

3.4 The $M/M/c$ Retrial Queue   73

and the derivatives of the partial generating functions follow from the property

$$\frac{d}{dz}F(a,b,c;f(z)) = \frac{ab}{c}f'(z)F(a+1,b+1,c+1;f(z)).$$

In order to understand how the generalized truncated models approximate the $M/M/c$ retrial queue, we next compare the main performance characteristics of the truncated process $\mathcal{X}^W$ and the generalized processes $\mathcal{X}^F$, $\mathcal{X}^{NR}$ and $\mathcal{X}^{AP}$. In particular, the following questions should be addressed rigorously:

(i) How to choose an appropriate truncated model?
(ii) How to choose the truncation level?

A good choice of the truncation level guarantees the calculation of any specific performance measure with a desired accuracy. However, it is clear that an appropriate level for a certain truncated model could be excessively large for another different truncated model. Thus, it seems reasonable to start giving an answer to question (i).

Comparing several approximations of any stochastic model, a significant observation is that the superiority of one approximate model depends mainly on the set of system parameters under consideration. For instance, a choice of the parameters leading to heavy load traffic could give support to the superiority of a certain approximation, whereas under light traffic another approximate model may provide a better performance. In what follows, we introduce two criteria for the comparative analysis of the approximate truncated models. For each criterion, we perform numerical experiments in order to establish the domain of parameters giving support to the superiority of each truncated model.

We concentrate on the blocking probability B and the mean number of customers in orbit $E[N]$. These performance characteristics of the truncated models converge to the corresponding characteristics of the $M/M/c$ retrial queue [44, 288, 604]. We propose to evaluate the accuracy through the speed of convergence. In this sense, a lesser level of truncation for reaching the true value of B and $E[N]$ means a faster convergence and, consequently, a better approximation. Accordingly to this spirit, Criterion I proposes to experiment with each truncated model and successively increase the level of truncation until the four first decimal digits are fitted.

Some numerical results for Criterion I are summarized in Tables 3.8 and 3.9 where we deal with a model with $c = 5$. In both tables we normalize the service rate ν to be equal to 1.0 and vary the retrial rate μ and the traffic intensity ρ. For each value of ρ in Table 3.8, we record the truncation levels L, M, N and K such that the associated truncated models (i.e., $\mathcal{X}^W$, $\mathcal{X}^F$, $\mathcal{X}^{NR}$ and $\mathcal{X}^{AP}$) fit the true value of the blocking probability up to its fourth decimal digit. The numbers in bold indicate the lowest truncation levels verifying the criterion. We notice that the truncation level for the process $\mathcal{X}^{NR}$ should be at least 1, whereas the rest of truncated models can take the level 0. In all cases, we observe that the fastest convergence always corresponds to the process

74   3 Limiting Distribution of the System State

Table 3.8. The blocking probability. Criterion I

ρ		$\mu = 0.05$	$\mu = 0.5$	$\mu = 1.0$	$\mu = 2.5$	$\mu = 5.0$
0.2	L	1	2	1	2	2
	M	2	2	5	2	1
	N	1	2	1	1	1
	K	1	0	4	0	0
	B	0.00311	0.00322	0.00329	0.00343	0.00355
0.4	L	10	7	13	7	7
	M	12	8	6	6	7
	N	6	5	10	4	3
	K	9	6	4	4	4
	B	0.04216	0.04488	0.04690	0.05036	0.05309
0.6	L	38	22	18	16	17
	M	39	16	14	13	10
	N	**26**	15	11	9	9
	K	28	10	9	7	4
	B	0.16256	0.17421	0.18275	0.19727	0.20863
0.8	L	136	51	47	39	38
	M	133	48	35	35	28
	N	**100**	31	28	20	18
	K	113	**27**	**20**	8	3
	B	0.41806	0.43948	0.45534	0.48238	0.50347

$\mathcal{X}^{NR}$ and/or $\mathcal{X}^{AP}$. Hence, we may discard the consideration of the models $\mathcal{X}^W$ and $\mathcal{X}^F$. In the light of the numerical data, we recommend to employ the truncated model $\mathcal{X}^{NR}$ when $\rho \in \{0.6, 0.8\}$ and $\mu < 0.5$. In contrast, the process $\mathcal{X}^{AP}$ seems to be superior in the domain $\rho \in \{0.6, 0.8\}$ and $\mu \geq 0.5$. If $\rho \in \{0, 2, 0.4\}$, the truncation levels are similar and moderately small for both processes $\mathcal{X}^{NR}$ and $\mathcal{X}^{AP}$.

Similar conclusions can be derived from Table 3.9, where the mean number of customers in orbit is computed with the help of the semi-explicit formula (2.5). Since $C(t)$ takes values on the finite set $\{0, ..., c\}$, we expect a better approximation when we calculate $E[N]$ through expression (2.5) rather than by using its direct definition.

Artalejo and Pozo [85] propose Criterion II which employs relative errors to polish Criterion I. Firstly, the true value of B is computed by choosing a sufficiently large truncation level in one of the three generalized truncated models. For instance, we experiment with the model $\mathcal{X}^{AP}$ and successively increase the truncation level until finding the first integer T^* such that $|1 - B^{AP}(T^* - 1)/B^{AP}(T^*)| < 10^{-8}$. Then, we take $B = B^{AP}(T^*)$; that is, $B^{AP}(T^*)$ is chosen as the true value of B. In a second step, the first truncation level M^* satisfying $|1 - B^F(M^*)/B| < 10^{-4}$ is determined. The same rule is applied to processes $\mathcal{X}^{NR}$ and $\mathcal{X}^{AP}$ in order to get the critical levels N^* and K^*, respectively.

3.4 The $M/M/c$ Retrial Queue 75

Table 3.9. The mean number of customers in orbit. Criterion I

ρ		$\mu = 0.05$	$\mu = 0.5$	$\mu = 1.0$	$\mu = 2.5$	$\mu = 5.0$
0.2	L	5	3	3	3	3
	M	5	3	4	4	2
	N	3	2	2	2	2
	K	5	3	3	3	1
	$E[N]$	0.06344	0.00734	0.00420	0.00230	0.00165
0.4	L	16	11	10	9	9
	M	19	9	7	6	5
	N	12	8	6	5	4
	K	19	8	6	5	4
	$E[N]$	1.79330	0.22276	0.13446	0.08004	0.06095
0.6	L	47	22	28	17	16
	M	49	20	16	14	12
	N	43	16	21	11	9
	K	46	17	12	10	8
	$E[N]$	11.92468	1.56465	0.98199	0.62264	0.49603
0.8	L	163	57	47	42	47
	M	161	52	41	32	35
	N	143	47	43	32	22
	K	140	39	28	20	20
	$E[N]$	59.25286	8.13197	5.26139	3.50158	2.88891

Table 3.10. The blocking probability. Criterion II

ρ		$\mu = 0.01$	$\mu = 0.1$	$\mu = 1.0$	$\mu = 10.0$	$\mu = 100.0$
0.1	M^*	4	4	3	3	2
	N^*	1	2	3	2	2
	K^*	3	3	3	2	0
0.3	M^*	14	8	7	5	3
	N^*	7	5	5	4	3
	K^*	12	7	5	3	0
0.6	M^*	109	31	16	11	7
	N^*	83	23	13	8	5
	K^*	90	23	11	4	0
0.8	M^*	419	88	36	21	12
	N^*	351	68	28	16	10
	K^*	398	75	20	2	0

Table 3.10 lists the critical values for the three generalized truncated models, in the case $c = 5$. The values correspond to the traffic intensities $\rho = 0.1$, 0.3, 0.6 and 0.8, and the retrial rates $\mu = 0.01$, 0.1, 1.0, 10.0 and 100.0. We present three numbers per value of ρ which correspond, in decreasing order, to M^*, N^* and K^*, respectively. Again, the numbers written in bold give the lowest truncation level.

76   3 Limiting Distribution of the System State

Table 3.11. The mean number of customers in orbit. Criterion II

ρ		$\mu = 0.01$	$\mu = 0.1$	$\mu = 1.0$	$\mu = 10.0$	$\mu = 100.0$
0.1	M^*	4	4	4	3	3
	N^*	2	3	4	3	2
	K^*	4	4	4	3	1
0.3	M^*	15	9	7	5	4
	N^*	9	6	6	5	3
	K^*	14	8	6	4	1
0.6	M^*	108	30	16	11	7
	N^*	84	23	12	8	5
	K^*	103	27	12	6	1
0.8	M^*	410	83	33	20	12
	N^*	344	66	26	15	9
	K^*	349	64	21	7	0

The numerical results show that the lowest rate of convergence is always associated with the process $\mathcal{X}^F$. The comparison between $\mathcal{X}^{NR}$ and $\mathcal{X}^{AP}$ depends on the retrial rate μ. Specifically, we may remark that the process $\mathcal{X}^{AP}$ provides a faster convergence when $\mu \geq 1$. The same conclusions are obtained from Table 3.11 for the mean number of customers in orbit.

Although the generalized truncated model $\mathcal{X}^{NR}$ provides the best approximation when $\mu < 1$, we propose the use of the model $\mathcal{X}^{AP}$. The reason is that running times are significantly lower when we employ $\mathcal{X}^{AP}$; for example, the running time for the matrix-geometric procedure involved in $\mathcal{X}^{NR}$ takes more than 19 minutes to calculate the main performance descriptors for a model with $c = 5$, $\rho = 0.8$, $\mu = 0.05$ and $N = 144$, while running time for the process $\mathcal{X}^{AP}$ is 3 seconds.

Preceding numerical examples in Tables 3.8-3.11 provide a comparative analysis of the performance obtained by employing the approximate models $\mathcal{X}^W$, $\mathcal{X}^F$, $\mathcal{X}^{NR}$ and $\mathcal{X}^{AP}$. Extending the comparison to the truncated model $\mathcal{X}^S$ presents some difficulties. The reason is that the process $\mathcal{X}^S$ relies in a more elaborated construction of the state space $\mathcal{S}^S$, where the role of level of truncation is replaced by the border of $\mathcal{S}^S$ (see Figure 3.1).

The next example is an attempt for comparing the models $\mathcal{X}^S$ and $\mathcal{X}^{AP}$. In Table 3.12, we deal with the expected number of customers in orbit for an $M/M/c$ retrial queue with $c = 50$ servers, service rate $\nu = 1.0$ and retrial rate $\mu = 10.0$. The arrival rate takes the values $\lambda = 40.0$, 45.0 and 49.0, which allows us to manage traffic intensities in the domain $[0.8, 0.98]$. For each value of λ, Table 3.12 lists the following: the true value of $E[N]$ is computed with an appropriate precision of $\varepsilon = 10^{-8}$ (see description of Criterion II), the approximate mean value $E\left[N^S\right]$ comes from Table 2 in [643, Section 7], and $E\left[N^{AP}\right]$ is evaluated accordingly to Criterion II with $\varepsilon = 10^{-5}$. In the light of the relative errors defined by

Table 3.12. Comparison of $E\left[N^S\right]$ and $E\left[N^{AP}\right]$

λ	$E[N]$	$E\left[N^S\right]$	$E\left[N^{AP}\right]$	R_S	R_{AP}
40.0	0.49371	0.49369	0.49371	3.4×10^{-5}	9.1×10^{-6}
45.0	4.10592	4.10588	4.10588	1.0×10^{-5}	9.0×10^{-6}
49.0	46.85110	46.85089	46.85063	4.3×10^{-6}	9.9×10^{-6}

$$R_S = \left|1 - \frac{E\left[N^S\right]}{E[N]}\right| \quad \text{and} \quad R_{AP} = \left|1 - \frac{E\left[N^{AP}\right]}{E[N]}\right|,$$

we conclude that both approximate models offer a similar performance, but the implementation related to $\mathcal{X}^S$ appears to be more sophisticated.

3.4.3 The *RTA* Approximation

In preceding Subsections 3.4.1 and 3.4.2 we approximated the limiting characteristics of the $M/M/c$ retrial queue by means of a variety of truncated models. In such a context, the key point was to deal with tractable approximating queueing models defined over simpler truncated state spaces. In this subsection, we present a different treatment where an approximating assumption is assumed in order to work with a simplified model.

We will approximate the limiting distribution of the number of busy servers defined as

$$P_{i\cdot} = \sum_{j=0}^{\infty} P_{ij}, \quad 0 \leq i \leq c,$$

as well as some characteristics of the quality of service like the mean number of customers in orbit $E[N]$ and the mean number of busy servers $E[C]$.

Our starting point consists of some balance equations obtained by equating flows on the scalar version of the Kolmogorov equations (2.1). First, we equate the flow between adjacent states. Then, we obtain

$$(\lambda + \mu N_i)P_{i\cdot} = (i+1)\nu P_{i+1,\cdot}, \quad 0 \leq i \leq c-1, \tag{3.83}$$

where $N_i P_{i\cdot} = \sum_{j=0}^{\infty} j P_{ij}$, so N_i denotes the conditional expected number of customers in orbit, given that there are i customers at the service facility.

On the other hand, by equating the rates that customers enter and leave the level j of the orbit, we get

$$(j+1)\mu \sum_{i=0}^{c-1} P_{i,j+1} = \lambda P_{cj}, \quad j \geq 0. \tag{3.84}$$

Summing equation (3.84) over j, we obtain

$$\mu \sum_{i=0}^{c} N_i P_{i\cdot} = (\lambda + \mu N_c) P_{c\cdot}. \tag{3.85}$$

The retrial see time averages (RTA) approximation due to Greenberg and Wolff [330, 689] finds a motivation in the well-known $PASTA$ property. Roughly speaking, $PASTA$ says that if the arrival process is Poisson, then the steady state probability that an arriving customer sees the system in state i is the same as the limiting probability that the system is in state i. Now the point is to translate the idea to understand what is happening at the epoch when a customer in orbit completes a repeated attempt.

If the retrial rate μ is small, then the system will have time to approach the steady state when retrials occur, and we expect that the probability that a retrial sees i customers at the service facility would be close to probability $P_{i\cdot}$. Making concrete this idea, we define O_i as the proportion of returning customers who find i customers receiving service; that is, we have that

$$O_i = \frac{\mu \sum_{j=0}^{\infty} j P_{ij}}{\mu \sum_{k=0}^{c} \sum_{j=0}^{\infty} j P_{kj}} = \frac{N_i P_{i\cdot}}{E[N]}, \quad 0 \leq i \leq c. \tag{3.86}$$

The RTA assumption means that

$$O_i = P_{i\cdot}, \quad 0 \leq i \leq c. \tag{3.87}$$

It is obvious from (3.86) that the approximating assumption (3.87) amounts to

$$N_i = E[N], \quad 0 \leq i \leq c.$$

Let $\widehat{E[N]}$ be the approximation of $E[N]$. Now an appeal to (3.83) yields

$$\hat{P}_{i\cdot} = \frac{(\lambda + \mu \widehat{E[N]})^i}{i! \nu^i} \hat{P}_{0\cdot}, \quad 1 \leq i \leq c. \tag{3.88}$$

Furthermore, from the normalizing condition $\sum_{i=0}^{c} \hat{P}_{i\cdot} = 1$, we have that

$$\hat{P}_{0\cdot} = \left(1 + \sum_{i=1}^{c} \frac{(\lambda + \mu \widehat{E[N]})^i}{i! \nu^i}\right)^{-1}.$$

For each positive value $\widehat{E[N]}$, we have an approximate distribution $\{\hat{P}_{i\cdot}; 0 \leq i \leq c\}$. In order to determine a unique approximation, we substitute expression (3.88) for $\hat{P}_{i\cdot}$ into (3.85), so we get

$$\hat{P}_{c\cdot} = \frac{\mu \widehat{E[N]}}{\lambda + \mu \widehat{E[N]}}. \tag{3.89}$$

Then, by combining (3.88) and (3.89), we reduce to the polynomial equation

3.4 The $M/M/c$ Retrial Queue

Table 3.13. *RTA approximations of $P_{0.}$ and B for queues with $(\nu, c) = (1.0, 5)$*

ρ		RTA approximation	$\mu = 0.05$	$\mu = 0.2$	$\mu = 1.0$	$\mu = 10.0$	$\mu = \infty$
0.2	$P_{0.}$	0.36695	0.36701	0.36715	0.36750	0.36778	0.36781
	B	0.00310	0.00311	0.00315	0.00329	0.00365	0.00383
0.4	$P_{0.}$	0.12655	0.12702	0.12816	0.13106	0.13391	0.13432
	B	0.04177	0.04216	0.04320	0.04690	0.05544	0.05970
0.6	$P_{0.}$	0.03305	0.03375	0.03546	0.04015	0.04566	0.04664
	B	0.16086	0.16256	0.16707	0.18275	0.21838	0.23615
0.8	$P_{0.}$	0.00333	0.00362	0.00444	0.00729	0.01191	0.01298
	B	0.41499	0.41806	0.42630	0.45534	0.52146	0.55411

$$\mu \widehat{E[N]} \left(1 + \sum_{i=1}^{c} \frac{(\lambda + \mu \widehat{E[N]})^i}{i! \nu^i} \right) = \frac{(\lambda + \mu \widehat{E[N]})^{c+1}}{c! \nu^c}. \qquad (3.90)$$

It can be shown [52] that the polynomial (3.90) has a single root in $(0, \infty)$. Note also that the *RTA* approximation does not depend on the retrial rate μ.

Greenberg and Wolff [330] consider the *RTA* approximation for a more general multiserver retrial queue with non-persistent customers. In that context, they prove that the approximate mean number of busy servers results in an upper bound of the true value $E[C]$. If we reduce to the $M/M/c$ retrial queue, then it can be easily verified from (3.88) and (3.89) that the approximation becomes exact; that is, we have

$$E[C] = \sum_{i=0}^{c} i P_{i.} = \sum_{i=0}^{c} i \hat{P}_{i.} = \frac{\lambda}{\nu}.$$

In Table 3.13, we compare the true blocking probability $B = P_{c.}$ and the probability that all servers are idle $P_{0.}$, and their *RTA* approximations for a model with $c = 5$, $\nu = 1.0$ and several values of λ and μ. The column corresponding to $\mu = \infty$ lists the descriptor in the standard $M/M/c$ queue. For later use, we let $\hat{B}$ denote the *RTA* approximation of B.

In Table 3.14, we present results for comparing the true mean number of customers in orbit $E[N]$ and two *RTA* estimations. The first estimation of $E[N]$ is obtained by dividing the positive root of (3.90) by μ. The second estimation is evaluated from the semi-explicit formula for $E[N]$ in (2.5) once the value of $Var(C)$ is replaced by the variance of the approximate distribution $\{\hat{P}_{i.}; 0 \leq i \leq c\}$ in (3.88). It is easy to see that such an estimation has the form

$$\widehat{E[N]}_{se} = \widehat{E[N]} \left(1 + \frac{\mu}{\nu} \right).$$

The results in Tables 3.13 and 3.14 exhibit the expected behavior; that is, the *RTA* approximation gives a better estimation as far as μ becomes smaller.

80   3 Limiting Distribution of the System State

Table 3.14. RTA approximations of $E[N]$ for queues with $(\nu, c) = (1.0, 5)$

ρ		$\mu = 0.05$	$\mu = 0.2$	$\mu = 1.0$	$\mu = 10.0$	$\mu = \infty$
0.2	$E[N]$	0.06344	0.01670	0.00420	0.00131	0.00095
	$\widehat{E[N]}$	0.06231	0.01557	0.00311	0.00031	—
	$\widehat{E[N]}_{se}$	0.06542	0.01869	0.00623	0.00342	—
0.4	$E[N]$	1.79330	0.48510	0.13446	0.05085	0.03980
	$\widehat{E[N]}$	1.74383	0.43595	0.08719	0.00871	—
	$\widehat{E[N]}_{se}$	1.83102	0.52315	0.17438	0.09591	—
0.6	$E[N]$	11.92468	3.29534	0.98199	0.42870	0.35422
	$\widehat{E[N]}$	11.50216	2.87554	0.57510	0.05751	—
	$\widehat{E[N]}_{se}$	12.07727	3.45064	1.15021	0.63261	—
0.8	$E[N]$	59.25286	16.67209	5.26139	2.56692	2.21645
	$\widehat{E[N]}$	56.75078	14.18769	2.83753	0.28375	—
	$\widehat{E[N]}_{se}$	59.58832	17.02523	5.67507	3.12129	—

As the reader may see, the approximation is very good when μ is close to 0. In our numerical examples, the true performance descriptors $P_{0\cdot}$, B and $E[N]$ have been computed following the procedure described in Subsection 3.4.2.

We shall return to the relevance of $\hat{B}$, $\widehat{E[N]}$ and $\widehat{E[N]}_{se}$ in estimating B and $E[N]$ at Subsection 3.4.5.

3.4.4 The Fredericks and Reisner Approximation

In contrast to the RTA assumption, Fredericks and Reisner [301] suggest an approximation assumption, denoted from now on by FR assumption, which explicitly depends on the retrial rate μ. Specifically, they rewrite (3.83) as

$$\lambda_i P_{i\cdot} = (i+1)\nu P_{i+1,\cdot}, \quad 0 \leq i \leq c-1, \tag{3.91}$$

with $\lambda_i = \lambda + \mu N_i$, which resembles the recursive formula for the state probabilities on an one-dimensional birth-and-death process. However, care must be taken in this assertion because λ_i corresponds to the average birth rate when there are i customers at the service facility but its computation also depends on the number of customers in orbit.

Clearly, to obtain the probabilities $P_{i\cdot}$ from (3.91) it would be necessary to determine the values of λ_i exactly. Since this can only be done by evaluating the two-dimensional limiting probabilities $\{P_{ij}; (i,j) \in S\}$, which is indeed the problem to avoid, Fredericks and Reisner [301] turn to approximations for the unknown values of λ_i or, similarly, for the rates $\lambda'_i = \mu N_i$.

Note that (3.85) and (3.86) yield

$$\lambda'_i P_{i\cdot} = (\lambda + \lambda'_c) P_{c\cdot} O_i, \quad 0 \leq i \leq c. \tag{3.92}$$

To determine the proportion O_i of returning customers who find i customers receiving service, we need the transient probabilities

$$Q_{mn}(t) = P\left(C(t) = n | C(0) = m\right), \quad 0 \leq m, n \leq c,$$

where such probabilities satisfy the boundary conditions $Q_{mn}(0) = \delta_{mn}$. Although $Q_{mn}(t)$ concerns only with the marginal distribution of the service facility, its study cannot ignore the dependence on the unknown initial distribution of the number of customers in orbit.

Instead of the exact value of O_i, Fredericks and Reisner consider $\mu Q_{ci}^*(\mu)$, where $Q_{ci}^*(\mu)$ is the Laplace-Stieltjes transform of $Q_{ci}(t)$ at point μ. To motivate this assumption, suppose that at time 0, either a primary customer or a returning customer finds the service facility full, which occurs at rate $(\lambda + \lambda_c')P_{c\cdot}$, and consequently joins the orbit. This customer will return after an exponentially distributed amount of time with intensity μ, which does not depend on the transient behavior of the service facility. Then, O_i is approximated by

$$\mu Q_{ci}^*(\mu) = \int_0^\infty Q_{ci}(t) \mu e^{-\mu t} dt, \quad 0 \leq i \leq c.$$

From (3.92), the above assumption amounts to the estimation of λ_i by

$$\tilde{\lambda}_i = \lambda + \tilde{\lambda}_c \frac{\tilde{P}_{c\cdot}}{\tilde{P}_{i\cdot}} \mu Q_{ci}^*(\mu), \quad 0 \leq i \leq c, \tag{3.93}$$

where $\tilde{\lambda}_c$ can be directly evaluated as

$$\tilde{\lambda}_c = \frac{\lambda}{1 - \mu Q_{cc}^*(\mu)}. \tag{3.94}$$

At this point, we show how it is more beneficial the use of recursive formulas for the unknown values of $\tilde{\lambda}_i$ instead of attempting to solve (3.93) and (3.94). Our main tool is the transient analysis of the birth-and-death process defined through (3.91).

Theorem 3.12. *The rates $\tilde{\lambda}_i$ satisfy*

$$\tilde{\lambda}_{i+1} = \lambda + \left(1 + \frac{\mu}{\tilde{\lambda}_i}\right)(\tilde{\lambda}_i - \lambda) + i\nu\left(1 - \frac{\tilde{\lambda}_{i-1}}{\tilde{\lambda}_i}\right), \quad 0 \leq i \leq c-1, \tag{3.95}$$

$$\tilde{\lambda}_c = \frac{\lambda\mu}{c\nu} + \tilde{\lambda}_{c-1}. \tag{3.96}$$

Proof. For ease, we first derive a recursion for the auxiliary values

$$\delta_i = \frac{\tilde{\lambda}_i - \lambda}{\tilde{\lambda}_{i-1} - \lambda}, \quad 1 \leq i \leq c.$$

From (3.93), we readily find

$$\delta_i = \frac{i\nu Q_{ci}^*(\mu)}{\tilde{\lambda}_{i-1} Q_{c,i-1}^*(\mu)}, \quad 1 \leq i \leq c. \tag{3.97}$$

Expressions for the Laplace-Stieltjes transforms $Q_{c,i-1}^*(\mu)$ and $Q_{ci}^*(\mu)$ in (3.97) follow from the derivation of the transient analysis given by Riordan [581, Section 5.3] for the birth-and-death process with birth and death rates defined by $\tilde{\lambda}_i$ and $i\nu$, respectively, and consequently satisfies

$$det(\mathbf{D}) Q_{ci}^*(\mu) = \nu^{c-i} \prod_{k=1}^{c-i} (i+k) D_i, \quad 0 \leq i \leq c, \tag{3.98}$$

where D_i is evaluated as

$$D_0 = 1,$$
$$D_1 = \tilde{\lambda}_0 + \mu,$$
$$D_i = (\tilde{\lambda}_{i-1} + (i-1)\nu + \mu) D_{i-1} - \tilde{\lambda}_{i-2}(i-1)\nu D_{i-2}, \quad 2 \leq i \leq c,$$

and $det(\mathbf{D}) = (c\nu + \mu) D_c - \tilde{\lambda}_{c-1} c\nu D_{c-1}$. We remark here that $\mathbf{D}$ is the square matrix of order $c+1$ governing the usual system of ordinary differential equations for the transient solution of the birth-and-death process.

By introducing (3.98) into (3.97), it is easily seen that

$$\delta_1 = 1 + \frac{\mu}{\tilde{\lambda}_0}, \tag{3.99}$$

$$\delta_{i+1} = \frac{(\tilde{\lambda}_i + i\nu + \mu)\delta_i - i\nu}{\tilde{\lambda}_i \delta_i}, \quad 1 \leq i \leq c-1, \tag{3.100}$$

after some straightforward algebra. Furthermore, for $i = c$, (3.98) and (3.100) yield

$$Q_{cc}^*(\mu) = \frac{\delta_c}{(c\nu + \mu)\delta_c - c\nu}. \tag{3.101}$$

Then, an appeal to the definition of δ_i leads directly to (3.95), by (3.99) and (3.100). Formula (3.96) is easily derived by equating the expressions for $Q_{cc}^*(\mu)$ in (3.94) and (3.101). □

From Theorem 3.12, it becomes easy to estimate the unknowns $\tilde{\lambda}_i$ by solving the system of $c+1$ non-linear equations (3.95) and (3.96) (see e.g. [206]).

Once the values of $\tilde{\lambda}_i$ are evaluated, the approximation of the limiting distribution $\{P_i; 0 \leq i \leq c\}$ is computed from (3.91) as

$$\tilde{P}_i = \nu^{c-i} \prod_{k=1}^{c-i} \frac{i+k}{\tilde{\lambda}_{i+k-1}} \tilde{P}_{c\cdot}, \quad 0 \leq i \leq c-1, \tag{3.102}$$

3.4 The $M/M/c$ Retrial Queue   83

Table 3.15. FR approximations of B and $E[N]$ for queues with $(\nu, c) = (1.0, 5)$

ρ		$\mu = 0.05$	$\mu = 0.2$	$\mu = 1.0$	$\mu = 10.0$	$\mu = \infty$
0.2	B	0.00311	0.00315	0.00329	0.00365	0.00383
	$\tilde{B}$	0.00311	0.00315	0.00328	0.00363	—
	$E[N]$	0.06344	0.01670	0.00420	0.00131	0.00095
	$\widetilde{E[N]}$	0.06341	0.01665	0.00411	0.00115	—
	$\widetilde{E[N]}_{se}$	0.06361	0.01685	0.00431	0.00133	—
0.4	B	0.04216	0.04320	0.04690	0.05544	0.05970
	$\tilde{B}$	0.04207	0.04289	0.04601	0.05452	—
	$E[N]$	1.79330	0.48510	0.13446	0.05085	0.03980
	$\widetilde{E[N]}$	1.78480	0.47605	0.12400	0.03734	—
	$\widetilde{E[N]}_{se}$	1.80321	0.49395	0.14047	0.05257	—
0.6	B	0.16256	0.16707	0.18275	0.21838	0.23615
	$\tilde{B}$	0.16176	0.16432	0.17511	0.21059	—
	$E[N]$	11.92468	3.29534	0.98199	0.42870	0.35422
	$\widetilde{E[N]}$	11.74806	3.11921	0.80775	0.24289	—
	$\widetilde{E[N]}_{se}$	11.99957	3.36575	1.03695	0.44780	—
0.8	B	0.41806	0.42630	0.45534	0.52146	0.55411
	$\tilde{B}$	0.41578	0.41814	0.42990	0.49123	—
	$E[N]$	59.25286	16.67209	5.26139	2.56692	2.21645
	$\widetilde{E[N]}$	57.45191	14.89036	3.54470	0.92977	—
	$\widetilde{E[N]}_{se}$	59.50391	16.92924	5.52130	2.69934	—

where

$$\tilde{P}_{c\cdot} = \left(1 + \sum_{i=0}^{c-1} \nu^{c-i} \prod_{k=1}^{c-i} \frac{i+k}{\lambda_{i+k-1}}\right)^{-1}, \qquad (3.103)$$

and an estimation of the mean number of customers in orbit is obtained as

$$\widetilde{E[N]} = \frac{\tilde{\lambda}_c \tilde{P}_{c\cdot}}{\mu}.$$

An alternative estimation, denoted by $\widetilde{E[N]}_{se}$, is derived by replacing $Var(C)$ in (2.5) by the variance of $\{\tilde{P}_{i\cdot}; 0 \leq i \leq c\}$ computed from (3.102) and (3.103).

Similarly to the RTA approximation, straightforward algebra based on (3.91)-(3.94) allows us to show that the FR assumption leads to

$$E[C] = \sum_{i=0}^{c} i P_{i\cdot} = \sum_{i=0}^{c} i \tilde{P}_{i\cdot} = \frac{\lambda}{\nu}.$$

Next, we obtain some numerical results. As in Subsection 3.4.3, we consider that $c = 5$ and $\nu = 1.0$. In Table 3.15, we compare the true values of the

84   3 Limiting Distribution of the System State

blocking probability B and the mean value $E[N]$, with their FR estimations for several values of λ and μ. It is observed that the FR estimations of B and $E[N]$ behave very well when μ becomes small and ρ is moderately small.

It should be noted that a comparison of Tables 3.13-3.15 shows that, for both descriptors B and $E[N]$, the RTA estimations $\hat{B}$ and $\widehat{E[N]}$ seem to be lower bounds of the FR estimations $\tilde{B}$ and $\widetilde{E[N]}$, which also turn into lower bounds of the true descriptors, irrespective of the values of μ and ρ. Hence, our numerical results make clear the superiority of the FR assumption over the RTA assumption. Results in Table 3.15 also show that the best choice between $\widetilde{E[N]}$ and $\widetilde{E[N]}_{se}$ depends on the values of the system parameters.

3.4.5 Approximations by Interpolation

The notion of approximation by interpolation was introduced by Riordan [581, Section 5.2] to reduce the estimation of performance characteristics in the $M/M/c$ retrial queue to the extreme cases $\mu \to 0$ and $\mu \to \infty$.

The case $\mu \to 0$ is the one in which intervals between repeated attempts are so long that the retrial queue may be regarded as an $M/M/c/0$ loss system with input corresponding to a Poisson process with rate $\Lambda = \lambda + r$, provided that the Poisson processes of primary customers (with rate λ) and of repeated attempts (with unknown rate r) are assumed to be mutually independent. To determine r, we equate the mean number of busy servers at the $M/M/c$ retrial queue to the mean queue length at the approximating $M/M/c/0$ loss system, which is the mean of a truncated Poisson distribution of parameter Λ. Then, it follows that the value of r satisfies

$$\Lambda \left(\frac{\Lambda}{\nu}\right)^c \frac{1}{c!} = r \sum_{i=0}^{c} \left(\frac{\Lambda}{\nu}\right)^i \frac{1}{i!}, \quad (3.104)$$

or equivalently

$$\lambda \left(\frac{\Lambda}{\nu}\right)^c \frac{1}{c!} = r \sum_{i=0}^{c-1} \left(\frac{\Lambda}{\nu}\right)^i \frac{1}{i!}. \quad (3.105)$$

Equations (3.104) and (3.105) have two immediate consequences. First, by (3.104), we notice that r is the unique zero in $(0, \infty)$ of the polynomial function (3.90). As a result, we derive $r = \mu \widehat{E[N]}$, which connects the RTA assumption to the approximation of the $M/M/c$ retrial queue by the $M/M/c/0$ loss system with the increased arrival rate $\Lambda = \lambda + \mu \widehat{E[N]}$. On the other hand, from (3.105) and Subsection 2.2.1, the additional arrival rate r corresponds to the limit value $\lim_{\mu \to 0} \mu E[N]$, and therefore can be seen as the rate associated with a load formed by sources of repeated attempts. A rigorous proof of this limit result can be found in [288, Subsection 2.7.2].

3.4 The $M/M/c$ Retrial Queue

In the case $\mu \to \infty$, intervals between repeated attempts tend to zero and the performance measures in the $M/M/c$ retrial queue approach the corresponding measures in the standard $M/M/c$ queue. The reader is directed to our preceding Subsection 2.2.1 for the formal mathematical translation of this asymptotic behavior.

In estimating the blocking probability B, these two extreme cases may be combined into a single approximate formula, namely

$$I_1(B) = \frac{\nu}{\nu+\mu}\hat{B} + \frac{\mu}{\nu+\mu}B_\infty, \qquad (3.106)$$

where B_∞ is the blocking probability in the standard $M/M/c$ queue.

For the mean value $E[N]$, we can propose the approximate value

$$I_1(N) = \widehat{E[N]} + E[N_\infty], \qquad (3.107)$$

where $E[N_\infty] = \rho(1-\rho)^{-1}B_\infty$. To motivate the right-hand side of (3.107), note that we may estimate $Var(C)$ by interpolation through

$$Var(C) \approx \frac{\nu}{\nu+\mu}\widehat{Var(C)} + \frac{\mu}{\nu+\mu}Var(C_\infty), \qquad (3.108)$$

where $\widehat{Var(C)}$ is the RTA estimation of $Var(C)$ and $Var(C_\infty)$ denotes the variance of the number of busy servers in the standard $M/M/c$ queue; that is,

$$\widehat{Var(C)} = \frac{\lambda}{\nu} - \frac{\mu\widehat{E[N]}}{\nu}\left(c - \frac{\lambda}{\nu}\right), \qquad (3.109)$$

$$Var(C_\infty) = \frac{\lambda}{\nu}(1 - B_\infty). \qquad (3.110)$$

Hence, (3.109) and (3.110) turn (2.5) into the approximation (3.107) for $E[N]$.

From the basic approximations (3.106) and (3.107), subsequent approximations of B and $E[N]$ by interpolation can be built. For the blocking probability, we can deal with

$$I_2(B) = \frac{(c\nu - \lambda)\hat{B} + \mu(1-\hat{B})B_\infty}{c\nu - \lambda + \mu(1-\hat{B})}, \qquad (3.111)$$

$$I_3(B) = \frac{\nu}{\nu+\mu}\tilde{B} + \frac{\mu}{\nu+\mu}B_\infty, \qquad (3.112)$$

$$I_4(B) = \frac{(c\nu - \lambda)\tilde{B} + \mu(1-\tilde{B})B_\infty}{c\nu - \lambda + \mu(1-\tilde{B})}. \qquad (3.113)$$

To propose (3.111), recall that $\mu E[N]$ is given by $(\lambda + \mu N_c)P_c$, by (3.85), so that an appeal to the limit behavior of $\mu E[N]$ as $\mu \to 0$ implies $\lim_{\mu \to 0} \mu N_c = r$. This allows us, by analogy with (3.107), to estimate N_c by means of

86   3 Limiting Distribution of the System State

$$N_c \approx \widehat{E[N]} + \frac{\rho}{1-\rho},$$

since $\lim_{\mu \to \infty} N_c = \rho/(1-\rho)$. By introducing this estimation of N_c into (3.85), it is readily seen that the right-hand side of (3.111) follows from the estimation of $E[N]$ through $I_1(N)$. Now, $I_3(B)$ and $I_4(B)$ in (3.112) and (3.113) are proposed by similarity to (3.106) and (3.111) by replacing the *RTA* estimation $\hat{B}$ by its *FR* counterpart $\tilde{B}$.

If we turn the attention to the estimation of $E[N]$, then a second approximation is derived in the light of (3.106) as

$$I_2(N) = \frac{\nu}{\nu+\mu}\widehat{E[N]} + \frac{\mu}{\nu+\mu}E[N_\infty]. \tag{3.114}$$

Furthermore, by substituting $\widehat{E[N]}$ in (3.114) by $\widehat{E[N]}_{se}$, we may obtain

$$I_3(N) = \widehat{E[N]} + \frac{\mu}{\nu+\mu}E[N_\infty]. \tag{3.115}$$

Replacing the role of *RTA* estimations in the construction of expressions (3.107), (3.114) and (3.115) by their *FR* equivalents leads to

$$I_4(N) = \frac{(\nu+\mu)(\lambda - \nu\bar{V})}{\mu(c\nu - \lambda)}, \tag{3.116}$$

$$I_5(N) = \frac{\nu}{\nu+\mu}\widetilde{E[N]} + \frac{\mu}{\nu+\mu}E[N_\infty], \tag{3.117}$$

$$I_6(N) = \frac{\nu}{\nu+\mu}\widetilde{E[N]}_{se} + \frac{\mu}{\nu+\mu}E[N_\infty], \tag{3.118}$$

where $\bar{V}$ represents the *FR* approximation of $Var(C)$ by interpolation; that is,

$$\bar{V} = \frac{\nu}{\nu+\mu}\widetilde{Var(C)} + \frac{\mu}{\nu+\mu}Var(C_\infty).$$

At this point, it may be appropriate to make some comments. Firstly, we recall that Falin and Templeton [288, page 170] notice numerically that $I_1(B)$ is less accurate than $I_2(B)$. In fact, $I_1(B)$ is observed to be less accurate than the *RTA* estimation $\hat{B}$ for particular values of the system parameters. Besides, our numerical work allows us to note the set of inequalities

$$\hat{B} < \tilde{B} < B < I_4(B) < I_3(B),$$

which gives a reasonable argument to reduce the estimation of B to a choice among $\tilde{B}$, $I_2(B)$ and $I_4(B)$.

We now turn our attention to the estimation of $E[N]$ and observe that a simple algebra yields $I_2(N) < I_3(N) < I_1(N)$. On the other hand, our numerical experiments show that $I_1(N) < E[N]$ and that $\widehat{E[N]}_{se}$ and the

Table 3.16. $I_2(B)$ and $I_4(B)$ for queues with $(\nu, c) = (1.0, 5)$

ρ		$\mu = 0.05$	$\mu = 0.2$	$\mu = 1.0$	$\mu = 10.0$
0.2	$I_2(B)$	0.00311	0.00314	0.00325	0.00362
	$I_4(B)$	0.00312	0.00318	0.00339	0.00377
0.4	$I_2(B)$	0.04205	0.04285	0.04611	0.05542
	$I_4(B)$	0.04235	0.04390	0.04931	0.05845
0.6	$I_2(B)$	0.16241	0.16669	0.18311	0.22166
	$I_4(B)$	0.16329	0.16986	0.19293	0.23098
0.8	$I_2(B)$	0.41894	0.42956	0.46634	0.53380
	$I_4(B)$	0.41971	0.43231	0.47500	0.54378

Table 3.17. $I_1(N)$ for queues with $(\nu, c) = (1.0, 5)$

ρ	$\mu = 0.05$	$\mu = 0.2$	$\mu = 1.0$	$\mu = 10.0$
0.2	0.06326	0.01653	0.00407	0.00126
0.4	1.78363	0.47575	0.12699	0.04852
0.6	11.85639	3.22976	0.92933	0.41173
0.8	58.96723	16.40414	5.05398	2.50020

FR estimations by interpolation (3.116)-(3.118) can be discarded. Thus, the problem reduces to compare $\widehat{E[N]}$, $\widehat{E[N]}_{se}$ and $I_1(N)$.

According to this, Table 3.16 lists values of $I_2(B)$ and $I_4(B)$ completing the numerical example considered in Table 3.15. As the reader may see, we prefer $\tilde{B}$ and $I_2(B)$ to $I_4(B)$. In particular, the FR estimation $\tilde{B}$ is higher accurate for all values of μ as ρ is small. $I_2(B)$ exhibits the best behavior when ρ increases, irrespective of the values of μ.

In the case of $E[N]$, Table 3.17 gives values for $I_1(N)$, whereas values for $E[N]$, $\widehat{E[N]}$ and $\widehat{E[N]}_{se}$ are reported in Table 3.15. In a nutshell, the choice among them depends on ρ and μ in a more complex manner than in the analysis for B. For example, $\widehat{E[N]}$ seems to be higher accurate for smaller values of ρ and μ, but $\widehat{E[N]}_{se}$ and $I_1(N)$ appear to be better than $\widehat{E[N]}$ when we deal with larger values of ρ.

3.5 A Multiserver Retrial Queue with Finite Population

The present section focuses on a queueing model with finite population of size M where the input process is characterized by the fact that each idle source generates requests independently and with the same exponentially distributed inter-arrival time of rate α. This particular type of finite-source input is often called quasi-random input. Some examples of its applications in retrial queues were reported in Section 2.3.

88 3 Limiting Distribution of the System State

In Subsection 3.1.2 we showed how to compute the limiting probabilities of a single server retrial queue with finite population and general service times. We now turn our attention to the multiserver case. The description of the service and retrial mechanisms agrees with the formulation given for the main multiserver retrial queue of type $M/M/c$. However, as a result of dealing with a finite population model, the arrival rate becomes $\lambda_{ij} = (M - i - j)\alpha$, for $(i, j) \in \mathcal{S} = \{0, ..., c\} \times \{0, ..., M - c\}$.

If the population size M increases to infinity and the arrival rate α decreases to zero in such a way that the overall rate αM remains constant (i.e., $\alpha M = \lambda$), then it is easy to verify that the arrival process converges to a Poisson flow of rate λ. Therefore, despite of its intrinsic interest, the model with finite population can be employed to approximate the $M/M/c$ retrial queue.

Let $\{P_{ij}; (i,j) \in \mathcal{S}\}$ be the limiting distribution of the number of busy servers and the number of customers in orbit. Then, the Kolmogorov equations are as follows:

$$((M - i - j)\alpha + i\nu + j\mu)P_{ij} = (M - i + 1 - j)\alpha P_{i-1,j} + (j+1)\mu P_{i-1,j+1}$$
$$+ (i+1)\nu P_{i+1,j}, \quad 0 \le i \le c-1, 0 \le j \le M-c-1, \quad (3.119)$$

$$((M - c - j)\alpha + c\nu)P_{cj} = (M - c + 1 - j)\alpha P_{c-1,j} + (j+1)\mu P_{c-1,j+1}$$
$$+ (M - c - j + 1)\alpha P_{c,j-1}, \quad 0 \le j \le M-c-1, \quad (3.120)$$

$$((c - i)\alpha + i\nu + (M - c)\mu)P_{i,M-c} = (c - i + 1)\alpha P_{i-1,M-c}$$
$$+ (i+1)\nu P_{i+1,M-c}, \quad 0 \le i \le c-1, \quad (3.121)$$

$$c\nu P_{c,M-c} = \alpha P_{c-1,M-c} + \alpha P_{c,M-c-1}. \quad (3.122)$$

We notice that $P_{ij} = 0$, for all $(i,j) \notin \mathcal{S}$.

Of course, the set of equations (3.119)-(3.122) is also subject to the normalization condition $\sum_{(i,j) \in \mathcal{S}} P_{ij} = 1$. The structure of the above equations is very similar to the structure of (3.65)-(3.68) for the truncated model $\mathcal{X}^W$ studied in Subsection 3.4.1. Thus, it can be solved with the help of a recursive scheme similar to that described in Theorem 3.10. The main difference concerns with the use of rates λ_{ij} instead of λ and some small changes in formulas (3.70) and (3.71), and in the definition of coefficients α_{ij} and β_{ij}. The interested reader may find the detailed algorithm in [288, Subsection 4.3.2].

Next we describe an alternative treatment due to Kornyshev [420]. Such an approach relies on an original transformation of the initial set of Kolmogorov equations. After some algebra, the transformation can be exploited for computing the main performance measures, including even limiting probabilities. From $\{P_{ij}; (i,j) \in \mathcal{S}\}$ we define the following series of transformations:

$$P_{ij}(q) = \frac{(j+1)P_{i,j+1}(q-1)}{\sum_{i=0}^{c}\sum_{j=0}^{M-c-q+1} jP_{ij}(q-1)}, \quad (3.123)$$

for $0 \le i \le c$, $0 \le j \le M - c - q$ and $1 \le q \le M - c$. We assume $P_{ij}(0)$ as being identically equal to P_{ij}.

3.5 A Multiserver Retrial Queue with Finite Population

We also consider some balance equations. First, we equate the flow rates into and out of the subset $\{(k,j); k \leq i\}$ to get

$$(M - i - j)\alpha P_{ij} + j\mu \sum_{k=0}^{i} P_{kj} = (i+1)\nu P_{i+1,j}$$

$$+ (1 - \delta_{j,M-c})(j+1)\mu \sum_{k=0}^{i-1} P_{k,j+1}, \quad (3.124)$$

for $0 \leq i \leq c-1$ and $0 \leq j \leq M-c$. Equating the crossing flow into the level j of the orbit, we obtain

$$(j+1)\mu \sum_{i=0}^{c-1} P_{i,j+1} = (M - c - j)\alpha P_{cj}, \quad 0 \leq j \leq M - c - 1. \quad (3.125)$$

Finally, we equate the flows into and out of the subset $\{(i,j); i + j = k\}$ and add the resulting equations over $k = 0, ..., M - 1$ in order to get the following expression for the mean rate of generation of primary arrivals:

$$\bar{\lambda} = \alpha \sum_{i=0}^{c} \sum_{j=0}^{M-c} (M - i - j) P_{ij} = \nu \sum_{i=0}^{c} i P_{i..}$$

Multiplying equations (3.124) and (3.125) by j, we obtain for each value of q

$$(i+1)\nu P_{i+1,j}(q) = (M - i - q)\alpha P_{ij}(q) + j(\mu - \alpha) P_{ij}(q)$$

$$+ q\mu \sum_{k=0}^{i} P_{kj}(q) - (1 - \delta_{j,M-c-q})(j+1)\mu \sum_{k=0}^{i-1} P_{k,j+1}(q)$$

$$+ j\mu \sum_{k=0}^{i-1} P_{kj}(q), \quad 0 \leq i \leq c - 1, \ 0 \leq j \leq M - c - q, \quad (3.126)$$

$$(M - c - j - q)\alpha P_{cj}(q) = (j+1)\mu \sum_{i=0}^{c-1} P_{i,j+1}(q), \ 0 \leq j \leq M - c - q - 1.$$

$$(3.127)$$

If we now define

$$P_{i.}(q) = \sum_{j=0}^{M-c-q} P_{ij}(q), \quad 0 \leq i \leq c, \ 0 \leq q \leq M - c,$$

then in the light of (3.123) and (3.127) we have

$$\sum_{j=0}^{M-c-q} j P_{ij}(q) = \frac{(M - c - q)\alpha P_{c.}(q)}{\mu - (\mu - \alpha) P_{c.}(q+1)} P_{i.}(q+1), \quad (3.128)$$

for $0 \leq i \leq c$ and $0 \leq q \leq M - c$.

After some algebraic manipulations over (3.126) and (3.128), we obtain the following relationship connecting two successive steps q and $q+1$:

$$(i+1)\nu P_{i+1,\cdot}(q) = (M-i-q)\alpha P_{i\cdot}(q) + q\mu \sum_{k=0}^{i} P_{k\cdot}(q)$$
$$+ \frac{(M-c-q)\alpha P_{c\cdot}(q)}{\frac{\mu}{\mu-\alpha} - P_{c\cdot}(q+1)} P_{i\cdot}(q+1), \qquad (3.129)$$

for $0 \leq i \leq c-1$ and $0 \leq q \leq M - c$. Finally, if we define

$$F_i(q) = \sum_{k=0}^{i} P_{k\cdot}(q), \quad 0 \leq i \leq c, \ 0 \leq q \leq M - c,$$

then (3.129) takes the form

$$(i+1)\nu F_{i+1}(q) - ((i+1)\nu + (M-i-q)\alpha + q\mu) F_i(q)$$
$$+ (1 - \delta_{0i})(M-i-q)\alpha F_{i-1}(q)$$
$$= \frac{(M-c-q)\alpha(1 - F_{c-1}(q))}{\frac{\alpha}{\mu-\alpha} + F_{c-1}(q+1)} (F_i(q+1) - (1-\delta_{0i})F_{i-1}(q+1)), \quad (3.130)$$

for $0 \leq i \leq c-1$ and $0 \leq q \leq M - c$.

The system of equations (3.130) can be recursively solved in decreasing order $q = M - c, ..., 0$. For each step q, (3.130) provides a system of c equations for c unknown values $F_i(q)$, for $0 \leq i \leq c-1$, since $F_c(q) = 1$, for $0 \leq q \leq M - c$. Then, we can simply compute $P_{i\cdot}(q)$ from the identity

$$P_{i\cdot}(q) = F_i(q) - (1-\delta_{0i})F_{i-1}(q), \quad 0 \leq i \leq c, \ 0 \leq q \leq M - c.$$

This allows us to get the blocking probability as $B = P_{c\cdot}(0)$. Similarly, $P_{c\cdot}(1)$ corresponds to the fraction of blocked repeated attempts defined as

$$B_r = \frac{\sum_{j=0}^{M-c} j P_{cj}}{\sum_{i=0}^{c} \sum_{j=0}^{M-c} j P_{ij}}.$$

Indeed, if we know B and B_r, then we can easily find many other quality characteristics. In particular, $E[N]$ and $\bar{\lambda}$ can be computed from the relationships

$$E[N] = \frac{(M-c)\alpha B}{\alpha B_r + \mu(1-B_r)},$$
$$\bar{\lambda} = \frac{\alpha \nu}{\alpha + \nu}(M - E[N]),$$

which can be readily proved.

Then, we may obtain the fraction of blocked primary arrivals and the mean waiting time by means of the following expressions:

$$B_a = \frac{\sum_{j=0}^{M-c}(M-c-j)P_{cj}}{\sum_{i=0}^{c}\sum_{j=0}^{M-c}(M-i-j)P_{ij}} = \frac{\mu(1-B_r)E[N]}{\bar{\lambda}},$$

$$E[W] = \frac{B_a}{\mu(1-B_r)}.$$

Our last objective is the computation of $\{P_{ij}; (i,j) \in \mathcal{S}\}$. Summing (3.128) over i, we find that

$$E[N(q)] = \sum_{i=0}^{c}\sum_{j=0}^{M-c-q} jP_{ij}(q) = \frac{(M-c-q)\alpha P_{c.}(q)}{\mu-(\mu-\alpha)P_{c.}(q+1)}, \quad 0 \leq q \leq M-c-1.$$

An iterative use of (3.123) reduces $P_{i.}(q)$ to the following expression in terms of the qth partial factorial moment:

$$M_q^i = \sum_{j=q}^{M-c} j(j-1)...(j-q+1)P_{ij} = P_{i.}(q)\prod_{k=0}^{q-1} E[N(k)],$$

for $0 \leq i \leq c$ and $1 \leq q \leq M-c$. Thus, noting that $M_0^i = P_{i.}(0)$, we complete a system of $M-c+1$ equations for the computation of the unknowns $\{P_{ij}; 0 \leq j \leq M-c\}$.

3.6 Bibliographical Notes

Tijms [662] provides an excellent treatment of the powerful regenerative approach for the computation of limiting probabilities of single server queues with Poisson input. The approach has been exploited in the retrial literature to study a variety of models [48, 69, 71, 365]. For alternative treatments based on supplementary variables or Markov renewal processes, see [288, 380] and [449, 450, 511], respectively.

If the service time distribution is of phase-type (see Subsection 7.1.4), the distribution of the Poisson events arriving during a service time is of discrete phase-type [544, Theorem 2.2.8] and can be recursively computed without numerical integration. This is helpful to evaluate a_j in explicit form; see formula (3.5).

The exposition of the maximum entropy approach for the $M/G/1$ retrial queue is based on the paper by Artalejo and Lopez-Herrero [94]. For applications to models with finite population, see the paper of Artalejo and Gómez-Corral [53].

The retrial process with non-exponential retrial times is a complex nonrenewal process. As a result, the investigation of the subject is very limited.

The first work reported on general retrial times [375] was found to be incorrect [268]. Another account of the subject is the treatment of multiserver retrial queues given by Pourbabai [563] where customers in orbit are replaced by a new independent input process whose parameters are approximated from those of the overflow process. As a related work, see also [482]. It should be noted that Yang et al. [700] provide $\rho < 1$ as a necessary condition for stability, but they do not prove that such a condition is also sufficient. See [211] for a necessary and sufficient condition for ergodicity of the $M/PH/1$ retrial queue with phase-type retrial times.

The reader is alerted to the fact that the inherent difficulty with general retrial times might be avoided by considering retrial queues where only one customer in orbit can make retrials; see e.g. [183] for the study of such a variant on the $M/M/1$ retrial queue with general retrial times, and [315] for the extension to the $M/G/1$ retrial queue. Among systems modelled with general retrial times we mention [427, 516].

The paper by Yang and Li [701] contributed to extend the study of retrial queues to discrete-time systems. Most of the existing papers deal with single server models and mainly follow the methodology introduced in [701]. The analysis of a multiserver discrete-time retrial queue with finite population can be found in [111]. For an exhaustive analysis of the $Geo/Geo/c$ retrial queue, we refer to [112].

Although the use of a truncated model to approximate the $M/M/c$ retrial queue was suggested by Wilkinson [685], the first formulation in terms of an algorithm is due to Jonin and Sedol [366]. The convergence of the limiting characteristics of the truncated models to the corresponding characteristics of the $M/M/c$ retrial queue is based on stochastic monotonicity results and on censored Markov chains [288, 604, 611]. A related treatment based on a synchronization method is given in [44].

The approximations by interpolation in Subsection 3.4.5 extend the analysis done in the book by Falin and Templeton [288]. For a related approximation taking into account peakedness and correlation between customers and repeated attempts, see the paper by Reeser [577].

The literature on approximations and variants of multiserver retrial queues is voluminous. The paper by Cohen [200] plays a central role on the theory of retrial queues. Such a paper provided a rigorous mathematical generalization of earlier accounts and was influential in promoting retrial queues in contraposition to classical loss models. Cohen's model takes into account not only retrials but the effect of the abandonments as well. The solution is given in terms of contour integrals. The work of Stepanov [622]-[643] is focused on the numerical analysis of multiserver retrial queues. Implementation of his methods is illustrated in a number of models with non-persistent customers, waiting positions, inner blocking, etc. Pearce [559] investigates a variant of Cohen's model with finite orbit capacity. Solutions are established by using sigma polynomials. The usefulness of continued fractions in solving recurrence relations arising in retrial queues has been shown in several papers [345, 557, 560].

Some decomposition results for multiserver queues with waiting positions and constant retrial policy are given in [574].

A few papers [306, 332, 507] deal with time-dependent multiserver models with retrials and abandonments. In this context, fluid and diffusion approximations are helpful both for analytical and computational purposes. A fluid approximation is also used by Aguir et al. [10] to investigate a retrial queue with balking and impatience operating in non-stationary regime. In [280] a diffusion approximation for the $M/M/c$ retrial queue can be found. Abramov [6] studies a multiserver model where the arrival process is a point process with strictly stationary and ergodic increments. The underlying methodology is close to a semimartingale decomposition of the queue length process; see also [7].

Shin [603] provides an algorithm based on the adaptive uniformization technique to find transient probabilities of a retrial queue with impatient customers. Yang et al. [698] extend the investigation to a retrial queue in which customers arrive according to a Coxian process instead of a Poisson stream.

A number of papers use multiserver retrial queues with finite population to model the customer behavior in cellular mobile networks [22, 24, 107, 665]. For the analysis of the busy period and the waiting time in models with finite population, the reader is referred to [288]-[290].

4
Busy Period

In this chapter, we study the length of the busy period, denoted by L, and related performance measures for the $M/G/1$ and $M/M/c$ retrial queues. This seems that the existing numerical inversion techniques would be inadequate to handle. For this reason, alternative approaches are needed, and Subsection 4.1.1 concentrates on methods based on the principle of maximum entropy and on the truncation of the orbit capacity for the $M/G/1$ retrial queue. Subsections 4.1.2 and 4.1.3 complete the analysis of the single server model by studying the recursive computation of the number of customers served and the maximum orbit size during a busy period. In Section 4.2, we consider the $M/M/c$ retrial queue. In particular, Subsection 4.2.1 discusses the length of a busy period and the computation of its moments. The algorithms presented involve the approximating models $\mathcal{X}^W$, $\mathcal{X}^F$ and $\mathcal{X}^{NR}$, which were introduced in Section 3.4. In Subsection 4.2.2, we briefly show how to extend the analysis to the number of customers served. The exact computation of the maximal queue length is presented in Subsection 4.2.3. Finally, bibliographical remarks are given in Section 4.3.

4.1 The $M/G/1$ Retrial Queue

4.1.1 The Length of the Busy Period

Our objective in this section is to overcome the difficulties inherent to the length of a busy period in the $M/G/1$ retrial queue, which were already remarked in Subsection 2.2.2. To this end, we consider two approaches, the first approach consisting in the use of the principle of maximum entropy and the second one based on a truncated retrial queue.

We begin with the first approach, where the *PME* summarized in Subsection 3.1.3 is used to get an estimation for the density function of L. Initially, we assume that the available information consists of the first and second moments of L, which are provided by the expressions (2.11) and (2.12). After that, we add one more constraint by using the value of the Laplace-Stieltjes transform $L^*(s)$ of L at a given point $s = s_0$, determined numerically by equation (2.10).

The methodology described in Subsection 3.1.3 yields maximum entropy densities $\hat{f}_2(x)$ and $\hat{f}_{2,1}(x)$, respectively. Their corresponding functional forms look as follows:

$$\hat{f}_2(x) = \exp\left\{-(\hat{\alpha}_0 + x\hat{\alpha}_1 + x^2\hat{\alpha}_2)\right\}, \quad x \in (0, T),$$
$$\hat{f}_{2,1}(x) = \exp\left\{-(\hat{\alpha}_0 + x\hat{\alpha}_1 + x^2\hat{\alpha}_2 + e^{-s_0 x}\hat{\alpha}_{2,1})\right\}, \quad x \in (0, T).$$

As a practical remark, we observe that the potential function F and the balanced function G, given in formulas (3.22) and (3.23) respectively, involve integrals defined over $(0, \infty)$. Thus, solving the minimization problem implies firstly the consideration of a truncated interval $(0, T)$. The upper bound T may be chosen with the help of Tchebychev's inequality, such as $P(L > T) \leq 10^{-2}$.

Note that the maximum entropy densities satisfy the given constraints, in particular the first two moments of $\hat{f}_2(x)$ and $\hat{f}_{2,1}(x)$ coincide with the ones of $f_L(x)$, and so does the Laplace-Stieltjes transform associated with $\hat{f}_{2,1}(x)$, at $s = s_0$, with $L^*(s_0)$. Thus, we propose to check the adequacy of the maximum entropy estimation by measuring the relative errors associated with their estimates for the Laplace-Stieltjes transform; that is, we consider

$$R_2(s) = \left|1 - \frac{L_2^*(s)}{L^*(s)}\right| \quad \text{and} \quad R_{2,1}(s) = \left|1 - \frac{L_{2,1}^*(s)}{L^*(s)}\right|,$$

where $L_2^*(s) = \int_0^T e^{-sx}\hat{f}_2(x)dx$ and $L_{2,1}^*(s) = \int_0^T e^{-sx}\hat{f}_{2,1}(x)dx$.

Next we analyze the length of a busy period in an $M/E_3/1$ retrial queue with $\beta_1 = 1.0$. The arrival rate is chosen as $\lambda = 0.25, 0.5$ and 0.75. For a given value of λ, we assume that the retrial rate is $\mu = 2\lambda$.

Table 4.1 presents a comparison between the classical Laplace-Stieltjes transform of the busy period $L^*(s)$ and the Laplace-Stieltjes transform of the maximum entropy solution based on two moments $L_2^*(s)$. For most fixed values $s > 0$, we observe that both transforms become closer when the traffic intensity decreases.

We may improve the estimation by considering the maximum entropy solution $\hat{f}_{2,1}(x)$. For practical purposes, the choice of the point s_0 can be done by taking into account that the behavior of $f_L(x)$ and $L^*(s)$ near the boundaries of their domains is determined by the Tauberian relations

$$\lim_{s \to 0} sL^*(s) = \lim_{x \to \infty} f_L(x) \quad \text{and} \quad \lim_{s \to \infty} sL^*(s) = \lim_{x \to 0} f_L(x). \quad (4.1)$$

Consequently, smaller values of s_0 provide a better description of the tail behavior of $f_L(x)$, while a large value of s_0 describes better the behavior near the origin.

In Table 4.2 and Figure 4.1 we consider an $M/E_3/1$ retrial queue with $\beta_1 = 1.0$ and $\rho = 0.25$, and we use as constraint the value of $L^*(s)$ at the point

4.1 The $M/G/1$ Retrial Queue   97

Table 4.1. Comparing Laplace-Stieltjes transforms in an $M/E_3/1$ retrial queue

	$\rho = 0.25$		$\rho = 0.5$		$\rho = 0.75$	
s	$L^*(s)$	$L_2^*(s)$	$L^*(s)$	$L_2^*(s)$	$L^*(s)$	$L_2^*(s)$
0.05	0.89166	0.90632	0.84622	0.85323	0.76778	0.70380
0.1	0.83598	0.83324	0.76951	0.75077	0.67777	0.54867
0.25	0.69107	0.67587	0.61400	0.55568	0.53178	0.33193
0.5	0.53792	0.51711	0.46800	0.38979	0.40388	0.20069
1.0	0.35856	0.35355	0.30994	0.24492	0.26869	0.11219
1.5	0.25426	0.26907	0.22107	0.17874	0.19324	0.07787
3.0	0.11078	0.15700	0.09882	0.09880	0.08854	0.04061
4.5	0.05802	0.11090	0.05282	0.06828	0.04822	0.02747
6.0	0.03409	0.08574	0.03150	0.05217	0.02916	0.02075
10.0	0.01159	0.05343	0.01095	0.03202	0.01038	0.01256

Table 4.2. Relative errors in the $M/E_3/1$ retrial queue for $\rho = 0.25$

s	$L^*(s)$	$L_2^*(s)$	$R_2(s)$	$L_{2,1}^*(s)$	$R_{2,1}(s)$
0.05	0.89166	0.90632	0.01644	0.90586	0.01593
0.1	0.83598	0.83324	0.00327	0.83101	0.00593
0.25	0.69107	0.67587	0.02200	0.66439	0.03861
0.5	0.53792	0.51711	0.03868	0.48998	0.08912
1.0	0.35856	0.35355	0.01395	0.30709	0.14352
1.5	0.25426	0.26907	0.05823	0.21404	0.15821
3.0	0.11078	0.15700	0.41719	0.09887	0.10750
4.5	0.05802	0.11090	0.91138	0.05802	0.00000
6.0	0.03409	0.08574	1.51504	0.03871	0.13564
10.0	0.01159	0.05343	3.60810	0.01841	0.58837

$s_0 = 4.5$. We can observe in Table 4.2 that the relative errors are moderately small as s is close to zero. Nevertheless, we also observe that, for large values of s, relative errors are notably diminished when we employ $\hat{f}_{2,1}(x)$, instead of $\hat{f}_2(x)$. This fact and the boundary behavior given in (4.1) indicate that, near the origin, the classical density function is better described by $\hat{f}_{2,1}(x)$ than by $\hat{f}_2(x)$.

Figure 4.1 shows different shapes of the maximum entropy densities $\hat{f}_2(x)$ and $\hat{f}_{2,1}(x)$. Previous discussion permits to assert that the distribution of L near the origin should be better represented by $\hat{f}_{2,1}(x)$; that is, a bell-shaped function. Moreover, Figure 4.1 illustrates the importance of including information about the Laplace-Stieltjes transform because the mode of the busy period distribution is not captured unless a constraint on $L^*(s)$ is specified.

In a second numerical example, see Table 4.3 and Figure 4.2, we consider an $M/H_2/1$ retrial queue with mean service time $\beta_1 = 1.0$ and coefficient of variation $c_v = 1.25$. The parameters of the H_2 distribution cannot be uniquely

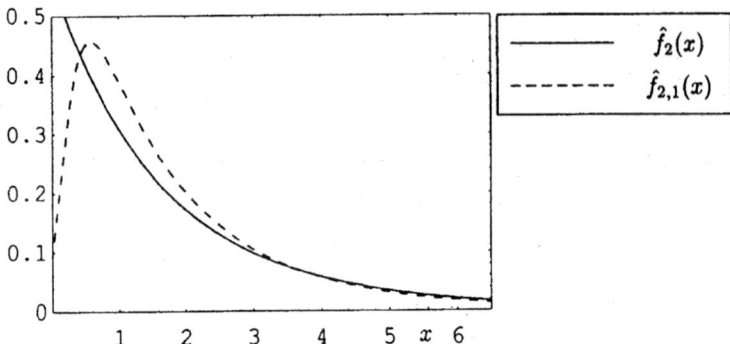

Fig. 4.1. Maximum entropy estimations in an $M/E_3/1$ retrial queue with $\rho = 0.25$

Table 4.3. Relative errors in the $M/H_2/1$ retrial queue for $\rho = 0.25$

s	$L^*(s)$	$L_2^*(s)$	$R_2(s)$	$L_{2,1}^*(s)$	$R_{2,1}(s)$
0.05	0.86545	0.82936	0.04170	0.83688	0.03301
0.1	0.80664	0.71491	0.11371	0.73749	0.08573
0.25	0.70620	0.50750	0.28135	0.57245	0.18939
0.5	0.60442	0.34316	0.43224	0.45084	0.25410
1.0	0.47820	0.20868	0.56361	0.35129	0.26539
1.5	0.39784	0.15000	0.62294	0.30388	0.23617
3.0	0.26601	0.08140	0.69398	0.23558	0.11438
4.5	0.20016	0.05586	0.72090	0.20016	0.00000
6.0	0.16049	0.04252	0.73503	0.17635	0.09887
10.0	0.10486	0.02598	0.75224	0.13675	0.30408

determined by fitting the above values unless an additional condition was assumed. Thus, we assume that the distribution has balanced means; that is, $p_1/\nu_1 = p_2/\nu_2$. The retrial rate is chosen as $\mu = \lambda/2$.

In Table 4.3 we compare the Laplace-Stieltjes transforms associated with the maximum entropy densities based on two moments and on two moments plus the value of the Laplace-Stieltjes transform at the point $s_0 = 4.5$. The entries for the relative errors $R_{2,1}(s)$ are smaller than the errors $R_2(s)$, thus showing the superiority of the estimation $\hat{f}_{2,1}(x)$.

In Figure 4.2, we plot the balanced function $G(\alpha_1, \alpha_2)$ in a neighborhood of the Lagrangian multipliers $(\hat{\alpha}_1, \hat{\alpha}_2)$. In agreement with the comments expressed in Subsection 3.1.3, the surface shows a rapid growth of G along some directions. The existence of a valley is also observed.

Although the mathematical formalism of the *PME* and underlying numerical techniques are common for both discrete and continuous distributions, the numerical effort to conduct the latter is considerably superior. More precisely, numerical implementation in a discrete case implies the estimation of a finite

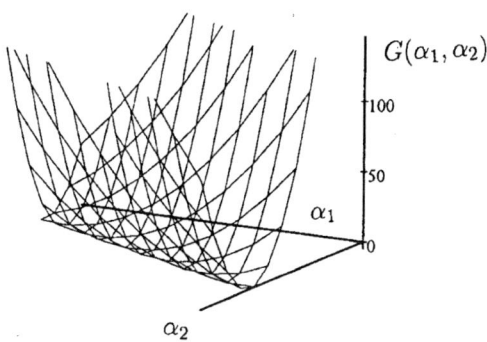

Fig. 4.2. The balanced function $G(\alpha_1, \alpha_2)$

set of probabilities which can be done in a personal computer after a few minutes run. Nevertheless, in a continuous situation, which is the case of L, the estimation of a density function defined over $(0, \infty)$ typically demands several hours of running time, and so often the computer program stops without converging to the optimal Lagrangian multipliers.

In what follows, we focus on a second approach consisting in placing a suitably chosen limit on the orbit capacity. Such an assumption yields a finite state space instead of the original infinite one, but it provides a simple and efficient procedure for the estimation of the density function of L.

We first investigate the length of a busy period in an $M/G/1$ retrial queue with finite capacity K of the retrial group. We employ the method of catastrophes which simplifies the probabilistic reasoning and leads readily to equations governing the dynamics of L.

Let us assume that L starts at the instant $t = 0$. We consider a catastrophe process independent of the functioning of our queueing system which generates catastrophes at a rate s according to a Poisson process. Suppose that, at the epoch of the kth service completion, there are j customers in orbit, no catastrophe occurred prior to that moment, and the busy period is still in progress. Let $P_j^{(k)}(s)$ denote the probability of that event. We also define $\Pi_k(s)$ as the probability that during the busy period, no catastrophe occurs and k customers are served.

These auxiliary probabilities play a key role to obtain the Laplace-Stieltjes transform $L^*(s)$ in the following.

Theorem 4.1. *The Laplace-Stieltjes transform $L^*(s)$ can be computed as follows:*

Step 1. Define the auxiliary quantities

$$a_j(s) = \int_0^\infty e^{-(s+\lambda)x} \frac{(\lambda x)^j}{j!} dB(x), \quad j \geq 0,$$

$$a_j^a(s) = \sum_{m=j}^{\infty} a_m(s), \quad 0 \leq j \leq K.$$

Step 2. Solve the system of linear equations

$$\phi_j(s) = a_j(s) + \sum_{n=1}^{j} \phi_n(s) \frac{\lambda}{s+\lambda+n\mu} a_{j-n}(s)$$

$$+ \sum_{n=1}^{j+1} \phi_n(s) \frac{n\mu}{s+\lambda+n\mu} a_{j-n+1}(s), \quad 0 \leq j \leq K-1, \quad (4.2)$$

$$\phi_K(s) = a_K^a(s) + \sum_{n=1}^{K} \phi_n(s) \frac{\lambda}{s+\lambda+n\mu} a_{K-n}^a(s)$$

$$+ \sum_{n=1}^{K} \phi_n(s) \frac{n\mu}{s+\lambda+n\mu} a_{K-n+1}^a(s). \quad (4.3)$$

Step 3. Calculate the Laplace-Stieltjes transform of L as $L^*(s) = \phi_0(s)$.

Proof. The probabilities $P_j^{(k)}(s)$ satisfy the following formulas:

$$P_j^{(1)}(s) = a_j(s), \quad 0 \leq j \leq K-1, \quad (4.4)$$

$$P_K^{(1)}(s) = a_K^a(s), \quad (4.5)$$

$$P_j^{(k)}(s) = \sum_{n=1}^{j} P_n^{(k-1)}(s) \frac{\lambda}{s+\lambda+n\mu} a_{j-n}(s)$$

$$+ \sum_{n=1}^{j+1} P_n^{(k-1)}(s) \frac{n\mu}{s+\lambda+n\mu} a_{j-n+1}(s), \quad 0 \leq j \leq K-1, \; k \geq 2, (4.6)$$

$$P_K^{(k)}(s) = \sum_{n=1}^{K} P_n^{(k-1)}(s) \frac{\lambda}{s+\lambda+n\mu} a_{K-n}^a(s)$$

$$+ \sum_{n=1}^{K} P_n^{(k-1)}(s) \frac{n\mu}{s+\lambda+n\mu} a_{K-n+1}^a(s), \quad k \geq 2, \quad (4.7)$$

where $a_j(s)$, for $0 \leq j \leq K-1$, and $a_j^a(s)$, for $0 \leq j \leq K$, were defined in Step 1.

To prove (4.4) we notice that the event under consideration takes place if no catastrophe occurs and j primary customers arrive during the service time. Equation (4.6) describes the motion between two successive service completion epochs. The term $\lambda/(s+\lambda+n\mu)$ (respectively, $n\mu/(s+\lambda+n\mu)$) indicates that the kth service time corresponds to a primary arrival (respectively, retrial customer). In the case $j = K$, we accumulate the probability of the blocked customers and use the same argument to get formulas (4.5) and (4.7).

It is clear that
$$\Pi_k(s) = P_0^{(k)}(s), \quad k \geq 1,$$
from which it follows that the Laplace-Stieltjes transform $L^*(s)$ can be obtained as
$$L^*(s) = \sum_{k=1}^{\infty} P_0^{(k)}(s).$$

We now extend the notation and define $\phi_j(s) = \sum_{k=1}^{\infty} P_j^{(k)}(s)$, for $0 \leq j \leq K$, so that $L^*(s) = \phi_0(s)$. Then, from (4.4)-(4.7), we have that $\phi_j(s)$, for $0 \leq j \leq K$, satisfy the system (4.2) and (4.3). □

We recall that the computation of $L^*(s)$, for complex values of s, is the key to use inversion algorithms (see e.g. [2] and [662, Appendix F]) for the numerical inversion of the density $f_L(x)$.

We now turn our attention to the computation of the moments
$$E[L^k] = (-1)^k \left.\frac{d^k}{ds^k} \phi_0(s)\right|_{s=0}, \quad k \geq 1.$$

We first introduce the notation
$$\phi_j^{(k)} = (-1)^k \left.\frac{d^k}{ds^k} \phi_j(s)\right|_{s=0}, \quad 0 \leq j \leq K,\ k \geq 1,$$
$$a_j^{(k)} = (-1)^k \left.\frac{d^k}{ds^k} a_j(s)\right|_{s=0}, \quad 0 \leq j \leq K-1,\ k \geq 1,$$
$$a_j^{a,(k)} = (-1)^k \left.\frac{d^k}{ds^k} a_j^a(s)\right|_{s=0}, \quad 0 \leq j \leq K,\ k \geq 1.$$

Then, the moments $E[L^k] = \phi_0^{(k)}$, for $k \geq 1$, can be recursively computed by solving the system of linear equations given in the following corollary.

Corollary 4.2. *For any fixed $k \geq 0$, the unknowns $\phi_j^{(k)}$, for $0 \leq j \leq K$, can be evaluated as the solution to*

$$\phi_j^{(k)} = a_j^{(k)} + \sum_{n=1}^{j} \frac{\lambda}{\lambda+n\mu} \sum_{l=0}^{k} \binom{k}{l} \phi_n^{(l)} \sum_{h=0}^{k-l} \binom{k-l}{h} \frac{h!}{(\lambda+n\mu)^h} a_{j-n}^{(k-l-h)}$$
$$+ \sum_{n=1}^{j+1} \frac{n\mu}{\lambda+n\mu} \sum_{l=0}^{k} \binom{k}{l} \phi_n^{(l)} \sum_{h=0}^{k-l} \binom{k-l}{h} \frac{h!}{(\lambda+n\mu)^h} a_{j-n+1}^{(k-l-h)},$$
$$0 \leq j \leq K-1, \quad (4.8)$$

$$\phi_K^{(k)} = a_K^{a,(k)} + \sum_{n=1}^{K} \frac{\lambda}{\lambda+n\mu} \sum_{l=0}^{k} \binom{k}{l} \phi_n^{(l)} \sum_{h=0}^{k-l} \binom{k-l}{h} \frac{h!}{(\lambda+n\mu)^h} a_{K-n}^{a,(k-l-h)}$$
$$+ \sum_{n=1}^{K} \frac{n\mu}{\lambda+n\mu} \sum_{l=0}^{k} \binom{k}{l} \phi_n^{(l)} \sum_{h=0}^{k-l} \binom{k-l}{h} \frac{h!}{(\lambda+n\mu)^h} a_{K-n+1}^{a,(k-l-h)}. \quad (4.9)$$

Table 4.4. Truncation thresholds for an $M/M/1$ retrial queue

ρ		$\mu = 0.05$	$\mu = 0.5$	$\mu = 1.0$	$\mu = 2.5$	$\mu = 5.0$
0.2	K	10	7	6	6	6
	$E[L]$	10.25878	1.83351	1.53524	1.36257	1.30603
0.4	K	31	17	12	12	11
	$E[L]$	245.57257	3.77000	2.61126	2.02152	1.84046
0.6	K	88	29	24	22	21
	$E[L]$	248351.01989	10.84504	5.55359	3.52485	2.98428
0.8	K	277	79	66	57	54
	$E[L]$	953674316405.0037	80.82899	21.39936	9.21045	6.83565

Proof. It follows directly by appropriate differentiation of (4.2) and (4.3). □

It is immediately verified that

$$a_j(s) = \left(\frac{\lambda}{s+\lambda}\right)^j \int_0^\infty e^{-(s+\lambda)x} \frac{((s+\lambda)x)^j}{j!} dB(x), \quad j \geq 0, \quad (4.10)$$

$$a_j^a(s) = \beta(s) - \sum_{m=0}^{j-1} a_m(s), \quad 0 \leq j \leq K, \quad (4.11)$$

$$a_j^{(k)} = \frac{(j+k)!}{j!\lambda^k} a_{j+k}, \quad 0 \leq j \leq K-1, \ k \geq 1, \quad (4.12)$$

$$a_j^{a,(k)} = \beta_k - \sum_{m=0}^{j-1} a_m^{(k)}, \quad 0 \leq j \leq K, \ k \geq 1, \quad (4.13)$$

where the values of a_j were defined by (3.5).

In order to estimate the density function of L, the truncation level K may be determined as the first positive integer such that the first four decimal digits of $E[L]$ are fitted. According to this criterion, we need to increase successively the value of K and solve the system (4.8) and (4.9) for $k = 1$. Along this process, we compare $\phi_0^{(1)}$ (i.e., the expectation of L in the truncated retrial queue) and the value of $E[L]$ in the model with infinite orbit. The latter can be calculated numerically with the help of (2.11) and (3.3). Table 4.4 gives the resulting thresholds for the $M/M/1$ case with $\beta_1 = 1.0$ and several choices of ρ and μ. We notice that the mean value $E[L]$ increases as ρ increases and decreases as μ increases.

It is interesting to show the influence of the traffic intensity. This is done in Figure 4.3 for the truncated $M/M/1$ retrial queue with $\mu = 2.5$ and $\rho = 0.2$, 0.4, 0.6 and 0.8. We display the numerical inversion of the density $f_L(x)$ and notice that the curves have a decreasing shape with tails becoming heavier for increasing values of ρ. We also observe that $f_L(0) = 1.0$. This fact agrees with the Tauberian statement

$$f_L(0) = \lim_{s \to \infty} sL^*(s). \quad (4.14)$$

As $\lim_{s\to\infty} sa_j(s) = \delta_{0j} f_B(0)$ (here $f_B(x)$ denotes the service time density), equations (4.2) and (4.14) yield $f_L(0) = f_B(0)$.

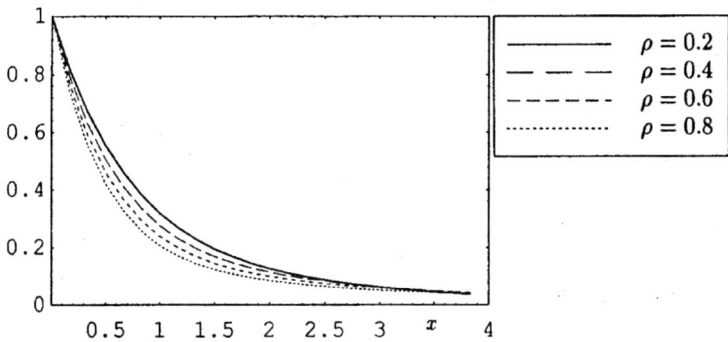

Fig. 4.3. Density functions of L as ρ varies

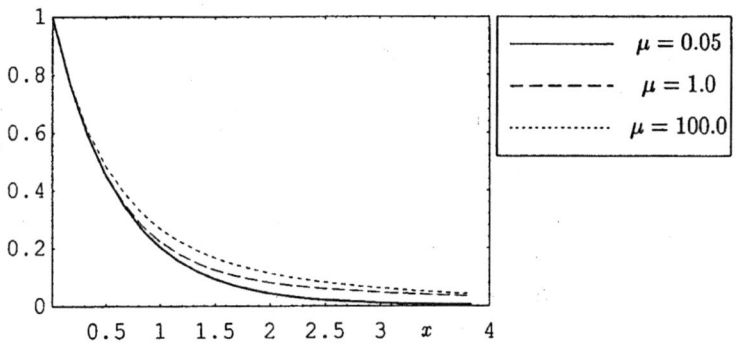

Fig. 4.4. Density functions of L as μ varies

In Figure 4.4, we take $\rho = 0.6$ and display the numerical inversion of the density $f_L(x)$ for the values of $\mu = 0.05, 1.0$ and 100.0. The curves have a decreasing shape with heavier tails when μ decreases.

We can also compare the shape of $f_L(x)$ for different choices of the service time distribution. In Figure 4.5, we consider exponential, E_4 and H_2 service times with $\rho = 0.4$ and $\beta_1 = \mu = 1.0$. We take the coefficient of variation of the H_2 law as 1.25. The cases exponential and H_2 exhibit a decreasing shape, whereas we observe a bell shape associated with E_4 case. Once more, we notice that the numerical inversion is consistent with the expression $f_L(0) = f_B(0)$.

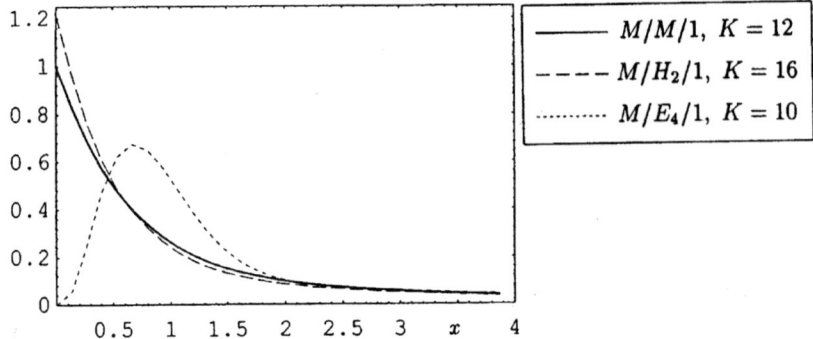

Fig. 4.5. The effect of the service time distribution on L

Table 4.5. $L^*(s)$ for an $M/E_3/1$ retrial queue as K increases

s	$K=1$	$K=3$	$K=5$	$K=7$	$K=\infty$
0.05	0.91420	0.90736	0.90719	0.90718	0.89166
0.1	0.84579	0.83731	0.83716	0.83716	0.83598
0.25	0.69783	0.69112	0.69107	0.69107	0.69107
0.5	0.54115	0.53793	0.53792	0.53792	0.53792
1.0	0.35938	0.35856	0.35856	0.35856	0.35856
1.5	0.25452	0.25426	0.25426	0.25426	0.25426
3.0	0.11080	0.11078	0.11078	0.11078	0.11078
4.5	0.05803	0.05803	0.05803	0.05803	0.05802
6.0	0.03412	0.03412	0.03412	0.03412	0.03409
10.0	0.01160	0.01160	0.01160	0.01160	0.01159

We finish this subsection by illustrating the convergence of the truncated retrial queue to the original $M/G/1$ retrial queue as $K \to \infty$. In Table 4.5, we take $\rho = 0.25$, $\mu = 0.5$ and E_3 service times with $\beta_1 = 1.0$. Since the distribution function of L in the $M/G/1$ retrial queue is unknown, we compare the value of the Laplace-Stieltjes transform $L^*(s)$ (given in column $K = \infty$) versus the Laplace-Stieltjes transform in the truncated models with $K = 1$, 3, 5 and 7 (here, the first four digits of $E[L]$ are fitted for $K = 7$). It is observed that the convergence is fast, thus showing that $K = 7$ provides a good approximation, for all values of s. Similar conclusions can be derived from Table 4.6 for the case of H_2 service times with $\beta_1 = 1.0$ and coefficient of variation $c_v = 1.25$. Further, $\rho = 0.25$ and $\mu = 0.125$.

From Tables 4.2, 4.3, 4.5 and 4.6, we may remark that this second approach based on a truncated model yields more accurate estimations for the Laplace-Stieltjes transform of L than those obtained from the use of the principle of maximum entropy.

4.1 The M/G/1 Retrial Queue 105

Table 4.6. $L^*(s)$ for an $M/H_2/1$ retrial queue as K increases

s	$K=1$	$K=4$	$K=8$	$K=12$	$K=\infty$
0.05	0.88192	0.86567	0.86545	0.86545	0.86545
0.1	0.81853	0.80670	0.80664	0.80664	0.80664
0.25	0.71061	0.70620	0.70620	0.70620	0.70620
0.5	0.60576	0.60442	0.60442	0.60442	0.60442
1.0	0.47847	0.47820	0.47820	0.47820	0.47820
1.5	0.39793	0.39785	0.39785	0.39785	0.39784
3.0	0.26602	0.26601	0.26601	0.26601	0.26601
4.5	0.20018	0.20018	0.20018	0.20018	0.20016
6.0	0.16054	0.16054	0.16054	0.16054	0.16049
10.0	0.10511	0.10511	0.10511	0.10511	0.10486

4.1.2 The Number of Customers Served

Our purpose here is to discuss the distribution of the number of customers served, I, in a busy period for the $M/G/1$ retrial queue. This characteristic is the natural discrete-valued counterpart of the length of the busy period studied in the preceding subsection. Indeed, the joint distribution of both random variables, L and I, can be derived in terms of a joint transform [288] which generalizes formula (2.10) for $L^*(s)$. In Subsection 2.2.2, we stressed the practical limitations of such a transform solution. Thus, we now focus on the algorithmic analysis of the number of customers served.

Previously, we give the following explicit expressions for the first two moments:

$$E[I] = \frac{1}{P_{00}}, \qquad (4.15)$$

$$E[I^2] = \frac{1}{P_{00}}\left(\frac{1-\rho^2+2\frac{\lambda}{\mu}\rho+\lambda^2\beta_2}{(1-\rho)^2} + \int_0^1 \frac{2\lambda}{\mu(\beta(\lambda-\lambda u)-u)}\right.$$

$$\left. \times \left(\frac{1}{1-\rho}\exp\left\{\frac{\lambda}{\mu}\int_u^1 \frac{1-\beta(\lambda-\lambda v)}{\beta(\lambda-\lambda v)-v}dv\right\} - \frac{(1-u)\beta(\lambda-\lambda u)}{\beta(\lambda-\lambda u)-u}\right)du\right). \qquad (4.16)$$

To derive (4.15), we notice that the amount of time in a cycle during which the server is busy, T_1, is related to the number of customers served through the formula

$$T_1 = \sum_{i=1}^{I} S_i,$$

where S_i is the service time of the ith customer served. Thus, we may apply Wald's identity to find $E[T_1] = \beta_1 E[I]$. On the other hand, the regenerative approach yields $E[T_1] = \rho(\lambda^{-1} + E[L])$. By combining these expressions and formula (2.11) for $E[L]$, we obtain (4.15).

The proof of (4.16) is based on the method of catastrophes; see [491]. In this paper, it is also shown that

$$P(I = 1) = a_0, \tag{4.17}$$

$$P(I = 2) = a_0 a_1 \frac{\mu}{\lambda + \mu}, \tag{4.18}$$

$$P(I = 3) = a_0^2 a_1 \frac{\lambda \mu}{(\lambda + \mu)^2} + a_0 a_1^2 \frac{\mu^2}{(\lambda + \mu)^2} + a_0^2 a_2 \frac{2\mu^2}{(\lambda + \mu)(\lambda + 2\mu)}. \tag{4.19}$$

Observe that the event $\{I = 1\}$ is fulfilled if and only if there are no primary arrivals during the first service time of the busy period, so (4.17) follows trivially. To prove (4.18) and (4.19), we consider all different possibilities for serving two and three customers, respectively, during a busy period. For example, the last term in (4.19) corresponds to the case with two primary customers arriving during the first service time. Then, both customers join the orbit and from there they go into the server and no other primary customer arrives during the busy period under consideration.

We next develop a recursive scheme to compute the probability mass function of I.

Let us observe the orbit state at service completion epochs and define $x_j^{(k)}$, for $1 \leq j \leq k$, as the conditional probability that, starting at a service completion epoch at which j customers are left behind, in the rest of the busy period up to k customers were served.

By conditioning on the number of primary customers arriving during the first service time, we have that

$$P(I = 1) = a_0,$$

$$P(I \leq k) = a_0 + \sum_{j=1}^{k-1} a_j x_j^{(k-1)}, \quad k \geq 2.$$

In the next theorem, we derive recursive equations for the probabilities $x_j^{(k)}$.

Theorem 4.3. *For any fixed $k \geq 1$, the probabilities $x_j^{(k)}$, for $1 \leq j \leq k$, can be computed as*

$$\mathbf{x}^k = \mathbf{T}_k \tilde{\mathbf{x}}^{k-1},$$

where $\mathbf{x}^k = (x_1^{(k)}, ..., x_k^{(k)})'$, $\tilde{\mathbf{x}}^0 = 1$, $\tilde{\mathbf{x}}^{k-1} = (1, \mathbf{x}^{k-1})'$, for $k \geq 2$, and $\mathbf{T}_k$ is the square matrix of order k with (i,j)th entry given by

$$t_{ij} = \begin{cases} \frac{j\mu}{\lambda + j\mu} a_0, & \text{if } 1 \leq i \leq k, \ j = i, \\ \frac{\lambda}{\lambda + i\mu} a_{j-i-1} + \frac{i\mu}{\lambda + i\mu} a_{j-i}, & \text{if } 1 \leq i < j \leq k, \\ 0, & \text{otherwise.} \end{cases}$$

4.1 The $M/G/1$ Retrial Queue 107

Proof. First, for $k = 1$, we observe that the preceding service completion left one customer in orbit. Besides, we notice that at most one more customer is served during the rest of the busy period if the customer in orbit gains the competition for the next service and, moreover, during that service time no other primary arrival occurs. We therefore have

$$x_1^{(1)} = \frac{\mu}{\lambda + \mu} a_0. \qquad (4.20)$$

For $k \geq 2$, a similar argument is used. We distinguish if the next customer served either is a primary arrival or comes from the orbit. Once again we condition on the number of arrivals occurring during that service time and so we find that

$$x_1^{(k)} = \frac{\mu}{\lambda + \mu} a_0 + \sum_{i=1}^{k-1} \frac{\lambda a_{i-1} + \mu a_i}{\lambda + \mu} x_i^{(k-1)},$$

$$x_j^{(k)} = \frac{j\mu}{\lambda + j\mu} a_0 x_{j-1}^{(k-1)} + \sum_{i=j}^{k-1} \frac{\lambda a_{i-j} + j\mu a_{i-j+1}}{\lambda + j\mu} x_i^{(k-1)}, \quad 2 \leq j \leq k-1,$$

$$x_k^{(k)} = \frac{k\mu}{\lambda + k\mu} a_0 x_{k-1}^{(k-1)}. \qquad (4.21)$$

For each $k \geq 1$, equations (4.20) and (4.21) can be written in matrix form, which leads to our statement. □

Figures 4.6 and 4.7 illustrate the effect of μ on the mean value and the standard deviation of I. We consider that service times are distributed according to E_2 and H_2 distributions with parameters chosen in such a manner that β_1 equals 1.0, and the coefficient of variation is 0.8 and 2.0, respectively. Each figure presents three pairs of curves corresponding to $\rho = 0.25$, 0.5 and 0.75.

In the light of the figures, similar conclusions can be inferred for both descriptors. They increase with increasing traffic intensity and decreasing retrial rate. For a fixed value of μ, we observe that the queue with E_2 service times exhibits less variability and higher expected number of customers served. Differences between both service distributions become more apparent for higher values of ρ.

The probabilities $P(I = k)$, for $1 \leq k \leq 6$, are displayed in Table 4.7 for several values of μ. Entries, from top to bottom, correspond to E_2, exponential and H_2 service times. We take $\lambda = 0.5$ and $\beta_1 = 1.0$. For all choices of μ, irrespectively of the service time distribution, we obtain a decreasing distribution giving a significant proportion of mass to the first point $k = 1$. In agreement with our expectations, the distribution becomes more sparse as far as μ decreases. Moreover, when $\mu \to \infty$, both $E[I]$ and $\sigma(I)$ are consistent with the corresponding values for the standard $M/G/1$ queue.

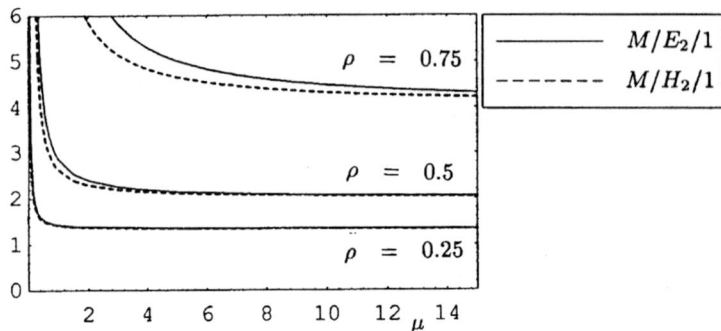

Fig. 4.6. $E[I]$ versus μ

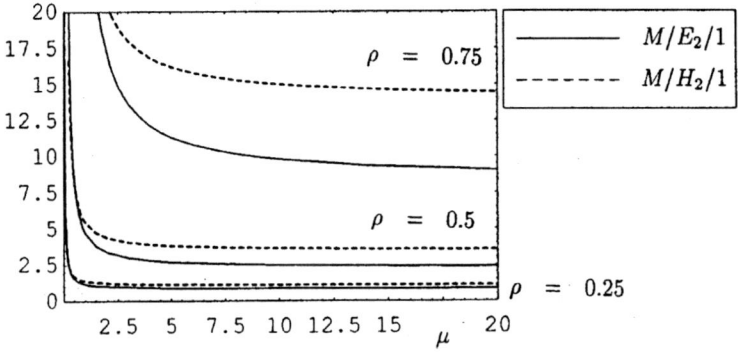

Fig. 4.7. $\sigma(I)$ versus μ

4.1.3 Maximal Queue Length in a Busy Period

In this subsection, we consider a variant of the preceding arguments in Subsection 4.1.2 to extend our study to the maximum orbit size, say N_{max}, during a busy period. We remark the interest in N_{max} as a descriptor clearly connected to the system congestion.

To determine the probability mass function of N_{max}, we have need of the embedded Markov chain at departure epochs. By defining $y_j^{(k)}$ as the conditional probability that, starting from the state j, this embedded Markov chain reaches the state 0 before hitting the state $k+1$, we readily derive

$$P(N_{max} = 0) = a_0,$$

$$P(N_{max} \leq k) = a_0 + \sum_{j=1}^{k} a_j y_j^{(k)}, \quad k \geq 1, \qquad (4.22)$$

4.1 The $M/G/1$ Retrial Queue   109

Table 4.7. Probability mass function of I, case $\rho = 0.5$

k		$\mu = 0.01$	$\mu = 0.1$	$\mu = 0.5$	$\mu = 2.0$	$\mu = 10.0$	$\mu = 100.0$
1	E_2	0.64724	0.64724	0.64724	0.64724	0.64724	0.64724
	Exp	0.66666	0.66666	0.66666	0.66666	0.66666	0.66666
	H_2	0.72727	0.72727	0.72727	0.72727	0.72727	0.72727
2	E_2	0.00313	0.02666	0.07999	0.12798	0.15236	0.15918
	Exp	0.00290	0.02469	0.07407	0.11851	0.14109	0.14741
	H_2	0.00251	0.02137	0.06411	0.10257	0.12211	0.12758
3	E_2	0.00202	0.01698	0.04932	0.06437	0.06996	0.07107
	Exp	0.00193	0.01619	0.04389	0.06028	0.06492	0.06576
	H_2	0.00182	0.01484	0.03779	0.04822	0.04935	0.04913
4	E_2	0.00133	0.01184	0.03176	0.03956	0.03959	0.03916
	Exp	0.00130	0.01150	0.03027	0.03719	0.03706	0.03663
	H_2	0.00133	0.01096	0.02558	0.02804	0.02614	0.02533
5	E_2	0.00088	0.00883	0.02385	0.02683	0.02492	0.02408
	Exp	0.00089	0.00868	0.02276	0.02532	0.02358	0.02285
	H_2	0.00098	0.00851	0.01882	0.01837	0.01614	0.01542
6	E_2	0.00059	0.00692	0.01891	0.01929	0.01673	0.01583
	Exp	0.00061	0.00686	0.01804	0.01828	0.01603	0.01526
	H_2	0.00073	0.00687	0.01462	0.01300	0.01100	0.01044

where the quantities a_j, for $j \geq 0$, are defined by (3.5). To prove (4.22), we note that $y_j^{(k)}$ can be interpreted as the conditional probability that the busy period in progress ends without exceeding k customers in orbit, given that we start at a departure epoch with j customers in orbit.

We now show that the probabilities $y_j^{(k)}$ in (4.22) can be evaluated as the solution to a set of linear equations.

Theorem 4.4. *For each fixed $k \geq 1$, the probabilities $y_j^{(k)}$, for $1 \leq j \leq k$, can be computed as the solution to the system*

$$\mathbf{y}^k = \mathbf{U}_k \mathbf{y}^k + \mathbf{c}, \tag{4.23}$$

where $\mathbf{y}^k = (y_1^{(k)}, ..., y_k^{(k)})'$, $\mathbf{U}_k$ is the square matrix of order k with (i,j)th entry defined by

$$u_{ij} = \begin{cases} \frac{\lambda}{\lambda+i\mu} a_{j-i} + \frac{i\mu}{\lambda+i\mu} a_{j-i+1}, & \text{if } 1 \leq i \leq j \leq k, \\ \frac{i\mu}{\lambda+i\mu} a_0, & \text{if } 2 \leq i \leq k,\ j = i-1, \\ 0, & \text{otherwise,} \end{cases}$$

and $\mathbf{c}$ is given by $(\frac{\mu}{\lambda+\mu} a_0, 0, ..., 0)'$.

Proof. We observe that, for $k = 1$, we can write down

$$y_1^{(1)} = \frac{\lambda}{\lambda+\mu} a_0 y_1^{(1)} + \frac{\mu}{\lambda+\mu}\left(a_0 + a_1 y_1^{(1)}\right), \tag{4.24}$$

from which it follows that

$$y_1^{(1)} = \frac{\mu a_0}{\lambda(1-a_0) + \mu(1-a_1)}.$$

Equation (4.24) is derived by noting that the previous departure left one customer in orbit. Then, by keeping in mind that the next customer to receive service may come from the orbit (which occurs with probability $\mu/(\lambda+\mu)$) or may be a primary arrival (which occurs with probability $\lambda/(\lambda+\mu)$), there are two cases to consider. In both cases, there can be at most one arrival during the corresponding service time.

Equations for $k \geq 2$ are derived similarly by distinguishing between the cases where the next customer to receive service is a primary arrival or a customer from the orbit. In either case, we must record the number of primary arrivals occurring during the subsequent service time so that, at its completion, the orbit size never reaches $k+1$ before the system empties out. This results in the linear equations

$$y_1^{(k)} = \frac{\mu}{\lambda+\mu} a_0 + \sum_{i=1}^{k}\left(\frac{\lambda}{\lambda+\mu}a_{i-1} + \frac{\mu}{\lambda+\mu}a_i\right) y_i^{(k)},$$

$$y_j^{(k)} = \frac{j\mu}{\lambda+j\mu} a_0 y_{j-1}^{(k)} + \sum_{i=j}^{k}\left(\frac{\lambda}{\lambda+j\mu}a_{i-j} + \frac{j\mu}{\lambda+j\mu}a_{i-j+1}\right) y_i^{(k)}, \; 2 \leq j \leq k.$$

(4.25)

In matrix form, (4.25) is written as claimed in (4.23). □

Since the matrix of coefficients in (4.23) is of $M/G/1$-type, (4.23) is amenable to be solved by Gaussian elimination. The mass function of N_{max} is then evaluated from (4.22) by substitution of the values of $y_j^{(k)}$.

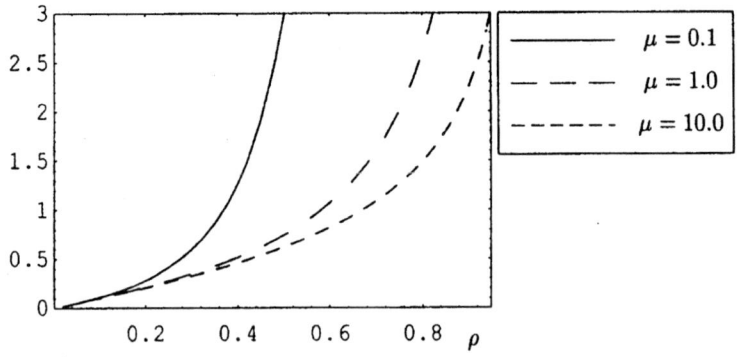

Fig. 4.8. $E[N_{max}]$ versus ρ

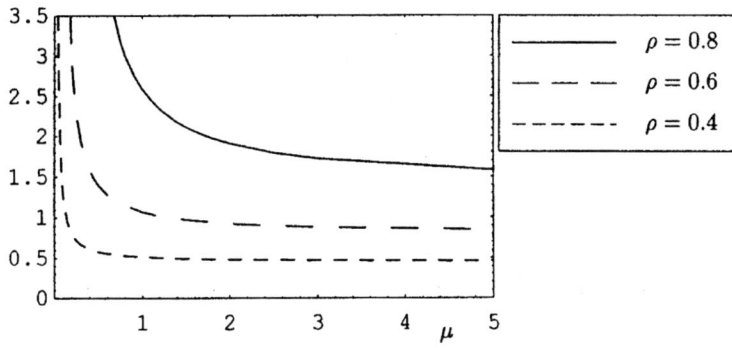

Fig. 4.9. $E[N_{max}]$ versus μ

Next we report numerical results on the behavior of N_{max} used in [493]. In a first example, we consider an $M/M/1$ retrial queue with mean service time $\beta_1 = 1.0$. In Figures 4.8 and 4.9, we display the expected value $E[N_{max}]$ as a function of ρ and μ, respectively. Curves in Figure 4.8 have increasing shapes. As is to be expected, the expected orbit occupancy increases with increasing values of the traffic intensity, irrespective of the value of μ. Curves in Figure 4.9 show that, for models with higher retrial rate, the orbit becomes less congested than for those having lower retrial rates.

In a second example, we assume that service times follow Erlang laws with $k = 2$ and 5 stages, and mean service time $\beta_1 = 1.0$. Table 4.8 lists values of $E[N_{max}]$ for several retrial and traffic intensities. Similarly to Figures 4.8 and 4.9, we observe that $E[N_{max}]$ increases with increasing values of ρ, but decreases with increasing values of μ. Also, we notice that the orbit is less congested when the Erlang law has more stages.

4.2 The $M/M/c$ Retrial Queue

In Subsections 3.4.1 and 3.4.2, we showed that the limiting probabilities of the $M/M/c$ retrial queue can be efficiently computed with the help of approximate truncated models. The aim of this section is to extend such an analysis to the busy period and related quantities; in particular, to the number of customers served, maximal queue length and moments. Thus, we develop algorithmic methods for studying the busy period of the models introduced by Wilkinson, Falin, and Neuts and Rao. We exploit that the process $\mathcal{X}^W$ is finite and the fact that the retrial rates of processes $\mathcal{X}^F$ and $\mathcal{X}^{NR}$ are homogeneous from a certain orbit level and up. However, we do not deal with the busy period of the process $\mathcal{X}^{AP}$ due to the non-homogeneity of its retrial rate.

Table 4.8. $E[N_{max}]$ versus μ and ρ for $M/E_k/1$ retrial queues

μ	k	$\rho = 0.2$	$\rho = 0.4$	$\rho = 0.6$	$\rho = 0.8$	$\rho = 0.9$
0.05	2	0.28150	1.31357	7.99605	51.48338	172.36573
	5	0.23659	0.73831	2.51308	15.43906	57.52662
0.1	2	0.24400	0.81234	3.07618	20.09706	72.89178
	5	0.22095	0.58402	1.45339	5.49134	18.75898
0.25	2	0.22128	0.58304	1.42930	5.10792	16.40142
	5	0.21135	0.49774	1.00600	2.41040	5.14011
0.5	2	0.21355	0.51397	1.07128	2.69802	6.00982
	5	0.20810	0.46934	0.87872	1.79296	3.13400
1.0	2	0.20963	0.48008	0.91692	1.92356	3.43629
	5	0.20646	0.45511	0.81796	1.53764	2.44046
2.5	2	0.20727	0.45979	0.83026	1.55798	2.44734
	5	0.20548	0.44655	0.78221	1.39838	2.09782
5.0	2	0.20647	0.45302	0.80217	1.44935	2.18413
	5	0.20515	0.44368	0.77039	1.35402	1.99395

4.2.1 The Length of the Busy Period

Denote the length of the busy period for the corresponding approximate models as L^W, L^F and L^{NR}. We also simplify the notation for the truncation levels so, in the sequel, K will denote the truncation parameter of any model under consideration.

We next develop algorithmic schemes for the computation of the Laplace-Stieltjes transform of the busy period. First, let us consider the model $\mathcal{X}^W$ and define L_{ij} as the first-passage time to the state $(0,0)$, starting from the state $(i,j) \in \mathcal{S}^W$. Then, the length of the busy period L^W corresponds to L_{10}. By employing a first-step argument, we get the following system for the Laplace-Stieltjes transforms $L_{ij}^*(s) = E[\exp\{-sL_{ij}\}]$:

$$L_{00}^*(s) = 1, \tag{4.26}$$

$$L_{ij}^*(s) = \frac{\lambda}{s+\lambda+i\nu+j\mu} L_{i+1,j}^*(s) + \frac{i\nu}{s+\lambda+i\nu+j\mu} L_{i-1,j}^*(s)$$
$$+ \frac{j\mu}{s+\lambda+i\nu+j\mu} L_{i+1,j-1}^*(s),$$
$$0 \le i \le c-1,\ 0 \le j \le K,\ (i,j) \ne (0,0), \tag{4.27}$$

$$L_{cj}^*(s) = \frac{\lambda}{s+\lambda+c\nu} L_{c,j+1}^*(s) + \frac{c\nu}{s+\lambda+c\nu} L_{c-1,j}^*(s),\quad 0 \le j \le K-1, \tag{4.28}$$

$$L_{cK}^*(s) = \frac{c\nu}{s+c\nu} L_{c-1,K}^*(s). \tag{4.29}$$

We observe that the set of equations (4.27), which is associated with the orbit level j, can be solved by exploiting its tridiagonal structure. More concretely, we may express any unknown $L_{ij}^*(s)$, for $0 \le i \le c-1$, in terms of

4.2 The M/M/c Retrial Queue

$L_{cj}^*(s)$ and elements $L_{i'j'}^*(s)$ of previously computed levels $j' = 0, ..., j-1$. Then, equations (4.28) and (4.29) allow us to eliminate several unknowns and relate $L_{cj}^*(s)$ and $L_{c,j+1}^*(s)$. Finally, we may use the boundary equation (4.26) to evaluate all transforms $L_{ij}^*(s)$ and, in particular, the desired Laplace-Stieltjes transform $L_W^*(s) = L_{10}^*(s)$. The algorithmic solution is summarized in the following theorem.

Theorem 4.5. *The Laplace-Stieltjes transform $L_W^*(s)$ can be computed as follows:*

Step 1. Define the coefficients b_{i0}, $\bar{D}_{i0}$, $\hat{F}_{i0}$ and $\hat{G}_{i0}$, for $1 \leq i \leq c-1$, recursively by the relations

$$b_{10} = \nu, \qquad b_{i0} = i\nu \frac{s + b_{i-1,0}}{s + \lambda + b_{i-1,0}}, \qquad 2 \leq i \leq c-1, \tag{4.30}$$

$$\bar{D}_{10} = \nu, \qquad \bar{D}_{i0} = i\nu \frac{\bar{D}_{i-1,0}}{s + \lambda + b_{i-1,0}}, \qquad 2 \leq i \leq c-1, \tag{4.31}$$

$$\hat{F}_{i0} = \frac{1}{\lambda} \sum_{k=i}^{c-1} \bar{D}_{k0} \prod_{n=i}^{k} \frac{\lambda}{s + \lambda + b_{n0}}, \qquad 1 \leq i \leq c-1, \tag{4.32}$$

$$\hat{G}_{i0} = \prod_{k=i}^{c-1} \frac{\lambda}{s + \lambda + b_{k0}}, \qquad 1 \leq i \leq c-1. \tag{4.33}$$

Step 2. For $j = 1, ..., K$, calculate sequentially the coefficients b_{ij}, G_{ij}, $H_{ij}^{(k)}$, $\bar{D}_{ij}$, $\tilde{D}_{ij}$, $\hat{F}_{ij}$ and $\hat{G}_{ij}$ given by

$$b_{0j} = 0, \qquad b_{ij} = i\nu \frac{s + j\mu + b_{i-1,j}}{s + \lambda + j\mu + b_{i-1,j}}, \qquad 1 \leq i \leq c-1,$$

$$G_{ij} = \prod_{k=i}^{c-1} \frac{\lambda}{s + \lambda + j\mu + b_{kj}}, \qquad 0 \leq i \leq c-1, \tag{4.34}$$

$$H_{ij}^{(k)} = \nu^{i-k+1} \prod_{n=k-1}^{i-1} \frac{n+1}{s + \lambda + j\mu + b_{nj}}, \qquad 0 \leq i \leq c-1, \ 1 \leq k \leq i+1,$$

$$\bar{D}_{ij} = j\mu \sum_{k=1}^{i+1} \left(\hat{F}_{k,j-1} + \hat{G}_{k,j-1} \frac{c\nu \hat{F}_{c-1,j-1}}{s + \lambda + c\nu(1 - \hat{G}_{c-1,j-1})} \right) H_{ij}^{(k)},$$
$$0 \leq i \leq c-1, \tag{4.35}$$

$$\tilde{D}_{ij} = \frac{\lambda j\mu}{s + \lambda + c\nu(1 - \hat{G}_{c-1,j-1})} \sum_{k=1}^{i+1} \hat{G}_{k,j-1} H_{ij}^{(k)}, \quad 0 \leq i \leq c-1, \tag{4.36}$$

$$\hat{F}_{ij} = \frac{1}{\lambda} \sum_{k=i}^{c-1} \bar{D}_{kj} \prod_{n=i}^{k} \frac{\lambda}{s + \lambda + j\mu + b_{nj}}, \qquad 0 \leq i \leq c-1, \tag{4.37}$$

$$\hat{G}_{ij} = G_{ij} + \frac{1}{\lambda} \sum_{k=i}^{c-1} \tilde{D}_{kj} \prod_{n=i}^{k} \frac{\lambda}{s + \lambda + j\mu + b_{nj}}, \qquad 0 \leq i \leq c-1. \tag{4.38}$$

Step 3W. Calculate $L^*_{cK}(s)$ as

$$L^*_{cK}(s) = \frac{c\nu \hat{F}_{c-1,K}}{s + c\nu(1 - \hat{G}_{c-1,K})}. \tag{4.39}$$

Step 4. Calculate $L^*_{c0}(s)$ by the formula

$$L^*_{c0}(s) = \prod_{j=0}^{K-1} \frac{\lambda}{s + \lambda + c\nu(1 - \hat{G}_{c-1,j})} L^*_{cK}(s)$$

$$+ c\nu \sum_{j=0}^{K-1} \frac{\lambda^j \hat{F}_{c-1,j}}{\prod_{n=0}^{j}(s + \lambda + c\nu(1 - \hat{G}_{c-1,n}))}. \tag{4.40}$$

Step 5. Calculate the Laplace-Stieltjes transform of L^W as

$$L^*_W(s) = \hat{F}_{10} + \hat{G}_{10} L^*_{c0}(s). \tag{4.41}$$

Proof. The set of equations (4.27), for $j = 0$, has the tridiagonal form

$$\alpha_{i0} L^*_{i-1,0}(s) + \beta_{i0} L^*_{i0}(s) + \gamma_{i0} L^*_{i+1,0}(s) = \omega_{i0}, \quad 1 \leq i \leq c - 1,$$

where $\alpha_{10} = 0$, $\alpha_{i0} = -i\nu$, for $2 \leq i \leq c - 1$, $\beta_{i0} = s + \lambda + i\nu$, for $1 \leq i \leq c - 1$, $\gamma_{i0} = -\lambda$, for $1 \leq i \leq c - 1$, $\omega_{10} = \nu$ and $\omega_{i0} = 0$, for $2 \leq i \leq c - 1$. This set of equations can be reduced to a diagonal formulation by using forward-elimination-backward-substitution method. Similarly to Theorem 3.10, we may introduce coefficients appropriately to avoid subtractions. We then observe that such a set of equations is reduced to

$$L^*_{i0}(s) = \frac{\bar{D}_{i0} + \lambda L^*_{i+1,0}(s)}{s + \lambda + b_{i0}}, \quad 1 \leq i \leq c - 1, \tag{4.42}$$

where b_{i0} and $\bar{D}_{i0}$ are given by (4.30) and (4.31). By iteration of (4.42), we easily find that

$$L^*_{i0}(s) = \hat{F}_{i0} + \hat{G}_{i0} L^*_{c0}(s), \quad 1 \leq i \leq c - 1, \tag{4.43}$$

where $\hat{F}_{i0}$ and $\hat{G}_{i0}$ are, respectively, given by (4.32) and (4.33). If we define $\hat{F}_{c0} = 0$ and $\hat{G}_{c0} = 1$, then (4.43) also holds for $i = c$.

For $j = 1, ..., K$, the set of equations (4.27) keeps a similar tridiagonal form, but now i varies from 0 to $c - 1$ and the coefficients are given by $\alpha_{ij} = -i\nu$, $\beta_{ij} = s + \lambda + i\nu + j\mu$, $\gamma_{ij} = -\lambda$ and $\omega_{ij} = j\mu L^*_{i+1,j-1}(s)$. By applying once more forward-elimination-backward-substitution, we obtain

$$L^*_{ij}(s) = F_{ij} + G_{ij} L^*_{cj}(s), \quad 0 \leq i \leq c - 1, \tag{4.44}$$

where G_{ij} is defined in (4.34) and

$$F_{ij} = \frac{1}{\lambda} \sum_{k=i}^{c-1} D_{kj} \prod_{n=i}^{k} \frac{\lambda}{s+\lambda+j\mu+b_{nj}}, \quad 0 \le i \le c-1, \qquad (4.45)$$

where $D_{0j} = j\mu L^*_{1,j-1}(s)$, $D_{ij} = j\mu L^*_{i+1,j-1}(s) + i\nu D_{i-1,j}(s+\lambda+j\mu+b_{i-1,j})^{-1}$, for $1 \le i \le c-1$.

For a given pair (i,j), the coefficient F_{ij} depends only on the transforms $L^*_{i'j'}(s)$ with $j' < j$, so (4.44) expresses $L^*_{ij}(s)$ in terms of $L^*_{cj}(s)$ and $L^*_{i'j'}(s)$, for $j' < j$. To remove the dependence on previous orbit levels, we iterate the definition of D_{i1} and find

$$D_{i1} = \mu \sum_{k=1}^{i+1} L^*_{k0}(s) H^{(k)}_{i1}, \quad 0 \le i \le c-1. \qquad (4.46)$$

Moreover, by combining (4.28), for $j = 0$, and (4.43), we get

$$L^*_{c0}(s) = \frac{c\nu \hat{F}_{c-1,0} + \lambda L^*_{c1}(s)}{s+\lambda+c\nu(1-\hat{G}_{c-1,0})}. \qquad (4.47)$$

By substituting (4.43) and (4.47) into (4.46), we obtain that $D_{i1} = \bar{D}_{i1} + \hat{D}_{i1} L^*_{c1}(s)$, where $\bar{D}_{i1}$ and $\hat{D}_{i1}$ are, respectively, given by (4.35) and (4.36). Then, by (4.44) and (4.45), we have that

$$L^*_{i1}(s) = \hat{F}_{i1} + \hat{G}_{i1} L^*_{c1}(s), \quad 0 \le i \le c-1, \qquad (4.48)$$

where $\hat{F}_{i1}$ and $\hat{G}_{i1}$ are defined by (4.37) and (4.38).

We conclude the proof by induction. Indeed, we may prove that

$$L^*_{ij}(s) = \hat{F}_{ij} + \hat{G}_{ij} L^*_{cj}(s), \quad 0 \le i \le c-1, \ 1 \le j \le K, \qquad (4.49)$$

$$L^*_{c,j-1}(s) = \frac{c\nu \hat{F}_{c-1,j-1} + \lambda L^*_{cj}(s)}{s+\lambda+c\nu(1-\hat{G}_{c-1,j-1})}, \quad 1 \le j \le K. \qquad (4.50)$$

Expressions (4.47) and (4.48) show the validity of (4.49) and (4.50) for $j = 1$. After assuming the inductive hypothesis, we can prove, in a similar manner to the case $j = 1$, that formulas (4.49) and (4.50) are also true for $j = k$. Finally, by replacing (4.29) in (4.49) for $j = K$, we derive

$$L^*_{c-1,K}(s) = \frac{(s+c\nu)\hat{F}_{c-1,K}}{s+c\nu(1-\hat{G}_{c-1,K})}$$

and, by using again (4.29), we get (4.39). By iterating (4.50) we obtain (4.40). Then, formula (4.43) for $i = 1$ yields (4.41), which completes the proof. □

We now focus on the approximate model $\mathcal{X}^F$. In this case, the set of equations governing the Laplace-Stieltjes transforms is essentially as (4.26)-(4.29) in Wilkinson's model, except that equation (4.28) is also valid for $j = K$ and (4.29) is replaced by

4 Busy Period

$$L_{cj}^*(s) = \frac{\lambda}{s+\lambda+c\nu}L_{c,j+1}^*(s) + \frac{c\nu}{s+\lambda+c\nu}L_{c,j-1}^*(s), \quad j \geq K+1.$$

The next theorem describes the necessary modifications in Theorem 4.5 to compute $L_F^*(s)$.

Theorem 4.6. *The Laplace-Stieltjes transform of L^F can be computed by following the steps of Theorem 4.5, with the exception of Step 3W, which is replaced by*

Step 3F. Calculate $L_{cK}^*(s)$ by the formula

$$L_{cK}^*(s) = \frac{c\nu \hat{F}_{c-1,K}}{s + \lambda(1-\Phi(s)) + c\nu(1-\hat{G}_{c-1,K})}, \quad (4.51)$$

where

$$\Phi(s) = \frac{s+\lambda+c\nu - \sqrt{(s+\lambda+c\nu)^2 - 4\lambda c\nu}}{2\lambda}. \quad (4.52)$$

Proof. Since the transforms $L_{ij}^*(s)$ for model $\mathcal{X}^F$ satisfy equations (4.26)-(4.28), we conclude that the proof given in Theorem 4.5 remains valid for Falin's model up to the establishment of (4.49) and (4.50). Indeed, (4.50) is valid for $j = K+1$, because of (4.28) for $j = K$.

We now observe that the model $\mathcal{X}^F$ evolves as the standard $M/M/1$ queue, with arrival rate λ and service rate $c\nu$, from the orbit level $K+1$ up. Thus, we have

$$L_{c,K+1}^*(s) = \Phi(s)L_{cK}^*(s), \quad (4.53)$$

where $\Phi(s)$ denotes the Laplace-Stieltjes transform of the busy period length of such a standard $M/M/1$ queue, which is given by (4.52). By substituting (4.53) in (4.50) for $j = K+1$, we readily obtain (4.51). This completes the proof. □

Finally, we concentrate on $\mathcal{X}^{NR}$. This model satisfies equations (4.26) and (4.27), and (4.28) is now valid for $j \geq 0$. In addition, we have the equations

$$L_{ij}^*(s) = \frac{\lambda}{s+\lambda+i\nu+K\mu}L_{i+1,j}^*(s) + \frac{i\nu}{s+\lambda+i\nu+K\mu}L_{i-1,j}^*(s)$$
$$+ \frac{K\mu}{s+\lambda+i\nu+K\mu}L_{i+1,j-1}^*(s), \quad 0 \leq i \leq c-1, \, j \geq K+1.$$

Since $\mathcal{X}^{NR}$ is a QBD process, the analysis of the Laplace-Stieltjes transforms $L_{ij}^*(s)$ is related to the so-called fundamental period (see details in [544, Section 3.3]). For completeness, we summarize some basic features here.

Let $\xi_{ii'}^{(n)}(x)$ be the conditional probability that, starting in $(i, K+n)$, the process $\mathcal{X}^{NR}$ reaches the set $\{(0,K), ..., (c,K)\}$ (i.e., the level K) for the first time no later than time x and does so by entering the state (i', K). Let $\boldsymbol{\Omega}^{(n)}(s)$ be a square matrix of order $c+1$ with (i, i')th element defined by the Laplace-Stieltjes transform

$$\omega_{ii'}^{(n)}(s) = \int_0^\infty e^{-sx} d\xi_{ii'}^{(n)}(x).$$

Again a first-step conditional argument is the key to show that $\boldsymbol{\Omega}^{(n)}(s) = \boldsymbol{\Omega}^n(s)$ and that the matrix $\boldsymbol{\Omega}(s)$ satisfies the quadratic matrix equation

$$\mathbf{X}(s) = (s\mathbf{I}_{c+1} - \mathbf{A}_1)^{-1}\mathbf{A}_2 + (s\mathbf{I}_{c+1} - \mathbf{A}_1)^{-1}\mathbf{A}_0\mathbf{X}^2(s), \qquad (4.54)$$

where $\mathbf{I}_{c+1}$ is the identity matrix of order $c+1$. The rate block matrices $\mathbf{A}_0$, $\mathbf{A}_1$ and $\mathbf{A}_2$ describe the homogeneous part of the process $\mathcal{X}^{NR}$; that is, in terms of the notation of Section 2.1, they are given by

$$\mathbf{A}_0 = \mathbf{A}_{K,K+1}, \quad \mathbf{A}_1 = \mathbf{A}_{KK} \quad \text{and} \quad \mathbf{A}_2 = \mathbf{A}_{K,K-1}.$$

Moreover, for every s, $\boldsymbol{\Omega}(s)$ is the minimal non-negative solution of the equation (4.54) and consequently can be computed as limit of the iterative scheme

$$\mathbf{X}_0(s) = \mathbf{0}_{(c+1)\times(c+1)},$$
$$\mathbf{X}_{n+1}(s) = (s\mathbf{I}_{c+1} - \mathbf{A}_1)^{-1}\mathbf{A}_2 + (s\mathbf{I}_{c+1} - \mathbf{A}_1)^{-1}\mathbf{A}_0\mathbf{X}_n^2(s), \quad n \geq 0, \quad (4.55)$$

where $\mathbf{0}_{p\times q}$ is the null matrix of dimension $p \times q$.

With these preliminaries in mind, we may now obtain an algorithmic procedure for computing $L_{NR}^*(s)$.

Theorem 4.7. *The Laplace-Stieltjes transform of L^{NR} can be computed by using the steps of Theorem 4.5 with Step 3W replaced by*

Step 3NR. Compute $\boldsymbol{\Omega}(s) = (\omega_{ii'}(s))$ by the iterative scheme (4.55). Then, calculate $L_{cK}^*(s)$ as

$$L_{cK}^*(s) = \frac{\lambda \sum_{i=0}^{c-1} \omega_{ci}(s)\hat{F}_{iK} + c\nu\hat{F}_{c-1,K}}{s + \lambda(1 - \omega_{cc}(s)) - \sum_{i=0}^{c-1} \omega_{ci}(s)\hat{G}_{iK} + c\nu(1 - \hat{G}_{c-1,K})}. \qquad (4.56)$$

Proof. The transforms $L_{ij}^*(s)$ in this model satisfy (4.26)-(4.28). Thus, the proof of Theorem 4.5 remains true up to the establishment of formulas (4.49) and (4.50). Combining (4.28), for $j = K$, and (4.49), for $(i,j) = (c-1, K)$, yields

$$\left(s + \lambda + c\nu\left(1 - \hat{G}_{c-1,K}\right)\right)L_{cK}^*(s) = c\nu\hat{F}_{c-1,K} + \lambda L_{c,K+1}^*(s). \qquad (4.57)$$

Conditioning on the state visited at the first entrance to level K results in

$$L_{c,K+1}^*(s) = \sum_{i=0}^{c}\omega_{ci}(s)L_{iK}^*(s). \qquad (4.58)$$

By combining (4.49) and (4.58), we may express $L_{c,K+1}^*(s)$ in terms of $L_{cK}^*(s)$. Finally, we solve (4.57) to derive the desired expression (4.56). □

Table 4.9. Truncation thresholds for an $M/M/5$ retrial queue

ρ	$\mu = 0.1$	$\mu = 1.0$	$\mu = 10.0$	$\mu = 100.0$
0.1	(2,1)	(1,0)	(1,0)	(1,0)
0.3	(6,6)	(3,4)	(2,2)	(1,0)
0.6	(25,29)	(11,14)	(6,8)	(4,4)
0.8	(81,94)	(30,37)	(17,21)	(10,12)

After having computed $L^*(s)$ by using any of the preceding procedures, we can obtain an accurate approximation for the density function of L by applying any of the well-known numerical inversion algorithms. We illustrate this for a queue with $c = 5$ and $\nu = 1.0$. Table 4.9 gives ordered pairs with the critical truncation levels K for the models $\mathcal{X}^{NR}$ and $\mathcal{X}^{F}$, respectively, which have been again obtained following Criterion II of Subsection 3.4.2. For example, the entry $(11, 14)$ means that the truncation level is 11 for the model $\mathcal{X}^{NR}$, but it is 14 in $\mathcal{X}^{F}$.

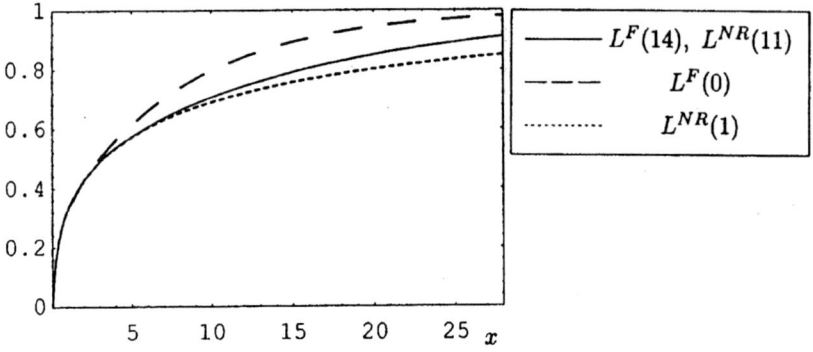

Fig. 4.10. Distribution functions of L^F and L^{NR}

In Figure 4.10, we fix $\rho = 0.6$ and $\mu = 1.0$ and display several distribution functions of L^F and L^{NR}. To this end, we appeal to the numerical inversion of the transforms $L_F^*(s)$ and $L_{NR}^*(s)$ computed from Theorems 4.6 and 4.7, respectively. The continuous curve represents simultaneously the distributions of L^F and L^{NR} calculated for the critical levels given in Table 4.9. Differences between both distribution functions are not visible because they do not differ until the fifth decimal digit.

At this point, we observe that

$$L^F \leq_d L \leq_d L^{NR}, \tag{4.59}$$
$$L^{NR}(K+1) \leq_d L^{NR}(K), \tag{4.60}$$
$$L^F(K) \leq_d L^F(K+1), \tag{4.61}$$

where $\theta \leq_d \eta$ denotes that θ is stochastically smaller in distribution than η. The extended notation $L^{NR}(K)$ remarks that we deal with $\mathcal{X}^{NR}$ with associated threshold K. For a proof of (4.59)-(4.61), we refer the reader to [44, 604].

In the light of (4.59), we see that the distribution function of L for the original $M/M/c$ retrial queue is placed between the distribution functions of L^{NR}, for $K = 11$, and L^F, for $K = 14$ (see Table 4.9). Since both curves differ in no more than 1.8×10^{-5}, they are graphically undistinguished, so the proposed approximations work notably well. On the other hand, the curves associated with the distribution functions of L^F, for $K = 0$, and L^{NR}, for $K = 1$, (i.e., the lowest possible thresholds for the approximate models) provide upper and lower bounds, respectively. As far as K increases, the distribution functions of $\mathcal{X}^{NR}$ approach the continuous curve from below, whereas the corresponding functions for $\mathcal{X}^F$ decrease to the same limiting curve. In what follows, we approximate L by L^{NR}.

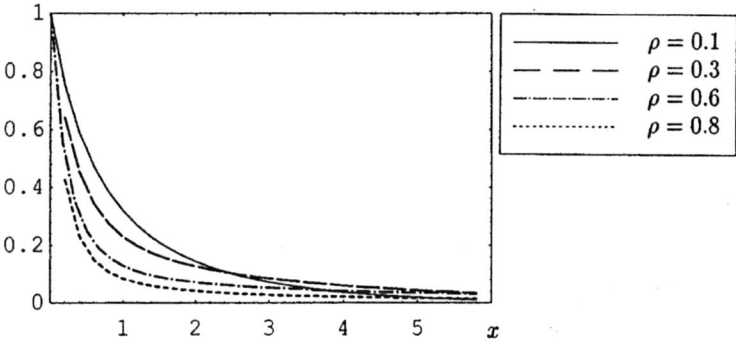

Fig. 4.11. Density functions of L as ρ varies

Figure 4.11 illustrates the influence of the traffic intensity ρ. We assume $\mu = 1.0$, in addition to the initial choices of c and ν. We observe the existence of one mode at the origin and that $f_L(0) = 1.0$. The latter is explained by the Tauberian relation

$$\lim_{s \to \infty} sL^*(s) = \lim_{x \to 0} f_L(x) = \nu.$$

We also observe that the tail of the distribution is heavier for larger values of the traffic load.

Figure 4.12 shows the effect of the retrial rate μ for the case $\rho = 0.6$. The three curves exhibit a decreasing shape and take the value 1.0 at the origin, but, for graphical convenience, densities are depicted on the domain where $f_L(x) < 0.55$. We notice that the retrial rate seems to be more influential at the tail of the distribution.

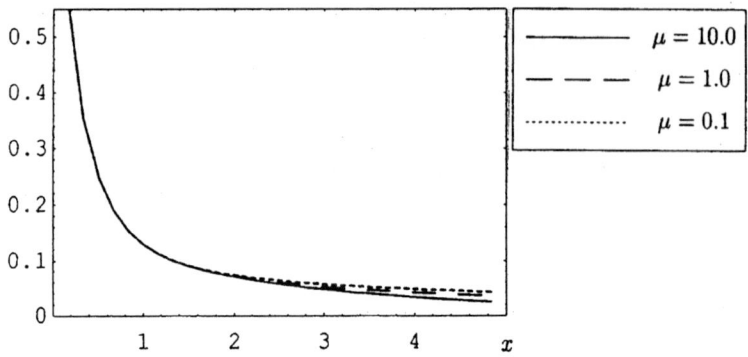

Fig. 4.12. Density functions of L as μ varies

In what follows, we analyze the recursive computation of the busy period moments for the three approximate models. We first multiply equations (4.26)-(4.29) by the corresponding denominators to express them in multiplicative form. Differentiating k times and letting $s = 0$, we obtain sets of equations amenable to be solved by the forward-elimination-backward-substitution method. We next present the algorithmic methods in full detail to be ready for implementation, but we omit the proof.

Let $L_{ij}^{(k)}$ be the kth moment of L_{ij}; that is,

$$L_{ij}^{(k)} = E[L_{ij}^k].$$

We then notice that $L_{ij}^{(0)} = 1$, for all (i, j), and that the kth moment of L, $L^{(k)} = E[L^k]$, is given by $L_{10}^{(k)}$.

Suppose that we have obtained $L_{ij}^{(n)}$ for a given approximate model and for all (i, j) with $0 \le i \le c$, $0 \le j \le K$ and $n < k$. Then, we may compute $L_{ij}^{(k)}$ according to the following recursive schemes.

Theorem 4.8. *The moments $L_{ij}^{(k)}$ for the model $\mathcal{X}^W$ can be recursively computed as follows:*

Step 1. Define the coefficients b_{i0}, $\bar{D}_{i0}$, $\hat{F}_{i0}$ and $\hat{G}_{i0}$, for $1 \le i \le c-1$, recursively by

$$b_{10} = \nu, \qquad b_{i0} = i\nu \frac{b_{i-1,0}}{\lambda + b_{i-1,0}}, \quad 2 \le i \le c-1,$$

$$\bar{D}_{10} = kL_{10}^{(k-1)}, \qquad \bar{D}_{i0} = kL_{i0}^{(k-1)} + i\nu \frac{\bar{D}_{i-1,0}}{\lambda + b_{i-1,0}}, \quad 2 \le i \le c-1,$$

$$\hat{F}_{i0} = \frac{1}{\lambda} \sum_{n=i}^{c-1} \bar{D}_{n0} \prod_{m=i}^{n} \frac{\lambda}{\lambda + b_{m0}}, \quad 1 \le i \le c-1,$$

4.2 The $M/M/c$ Retrial Queue

$$\hat{G}_{i0} = \prod_{n=i}^{c-1} \frac{\lambda}{\lambda + b_{n0}}, \quad 1 \leq i \leq c-1.$$

Step 2. For $j = 1, ..., K$, calculate sequentially the coefficients b_{ij}, G_{ij}, $H_{ij}^{(n)}$, $\bar{D}_{ij}$, $\tilde{D}_{ij}$, $\hat{F}_{ij}$ and $\hat{G}_{ij}$ given by

$$b_{0j} = 0, \quad b_{ij} = i\nu \frac{j\mu + b_{i-1,j}}{\lambda + j\mu + b_{i-1,j}}, \quad 1 \leq i \leq c-1,$$

$$G_{ij} = \prod_{n=i}^{c-1} \frac{\lambda}{\lambda + j\mu + b_{nj}}, \quad 0 \leq i \leq c-1,$$

$$H_{ij}^{(n)} = \nu^{i-n+1} \prod_{m=n-1}^{i-1} \frac{m+1}{\lambda + j\mu + b_{mj}}, \quad 0 \leq i \leq c-1,\ 1 \leq n \leq i+1,$$

$$\bar{D}_{ij} = j\mu \sum_{n=1}^{i+1} \left(\hat{F}_{n,j-1} + \hat{G}_{n,j-1} \frac{c\nu \hat{F}_{c-1,j-1} + kL_{c,j-1}^{(k-1)}}{\lambda + c\nu(1 - \hat{G}_{c-1,j-1})} \right) H_{ij}^{(n)}$$
$$+ k \sum_{n=1}^{i+1} L_{n-1,j}^{(k-1)} H_{ij}^{(n)}, \quad 0 \leq i \leq c-1,$$

$$\tilde{D}_{ij} = \frac{\lambda j\mu}{\lambda + c\nu(1 - \hat{G}_{c-1,j-1})} \sum_{n=1}^{i+1} \hat{G}_{n,j-1} H_{ij}^{(n)}, \quad 0 \leq i \leq c-1,$$

$$\hat{F}_{ij} = \frac{1}{\lambda} \sum_{n=i}^{c-1} \bar{D}_{nj} \prod_{m=i}^{n} \frac{\lambda}{\lambda + j\mu + b_{mj}}, \quad 0 \leq i \leq c-1,$$

$$\hat{G}_{ij} = G_{ij} + \frac{1}{\lambda} \sum_{n=i}^{c-1} \tilde{D}_{nj} \prod_{m=i}^{n} \frac{\lambda}{\lambda + j\mu + b_{mj}}, \quad 0 \leq i \leq c-1.$$

Step 3W. Calculate $L_{cK}^{(k)}$ by the formula

$$L_{cK}^{(k)} = \frac{c\nu \hat{F}_{c-1,K} + kL_{cK}^{(k-1)}}{c\nu(1 - \hat{G}_{c-1,K})}.$$

Step 4. Compute $L_{c,K-1}^{(k)}$, ..., $L_{c0}^{(k)}$ by the reversed recursive scheme

$$L_{c,j-1}^{(k)} = \frac{c\nu \hat{F}_{c-1,j-1} + \lambda L_{cj}^{(k)} + kL_{c,j-1}^{(k-1)}}{\lambda + c\nu(1 - \hat{G}_{c-1,j-1})}, \quad 1 \leq j \leq K,$$

which yields

$$L_{c0}^{(k)} = \prod_{j=0}^{K-1} \frac{\lambda}{\lambda + c\nu(1 - \hat{G}_{c-1,j})} L_{cK}^{(k)} + \sum_{j=0}^{K-1} \frac{\lambda^j (c\nu \hat{F}_{c-1,j} + kL_{cj}^{(k-1)})}{\prod_{n=0}^{j}(\lambda + c\nu(1 - \hat{G}_{c-1,n}))}.$$

Step 5. Calculate
$$L_{ij}^{(k)} = \hat{F}_{ij} + \hat{G}_{ij}L_{cj}^{(k)}, \quad 0 \leq i \leq c-1, \ 0 \leq j \leq K.$$

Then, the kth moment of L^W can be computed as
$$L_W^{(k)} = L_{10}^{(k)} = \hat{F}_{10} + \hat{G}_{10}L_{c0}^{(k)}.$$

Theorem 4.9. *The moments $L_{ij}^{(k)}$ for the model $\mathcal{X}^F$ can be computed by following the steps of Theorem 4.8 with Step 3W replaced by*

Step 3F. Calculate $L_{cK}^{(k)}$ as
$$L_{cK}^{(k)} = \frac{c\nu \hat{F}_{c-1,K} + \lambda \sum_{n=0}^{k-1} \binom{k}{n} L_{cK}^{(n)} \hat{\Phi}^{(k-n)} + kL_{cK}^{(k-1)}}{c\nu(1 - \hat{G}_{c-1,K})},$$

where $\hat{\Phi}^{(k)}$ denotes the kth moment of the busy period length in the standard $M/M/1$ queue with arrival and service rates given by λ and $c\mu$.

In case of $\mathcal{X}^{NR}$, we first compute the moment matrices
$$\hat{\boldsymbol{\Omega}}^{(n)} = (-1)^n \boldsymbol{\Omega}^{(n)}(0)$$

of the fundamental period for all $n < k$. Then, the matrix $\hat{\boldsymbol{\Omega}}^{(k)}$ is the unique solution of the linear system

$$\mathbf{A}_1 \hat{\boldsymbol{\Omega}}^{(k)} + \mathbf{A}_0 \left(\hat{\boldsymbol{\Omega}}^{(k)} \hat{\boldsymbol{\Omega}} + \hat{\boldsymbol{\Omega}} \hat{\boldsymbol{\Omega}}^{(k)} \right) = -k\hat{\boldsymbol{\Omega}}^{(k-1)} - \mathbf{A}_0 \sum_{n=1}^{k-1} \binom{k}{n} \hat{\boldsymbol{\Omega}}^{(n)} \hat{\boldsymbol{\Omega}}^{(k-n)}, \tag{4.62}$$

where $\hat{\boldsymbol{\Omega}} = \hat{\boldsymbol{\Omega}}^{(0)}$ can be computed by any well-known method; see [464].

Theorem 4.10. *The moments $L_{ij}^{(k)}$ for the model $\mathcal{X}^{NR}$ can be computed by following the steps of Theorem 4.8 with Step 3W replaced by*

Step 3NR. Compute $\hat{\boldsymbol{\Omega}}^{(k)} = (\hat{\omega}_{ii'}^{(k)})$ from formula (4.62). Then, calculate $L_{cK}^{(k)}$ by the formula

$$L_{cK}^{(k)} = \left(\lambda \left(1 - \hat{\omega}_{cc}^{(0)} - \sum_{i=0}^{c-1} \hat{G}_{iK} \hat{\omega}_{ci}^{(0)} \right) + c\nu \left(1 - \hat{G}_{c-1,K} \right) \right)^{-1}$$
$$\times \left(kL_{cK}^{(k-1)} + \lambda \sum_{i=0}^{c} \sum_{n=0}^{k-1} \binom{k}{n} L_{iK}^{(n)} \hat{\omega}_{ci}^{(k-n)} + \lambda \sum_{i=0}^{c-1} \hat{F}_{iK} \hat{\omega}_{ci}^{(0)} + c\nu \hat{F}_{c-1,K} \right).$$

To finish, in Table 4.10 we display the mean and the standard deviation of L for several choices of ρ and μ. Both measures increase with increasing values of ρ and decrease with increasing retrial rates.

Table 4.10. Moments of L

ρ		$\mu = 0.1$	$\mu = 1.0$	$\mu = 10.0$	$\mu = 100.0$
0.1	$E[L]$	1.29955	1.29745	1.29745	1.29745
	$\sigma(L)$	1.46188	1.44112	1.44112	1.44112
0.3	$E[L]$	2.89337	2.36130	2.32805	2.32589
	$\sigma(L)$	5.65751	3.01133	2.92673	2.92165
0.6	$E[L]$	183.14678	9.04420	6.96774	6.82736
	$\sigma(L)$	393.33588	13.43633	9.49789	9.24268
0.8	$E[L]$	891952.43694	51.12557	20.75163	19.16942
	$\sigma(L)$	1388817.68592	79.95126	29.85013	27.24057

4.2.2 The Number of Customers Served

In this subsection, we briefly discuss the analysis of the number of customers served during a busy period in the $M/M/c$ retrial queue. Let I_{ij} be the number of customers served in the first-passage time L_{ij}. Then, the number of customers served during a busy period, I, corresponds simply to I_{10}. By using a first-step argument, we obtain the equations governing the generating functions $I_{ij}(z) = E[z^{I_{ij}}]$, for $|z| \leq 1$. For instance, in the case of the model $\mathcal{X}^W$, we write down

$$I_{00}(z) = 1,$$

$$I_{ij}(z) = \frac{\lambda}{\lambda + i\nu + j\mu} I_{i+1,j}(z) + \frac{i\nu}{\lambda + i\nu + j\mu} z I_{i-1,j}(z)$$
$$+ \frac{j\mu}{\lambda + i\nu + j\mu} I_{i+1,j-1}(z), \ 0 \leq i \leq c-1, \ 0 \leq j \leq K, \ (i,j) \neq (0,0),$$

$$I_{cj}(z) = \frac{\lambda}{\lambda + c\nu} I_{c,j+1}(z) + \frac{c\nu}{\lambda + c\nu} z I_{c-1,j}(z), \quad 0 \leq j \leq K-1,$$

$$I_{cK}(z) = z I_{c-1,K}(z).$$

An algorithm for computing $I_{ij}(z)$ can be easily derived as a variant of Theorem 4.5. This is also true for the models $\mathcal{X}^F$ and $\mathcal{X}^{NR}$. More particularly, for $\mathcal{X}^{NR}$ some analogues for $\xi_{ii'}^{(n)}(x)$ and $\omega_{ii'}^{(n)}(s)$ are needed. Let $\sigma_{ii'}^{(n)}(m)$ be the conditional probability that, starting from the state $(i, K+n)$, the process $\mathcal{X}^{NR}$ reaches for the first time the level K entering the state (i', K) and during this time m customers were served. Define also the generating function

$$\chi_{ii'}^{(n)}(z) = \sum_{m=0}^{\infty} z^m \sigma_{ii'}^{(n)}(m),$$

which plays the role of $\omega_{ii'}^{(n)}(s)$. It should be pointed out that the results in [544, Section 3.3], about the number of left transitions in the fundamental period of a QBD process, are not applicable here. In the case of $\mathcal{X}^{NR}$, left

124   4 Busy Period

Table 4.11. Probability mass function $P(I = k)$

k	$\rho = 0.1$	$\rho = 0.3$	$\rho = 0.6$	$\rho = 0.8$
1	0.66666	0.40000	0.25000	0.20000
2	0.17777	0.13714	0.07500	0.05333
3	0.07788	0.08620	0.04500	0.02946
4	0.03799	0.06368	0.03342	0.02062
5	0.01926	0.05029	0.02747	0.01625
6	0.00988	0.04089	0.02364	0.01351
7	0.00509	0.03374	0.02085	0.01156
8	0.00262	0.02811	0.01867	0.01008
9	0.00135	0.02358	0.01693	0.00891
10	0.00070	0.01988	0.01550	0.00796
11	0.00036	0.01683	0.01432	0.00720
12	0.00018	0.01430	0.01331	0.00656
13	0.00009	0.01218	0.01246	0.00603
14	0.00004	0.01039	0.01172	0.00558
15	0.00002	0.00888	0.01108	0.00520

transitions correspond to successful retrials and not to service completions. However, a direct analysis can be carried out and, similarly to (4.54), it can be shown that $\mathbf{X}(z) = (\chi_{ii'}^{(1)}(z))$ satisfies the matrix equation

$$\mathbf{X}(z) = -\mathbf{A}_1^{-1}(z)\mathbf{A}_2 - \mathbf{A}_1^{-1}(z)\mathbf{A}_0\mathbf{X}^2(z),$$

where $\mathbf{A}_1(z)$ is defined from $\mathbf{A}_1$ multiplying its subdiagonal by z. Moreover, for every z, we can obtain $\mathbf{X}(z)$ as limit of the sequence $\mathbf{X}_n(z)$ defined by

$$\mathbf{X}_0(z) = \mathbf{0}_{(c+1)\times(c+1)},$$
$$\mathbf{X}_{n+1}(z) = -\mathbf{A}_1^{-1}(z)\mathbf{A}_2 - \mathbf{A}_1^{-1}(z)\mathbf{A}_0\mathbf{X}_n^2(z), \quad n \geq 0.$$

In Table 4.11, we fix $\mu = 1.0$ and display the probabilities $P(I = k)$, for $1 \leq k \leq 15$, in four cases corresponding to $\rho = 0.1, 0.3, 0.6$ and 0.8. We get a decreasing probability mass function with mode at $k = 1$. The probability of serving one customer is bigger for smaller values of ρ, agreeing with intuitive expectations.

Finally, Table 4.12 lists values of the mean and the standard deviation of I versus ρ and μ. Similar comments to those for Table 4.10 are also valid here. These two tables correct a bug in the code of paper [105]. From the tables, it can be verified the identity $E[I] = 1 + \lambda E[L]$.

4.2.3 Maximal Queue Length in a Busy Period

We now concentrate on the algorithmic analysis of the maximum queue length observed during a busy period L of the $M/M/c$ retrial queue.

4.2 The $M/M/c$ Retrial Queue   125

Table 4.12. Moments of I

ρ		$\mu = 0.1$	$\mu = 1.0$	$\mu = 10.0$	$\mu = 100.0$
0.1	$E[I]$	1.64977	1.64872	1.64872	1.64872
	$\sigma(I)$	1.24325	1.23258	1.23258	1.23258
0.3	$E[I]$	5.34006	4.54195	4.49207	4.48883
	$\sigma(I)$	9.60204	5.57724	5.41913	5.40937
0.6	$E[I]$	550.44034	28.13261	21.90322	21.48208
	$\sigma(I)$	1187.41401	43.04310	30.69749	29.88119
0.8	$E[I]$	3567810.74763	205.50229	84.00652	77.67769
	$\sigma(I)$	5555305.42817	326.43339	123.66801	112.99921

In addition to the assumption of dealing with a well-posed matrix structure, in the framework of two-dimensional Markov chains some typical requirements are the stationary regime and the system homogeneity. In the absence of such assumptions, the computation of the limiting distribution of the system state becomes extremely intricate or indeed intractable. In the sequel, we will show how the maximum queue length distribution overcomes these restrictions and can be efficiently computed.

Let N_{max} be the maximum number of retrial customers observed during a busy period. For each $k \geq 1$, we notice that $\{N_{max} < k\}$ amounts to the event that, starting from the state $(1,0)$, the process $\mathcal{X}$ hits the state $(0,0)$ before hitting (c,k). Hence, the computation of $P(N_{max} < k)$, for $k \geq 1$, reduces to finding the probability of absorption into a particular state for a finite Markov chain with two absorbing states, say $(0,0)$ and (c,k).

For a fixed k, let τ_{ij} be the probability of eventual absorption into $(0,0)$, starting from (i,j), for $0 \leq i \leq c$ and $0 \leq j \leq k-1$. By conditioning on the next state visited, we obtain

$$\tau_{00} = 1, \tag{4.63}$$

$$\tau_{ij} = \frac{\lambda}{\lambda + i\nu + j\mu}\tau_{i+1,j} + \frac{i\nu}{\lambda + i\nu + j\mu}\tau_{i-1,j} + \frac{j\mu}{\lambda + i\nu + j\mu}\tau_{i+1,j-1},$$
$$0 \leq i \leq c-1,\ 0 \leq j \leq k-1,\ (i,j) \neq (0,0), \tag{4.64}$$

$$\tau_{cj} = \frac{\lambda}{\lambda + c\nu}\tau_{c,j+1} + \frac{c\nu}{\lambda + c\nu}\tau_{c-1,j},\quad 0 \leq j \leq k-1, \tag{4.65}$$

$$\tau_{ck} = 0. \tag{4.66}$$

By setting $s = 0$ in the set of equations (4.26)-(4.29), we observe that the differences with equations (4.63)-(4.66) reduce to the boundary conditions. As a result, the algorithmic solution of (4.63)-(4.66) is parallel to Theorem 4.5. We thus omit repetitive details which can be found in [106].

In the next proposition, we summarize two explicit results. First, we reduce to the single server case and obtain an explicit formula for the distribution of N_{max}. In the second result, we obtain the relationship between the maximum number of customers in orbit and the maximum number of customers in the system, denoted by S_{max}.

Proposition 4.11. *(i) In the case $c = 1$, the distribution of N_{max} is given by*

$$P(N_{max} \leq k-1) = \frac{\sum_{m=0}^{k-1} \frac{m!}{\rho^m \prod_{l=1}^{m}(\frac{\lambda}{\mu}+l)}}{\rho + \sum_{m=0}^{k-1} \frac{m!}{\rho^m \prod_{l=1}^{m}(\frac{\lambda}{\mu}+l)}}, \quad k \geq 1. \qquad (4.67)$$

(ii) The distribution of S_{max} is given by

$$P(S_{max} \leq k) = \begin{cases} 1 - \left(\sum_{m=0}^{k} \frac{m!}{\rho^m}\right)^{-1}, & \text{if } 1 \leq k \leq c-1, \\ P(N_{max} \leq k-c), & \text{if } k \geq c. \end{cases} \qquad (4.68)$$

Proof. First, we set $c = 1$ and combine (4.64), for $i = 0$, and (4.65) in order to get

$$\left(\lambda + \frac{j\mu\nu}{\lambda + j\mu}\right)\tau_{1j} = \frac{j\mu\nu}{\lambda + j\mu}\tau_{1,j-1} + \lambda\tau_{1,j+1}, \quad 1 \leq j \leq k-1.$$

By setting $z_j = \tau_{1j} - \tau_{1,j-1}$, for $1 \leq j \leq k$, we obtain

$$\rho z_{j+1} = \frac{j}{\frac{\lambda}{\mu} + j} z_j, \quad 1 \leq j \leq k-1. \qquad (4.69)$$

By iterating (4.69), we get

$$\tau_{1,j+1} - \tau_{1j} = \frac{j!}{\rho^j \prod_{m=1}^{j}\left(\frac{\lambda}{\mu} + m\right)}(\tau_{11} - \tau_{10}), \quad 1 \leq j \leq k-1. \qquad (4.70)$$

Summing (4.70) for $j = 1, ..., k-1$ yields

$$\tau_{1k} - \tau_{11} = \sum_{j=1}^{k-1} \frac{j!}{\rho^j \prod_{m=1}^{j}\left(\frac{\lambda}{\mu} + m\right)}(\tau_{11} - \tau_{10}).$$

But $\tau_{1k} = 0$ and $\tau_{10} = (\lambda\tau_{11} + \nu)(\lambda + \nu)^{-1}$, so by eliminating τ_{11} and τ_{1k} we obtain the explicit formula (4.67).

We now turn the attention to statement (ii). Since the orbit size increases only through states at which all servers are busy, we easily find that

$$P(S_{max} \leq k) = P(N_{max} \leq k-c), \quad k \geq c.$$

On the other hand, if $S_{max} \leq c-1$, then there are no retrial customers during the busy period and consequently the problem can be viewed in terms of the standard $M/M/c$ queue modified to evolve in the set $\{(0,0), ..., (k +$

4.2 The $M/M/c$ Retrial Queue   127

Table 4.13. $P(N_{max} = k)$ as λ varies

k	$\rho = 0.2$	$\rho = 0.4$	$\rho = 0.6$	$\rho = 0.8$
0	0.99350	0.87500	0.61057	0.40465
1	—	0.06479	0.08446	0.03912
2	—	0.03315	0.07563	0.03732
3	—	0.01539	0.06397	0.03680
4	—	0.00674	0.05064	0.03672
5	—	—	0.03768	0.03668
6	—	—	0.02662	0.03646
7	—	—	0.01805	0.03592
8	—	—	0.01188	0.03499
9	—	—	0.00764	0.03365
10	—	—	0.00484	0.03191
11	—	—	—	0.02984
12	—	—	—	0.02751
⋮	⋮	⋮	⋮	⋮
27	—	—	—	0.00253
28	—	—	—	0.00207

$1, 0)\}$, for $1 \leq k \leq c - 1$. An appeal to well-known results for extreme values of a birth-and-death process [595] yields

$$P(S_{max} \leq k) = 1 - \left(\sum_{m=0}^{k} \frac{m!}{\rho^m}\right)^{-1}, \quad 1 \leq k \leq c - 1.$$

This completes the proof of expression (4.68). □

In order to study the influence of the system parameters on N_{max}, some numerical experiments are performed. In Table 4.13 we vary the arrival rate keeping other parameters constant, so we fix $c = 5$, $\nu = 1.0$, $\mu = 2.5$ while $\rho = 0.2, 0.4, 0.6$ and 0.8. Entries in the table give the probabilities $P(N_{max} = k)$ needed to reach the 99th percentile of N_{max}. It can be inferred that the 99th percentile increases as a function of ρ.

In a second example, we fix $\lambda = 2.5$ and vary the retrial rate on the domain $\mu = 0.08, 0.25, 0.5, 1.0, 10.0$ and 1000.0. The rest of parameters are chosen as in Table 4.13. Since the $M/M/c$ retrial queue approaches the standard $M/M/c$ queue, we obtain the entries for the last column by the formula

$$P(S^{\infty}_{max} \leq c + k) = 1 - \left(\sum_{m=0}^{c-1} \frac{m!}{(c\rho)^m} + \frac{c!}{c^c} \sum_{m=c}^{c+k} \frac{1}{\rho^m}\right)^{-1}, \quad k \geq 0.$$

We observe from Table 4.14 that, for $\mu = 0.08$, the distribution has two modes at $k = 0$ and $k = 8$. The distribution of N_{max} in the retrial queue appears to converge to the parallel distribution for the standard $M/M/c$ queue,

128   4 Busy Period

Table 4.14. $P(N_{max} = k)$ as μ varies

k	$\mu = 0.08$	$\mu = 0.25$	$\mu = 0.5$	$\mu = 1.0$	$\mu = 10.0$	$\mu = 1000.0$	$\mu = \infty$
0	0.74665	0.74665	0.74665	0.74665	0.74665	0.74665	0.74665
1	0.03602	0.06219	0.07507	0.08384	0.09490	0.09718	0.09721
2	0.02237	0.04676	0.05632	0.06159	0.06699	0.06778	0.06779
3	0.01810	0.03810	0.04196	0.04254	0.04154	0.04106	0.04105
4	0.01674	0.03092	0.02974	0.02739	0.02360	0.02278	0.02277
5	0.01662	0.02409	0.01988	0.01661	0.01272	0.01203	0.01202
6	0.01706	0.01776	0.01258	0.00962	0.00664	0.00618	0.00618
7	0.01759	0.01238	0.00762	0.00539	—	—	—
8	0.01784	0.00820	0.00446	—	—	—	—
9	0.01748	0.00520	—	—	—	—	—
10	0.01632	—	—	—	—	—	—
11	0.01439	—	—	—	—	—	—
12	0.01196	—	—	—	—	—	—
13	0.00937	—	—	—	—	—	—
14	0.00697	—	—	—	—	—	—
15	0.00495	—	—	—	—	—	—

Table 4.15. $P(N_{max} = k)$ as c varies

k	$c = 1$	$c = 3$	$c = 6$	$c = 12$	$c = 24$
0	0.66666	0.76923	0.71875	0.43422	0.10685
1	0.13333	0.07559	0.07960	0.09846	0.01157
2	0.06956	0.05018	0.06286	0.10998	0.02114
3	0.04472	0.03566	0.04788	0.10669	0.03761
4	0.03036	0.02507	0.03413	0.08868	0.06348
5	0.02054	0.01697	0.02272	0.06410	0.09786
6	0.01353	0.01098	0.01427	0.04155	0.13130
7	0.00861	0.00681	0.00856	0.02495	0.14660
8	0.00529	—	0.00496	0.01426	0.13341
9	—	—	—	0.00789	0.10052
10	—	—	—	—	0.06543
11	—	—	—	—	0.03860
12	—	—	—	—	0.02144
13	—	—	—	—	0.01151
14	—	—	—	—	0.00606

as $\mu \to \infty$. With increasing values of μ, the orbit size decreases and consequently smaller 99th percentiles are obtained.

Finally, in Table 4.15, we analyze the effect of the number c of available servers. The service rate is $\nu = 1.0$ and λ is varied in such a way that ρ fits the value 0.5. We also fix $\mu = 0.5$. In the cases $c = 12$ and 24, we notice the existence of two modes, the first at $k = 0$ whereas the second one appears to increase with increasing values of c.

4.3 Bibliographical Notes

A thorough treatment of the truncated approach for the length of a busy period of the $M/G/1$ retrial queue can be found in [109]. For an exhaustive numerical analysis of the approach based on the principle of maximum entropy, interested readers are referred to the papers [94, 494]. The recursive scheme for the computation of the distributions of I and N_{max} has its origin in [492] and [493], respectively.

Falin and Templeton [288, Section 1.6] investigate the joint distribution of the pair (L, I) in the $M/G/1$ retrial queue in terms of Laplace-Stieltjes transforms. Indeed, they deal with the more general notion of k-busy period; that is, a busy period starting when a primary arrival finds an idle server and $k - 1$ customers in orbit, and ending at the next departure epoch at which the system becomes empty. Thus, the standard busy period can be view as a 1-busy period. Their approach is based on the method of catastrophes and some results for bounded linear operators on a Banach space. Some bounds for the mean characteristics of the busy period were obtained by Falin [268]. The use of taboo probabilities provides another elegant alternative for the study of the busy period (for instance, see [75]). For a comprehensive overview of the methods of catastrophes and collective marks, we refer to the book by Kleinrock [398, Chapter 7].

An ordinary busy period can be decomposed into several orbit periods and intervals of competition between the remaining service time and the Poisson input process [55]. The notion of orbit busy period was introduced by Artalejo [47] as the period starting at the epoch when a primary customer arrives and finds the server busy and the orbit idle, and ending at the next epoch at which a repeated attempt finds the server idle and the orbit becomes empty.

Lopez-Herrero [491] uses the *PME* to estimate the distribution of the number of customers served when the available constraints are the four first probabilities and the two first moments.

Choo and Conolly [195] investigate some aspects of the busy period of the $M/M/1$ retrial queue. The approach developed in Section 4.2 for the $M/M/c$ retrial queue comes from some papers by Artalejo et al. [92, 105, 106]. For applications of the maximum queue length distribution to advanced multiserver models with retrials, we refer to the papers [98, 104]. The paper [104] discusses a retrial queue with impatience and customer balking and its applications to call center management. Artalejo and Chakravarthy [98] deal with a model with batch arrivals of positive and negative customers.

Extreme values of the queue length can be investigated following different approaches. Serfozo [594, 595] proposes an asymptotic analysis whereas Neuts [542] concentrates on the distribution of the maximum queue length during a busy period. In Subsections 4.1.3 and 4.2.3 we adopted the latter approach. The use of the asymptotic approach in the context of the $M/G/1$ retrial queue can be found in [318].

We carry out the numerical inversion of the Laplace-Stieltjes transforms by using the algorithms EULER and POST-WIDDER described in [2]. Both methods are variants of the Fourier series method but, as it is reported in [2], they provide different approaches to the inversion problem so they should be used in parallel to determine the desired accuracy by checking on each other.

As related work we mention a variety of papers dealing with the busy period analysis of retrial queues with finite population [289], constant retrial policy [184, 315], retrial queues with negative arrivals [62] and disasters [63, 70].

5
Waiting Time

In this chapter, we discuss the waiting time distribution for the $M/G/1$ and $M/M/c$ retrial queues under the natural assumption that customers are served in random order. In Subsection 5.1.1, maximum entropy, truncated and gamma approaches are discussed for the $M/G/1$ case. The numerical results give support to the truncated and gamma approximations, but the maximum entropy solution is itself of methodological interest. In Subsection 5.2.1, we provide a route for the computational investigation of the $M/M/c$ retrial queue with finite orbit capacity, which approximates satisfactorily the original infinite capacity model. Our objective in Subsections 5.1.2 and 5.2.2 is to study for both basic models the discrete counterpart of the waiting time, namely the number of retrials made by a tagged customer before entering into service.

5.1 The $M/G/1$ Retrial Queue

5.1.1 Waiting Time Distribution

In Subsection 4.1.1, we applied two approaches dealing with the principle of maximum entropy and truncated models to the estimation of the density function of a busy period. Our purpose here is to extend such an analysis to the waiting time distribution and make a numerical comparison between the resulting solutions.

We first present a maximum entropy analysis of the waiting time W based on the knowledge of $E[W]$, $E[W^2]$ and $W^*(s)$ at a given point $s = s_0$; see formulas (2.13)-(2.15). Since the definition of W excludes the service time, we notice that its probability distribution function $F_W(x)$ has a jump at $x = 0$ and is absolutely continuous in the interval $(0, \infty)$. The jump size at $x = 0$ corresponds to the probability that an arriving customer finds the server idle, thus being equal to $1 - \rho$. Therefore, we have

Table 5.1. Relative errors in the $M/M/1$ retrial queue for $\rho = 0.5$

s	$W^*(s)$	$W_2^*(s)$	$R_2(s)$	$W_{2,1}^*(s)$	$R_{2,1}(s)$
0.01	0.98548	0.98548	1.27×10^{-6}	0.98548	7.58×10^{-8}
0.05	0.93549	0.93559	0.00010	0.93548	9.56×10^{-6}
0.1	0.88632	0.88682	0.00057	0.88627	5.98×10^{-5}
0.25	0.78842	0.79120	0.00325	0.78842	0.00000
0.3	0.76562	0.76928	0.00478	0.76579	0.00022
0.6	0.67751	0.68605	0.01261	0.67987	0.00407
0.9	0.63035	0.64237	0.05078	0.63532	0.00788
1.2	0.60109	0.61536	0.02374	0.60823	0.01187
1.5	0.58130	0.59699	0.02699	0.59007	0.01508
10.0	0.50633	0.51763	0.02232	0.51545	0.01801
20.0	0.50193	0.50898	0.01405	0.50781	0.01171

$$\frac{dF_W(x)}{dx} = (1-\rho)u_0(x) + f_{W^c}(x), \quad x \geq 0,$$

where $u_0(x)$ is the unit impulse at the origin defined by

$$u_0(x) = \begin{cases} \infty, & \text{if } x = 0, \\ 0, & \text{if } x \neq 0. \end{cases}$$

Hence, the estimation of the distribution of W reduces to the density function $f_{W^c}(x)$ of the continuous contribution. In a first step, we assume that the available information consists of the moments $E[W]$ and $E[W^2]$. Later, we add the knowledge of $W^*(s)$ at the point $s_0 = 0.25$. Then, by following the general theory summarized in Subsection 3.1.3, we obtain the maximum entropy estimations $\hat{f}_2(x)$ and $\hat{f}_{2,1}(x)$.

In Table 5.1 we consider the $M/M/1$ retrial queue with $\lambda = 0.5$, $\beta_1 = 1.0$ and $\mu = 2.0$, so the traffic intensity is $\rho = 0.5$. We evaluate the accuracy of the maximum entropy estimations by comparing the classical Laplace-Stieltjes transform $W^*(s)$, which is obtained by numerical integration from (2.13), versus the maximum entropy versions $W_2^*(s)$ and $W_{2,1}^*(s)$, which are given by

$$W_2^*(s) = 1 - \rho + \int_0^T e^{-sx} \hat{f}_2(x) dx \quad \text{and} \quad W_{2,1}^*(s) = 1 - \rho + \int_0^T e^{-sx} \hat{f}_{2,1}(x) dx.$$

The entries $R_2(s)$ and $R_{2,1}(s)$ correspond to the relative errors which are defined as in Subsection 4.1.1 for the case of the busy period. We observe that the relative errors decrease when we employ $\hat{f}_{2,1}(x)$ rather than $\hat{f}_2(x)$. The upper bound T is 29.5, thus verifying $P(W > T) \leq 10^{-2}$.

For the same numerical example, in Figure 5.1, we display the maximum entropy densities $\hat{f}_1(x)$, $\hat{f}_2(x)$ and $\hat{f}_{2,1}(x)$, where $\hat{f}_1(x)$ is the estimation based on the knowledge of $E[W]$. In the light of the decreasing shape of the three densities, we conclude that all these solutions are close enough. However, at

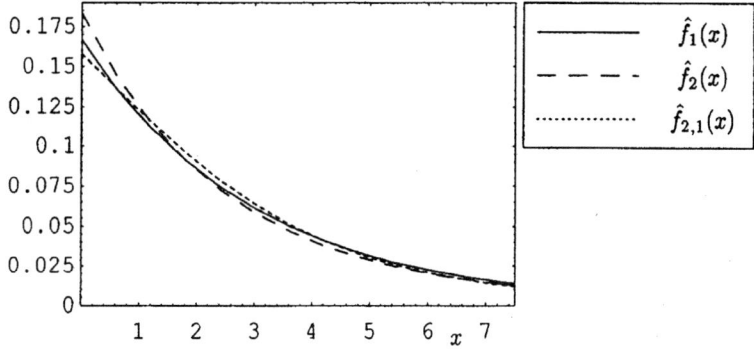

Fig. 5.1. Maximum entropy estimations in an $M/M/1$ retrial queue with $\rho = 0.5$

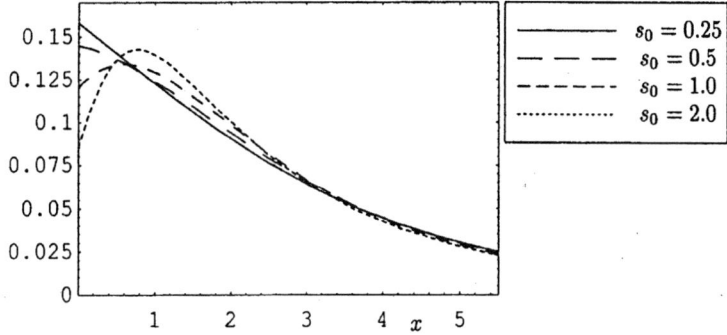

Fig. 5.2. The effect of the point s_0 on $\hat{f}_{2,1}(x)$

this point we remember the Tauberian relations (4.1), which give some light about the effect of the auxiliary point s_0. Accordingly, in Figure 5.2, we allow s_0 to take the values 0.25, 0.5, 1.0 and 2.0. With increasing values of s_0, we expect to get a better description of the behavior of $f_{W^c}(x)$ near the origin $x = 0$. Indeed, we observe that the density functions associated with the values 0.25 and 0.5 are decreasing functions whereas the density functions based on the values 1.0 and 2.0 exhibit a bell-shaped form.

To conclude our analysis based on maximum entropy, we plot in Figure 5.3 the potential function $F(\alpha_1, \alpha_2)$. The resulting surface is complementary to Figure 4.2, where we analyzed the length of a busy period and plotted the balanced function G. Once more we observe the existence of a long valley and asymptotic linearity when we leave a neighborhood of the Lagrangian multipliers. Since the maximum entropy formalism is independent of the service time distribution, we expect similar conclusions for other numerical examples.

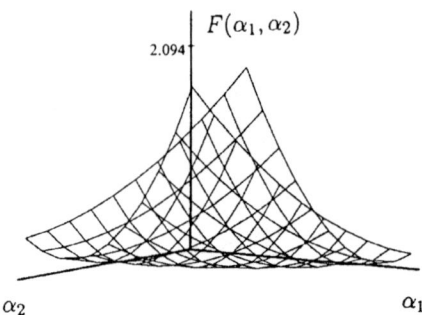

Fig. 5.3. The potential function $F(\alpha_1, \alpha_2)$

The next approach [108] uses a truncated model to estimate the waiting time distribution. To be concrete, similarly to the analysis of the busy period made in Subsection 4.1.1, we focus on the most natural and traditional approach consisting in assuming a suitably chosen truncation limit K in the orbit capacity.

For a fixed value of K, we describe the system state by means of the process $\{(C^K(t), N^K(t), \xi^K(t)); t \geq 0\}$. As usual, $C^K(t)$ and $N^K(t)$ denote the state of the server and the orbit size at time t, respectively. When $C^K(t) = 1$, the random variable $\xi^K(t)$ is defined as the elapsed service time of the customer in service. The limiting probabilities of the truncated retrial queue are denoted by P_{ij}^K, for $(i,j) \in \{0,1\} \times \{0, 1, ..., K\}$, and can be recursively computed from Theorem 3.2.

Let us assume that, at time $t = 0$, a primary customer finds the server busy and, consequently, he joins the orbit. We tag this blocked customer. Then, the system state becomes $(C^K(0+), N^K(0+), \xi^K(0+)) = (1, j, \tau)$, for any $j \geq 1$, where $N^K(0+) = j$ includes the tagged customer. Let $W_{1j\tau}$ be the waiting time in orbit of the tagged customer. Its Laplace-Stieltjes transform $W_{1j\tau}^*(s)$ satisfies the equation

$$W_{1j\tau}^*(s) = \sum_{n=j}^{K-1} a_{n-j,\tau}(s) W_{0n}^*(s) + a_{K-j,\tau}^a(s) W_{0K}^*(s), \quad 1 \leq j \leq K, \quad (5.1)$$

where

$$a_{j,\tau}(s) = \frac{1}{1 - B(\tau)} \int_\tau^\infty e^{-(s+\lambda)(x-\tau)} \frac{(\lambda(x-\tau))^j}{j!} dB(x), \ 0 \leq j \leq K-2, \ (5.2)$$

$$a_{j,\tau}^a(s) = \frac{1}{1 - B(\tau)} \int_\tau^\infty e^{-s(x-\tau)} \sum_{m=j}^\infty e^{-\lambda(x-\tau)} \frac{(\lambda(x-\tau))^m}{m!} dB(x)$$

$$= \frac{1}{1-B(\tau)} \int_\tau^\infty e^{-s(x-\tau)} dB(x) - \sum_{m=0}^{j-1} a_{m,\tau}(s), \quad 0 \le j \le K-1, \quad (5.3)$$

and $W_{0j}^*(s)$ denotes the Laplace-Stieltjes transform of the remaining waiting time, given that the service time just expired and the system state is $(0,j)$. We also define the Laplace-Stieltjes transform $W_{1j}^*(s)$ of the remaining waiting time of the tagged customer, given that a new service time starts and the system state is $(1,j)$.

Conditioning on the identity of the customer entering service (by distinguishing among primary arrival, tagged customer, and the rest of customers in orbit), we get

$$W_{0j}^*(s) = \frac{\lambda}{s+\lambda+j\mu} W_{1j}^*(s) + \frac{(j-1)\mu}{s+\lambda+j\mu} W_{1,j-1}^*(s) + \frac{\mu}{s+\lambda+j\mu},$$
$$1 \le j \le K. \quad (5.4)$$

Moreover, taking into account the number of primary arrivals during a service time results in

$$W_{1j}^*(s) = \sum_{n=j}^{K-1} a_{n-j}(s) W_{0n}^*(s) + a_{K-j}^a(s) W_{0K}^*(s), \quad 1 \le j \le K, \quad (5.5)$$

where the functions $a_j(s)$ and $a_j^a(s)$ were already defined in Subsection 4.1.1.

By solving the system of linear equations (5.4) and (5.5), we can obtain the value of $W_{ij}^*(s)$ for any given s. Then, we may use (5.1) to get the conditional Laplace-Stieltjes transform $W_{1j\tau}^*(s)$. Alternatively, it seems more interesting to investigate the unconditional version of which the Laplace-Stieltjes transform is given by

$$W^*(s) = 1 - \sum_{j=0}^{K-1} P_{1j}^K + \sum_{j=0}^{K-1} \int_0^\infty P_{1j\tau}^K W_{1,j+1,\tau}^*(s) d\tau, \quad (5.6)$$

where $P_{1j\tau}^K$ is the limiting probability of being in the state $(1,j,\tau)$. It is worth mentioning that, though the limiting probabilities P_{1j}^K can be efficiently computed, the probabilities $P_{1j\tau}^K$ have no simple analytical solution. Thus, we make an approximating assumption consisting in to replace the distribution of the residual service time that the tagged customer finds upon arrival by the distribution of the forward renewal service time

$$\hat{B}(x) = \frac{1}{\beta_1} \int_0^x (1-B(u)) du, \quad x \ge 0.$$

Formulas (5.1)-(5.3) and (5.6) are thus replaced by the equations summarized in the following theorem.

Theorem 5.1. *The approximation of the Laplace-Stieltjes transform $W^*(s)$ based on the use of the forward renewal service time yields*

$$W_e^*(s) = 1 - \sum_{j=0}^{K-1} P_{1j}^K + \sum_{j=0}^{K-1} P_{1j}^K W_{1,j+1}^{*e}(s), \qquad (5.7)$$

where

$$W_{1j}^{*e}(s) = \sum_{n=j}^{K-1} a_{n-j}^e(s) W_{0n}^*(s) + a_{K-j}^{a,e}(s) W_{0K}^*(s), \quad 1 \leq j \leq K, \qquad (5.8)$$

$$a_j^e(s) = \frac{1}{\beta_1} \left(\frac{\lambda}{s+\lambda}\right)^j \int_0^\infty e^{-(s+\lambda)x} \frac{((s+\lambda)x)^j}{j!} (1-B(x))dx,$$

$$0 \leq j \leq K-2, \quad (5.9)$$

$$a_j^{a,e}(s) = \frac{1}{\beta_1} \int_0^\infty e^{-sx} \sum_{m=j}^\infty e^{-\lambda x} \frac{(\lambda x)^m}{m!}(1-B(x))dx$$

$$= \frac{1-\beta(s)}{s\beta_1} - \sum_{m=0}^{j-1} a_m^e(s), \quad 0 \leq j \leq K-1. \qquad (5.10)$$

After solving the system (5.4) and (5.5), we may combine equations (5.7)-(5.10) and compute $W_e^*(s)$. This amounts to the use of algorithms for the numerical inversion of the waiting time density. For exponential service times, we have that $B(x) = \hat{B}(x)$, so that (5.1) reduces to (5.5) and the approximation results become exact, so $W_e^*(s) = W^*(s)$.

We remark that (5.7) consists of two contributions, the first is the discrete contribution $1 - \sum_{j=0}^{K-1} P_{1j}^K$ accounting for the probability of non-waiting and the second is the Laplace-Stieltjes transform of the continuous contribution with density $f_{W_c^e}(x)$ on the interval $(0, \infty)$. Note also that $\lim_{s\to\infty} sa_j^e(s) = \delta_{0j}\beta_1^{-1}$ allows us to derive the following Tauberian results:

$$f_{W_{ij}}(0) = \lim_{s\to\infty} sW_{ij}^*(s) = \delta_{0i}\mu, \quad i \in \{0,1\}, \ 1 \leq j \leq K,$$

$$f_{W_c^e}(0) = \lim_{s\to\infty} s\sum_{j=0}^{K-1} P_{1j}^K W_{1,j+1}^{*e}(s) = 0.$$

The next proposition provides a recursive scheme for computing first the conditional moments $W_{ij}^{(k)}$, and then the estimation $W_e^{(k)}$ of the unconditional kth moment.

Proposition 5.2. (i) *For any fixed $k \geq 1$, the moments $W_{ij}^{(k)}$, for $i \in \{0,1\}$ and $1 \leq j \leq K$, can be computed by solving the following equations:*

$$(\lambda + j\mu)W_{0j}^{(k)} - \lambda W_{1j}^{(k)} - (j-1)\mu W_{1,j-1}^{(k)} = kW_{0j}^{(k-1)}, \quad 1 \leq j \leq K, \quad (5.11)$$

$$W_{1j}^{(k)} = \sum_{l=0}^k \binom{k}{l} \left(\sum_{m=j}^{K-1} a_{m-j}^{(k-l)} W_{0m}^{(l)} + a_{K-j}^{a,(k-l)} W_{0K}^{(l)}\right), \quad 1 \leq j \leq K, \quad (5.12)$$

5.1 The M/G/1 Retrial Queue

Table 5.2. Truncation thresholds for an $M/M/1$ retrial queue

ρ		$\mu = 0.05$	$\mu = 0.5$	$\mu = 1.0$	$\mu = 2.5$	$\mu = 5.0$
0.2	K	11	7	7	7	7
	$E[W]$	5.25000	0.75000	0.50000	0.35000	0.30000
0.4	K	28	15	15	14	13
	$E[W]$	14.00000	2.00000	1.33333	0.93333	0.80000
0.6	K	66	30	27	25	25
	$E[W]$	31.50000	4.50000	3.00000	2.10000	1.80000
0.8	K	192	79	70	64	61
	$E[W]$	84.00000	12.00000	8.00000	5.60000	4.80000

where $a_j^{(k)}$ and $a_j^{a,(k)}$ follow from (4.10)-(4.13).

(ii) The estimation of the unconditional kth moment of W is given by

$$W_e^{(k)} = \sum_{j=0}^{K-1} P_{1j}^K \sum_{l=0}^{k} \binom{k}{l} \left(\sum_{n=j+1}^{K-1} a_{n-j-1}^{e,(k-l)} W_{0n}^{(l)} + a_{K-j-1}^{a,e,(k-l)} W_{0K}^{(l)} \right), \quad k \geq 1,$$

(5.13)

with

$$a_j^{e,(k)} = \frac{(j+k)!}{j! \lambda^k \beta_1} \hat{a}_{j+k}, \quad 0 \leq j \leq K-2, \; k \geq 1,$$

$$a_j^{a,e,(k)} = \frac{\beta_{k+1}}{(k+1)\beta_1} - \sum_{m=0}^{j-1} a_m^{e,(k)}, \quad 0 \leq j \leq K-1, \; k \geq 1,$$

where the values of $\hat{a}_j$ were already defined in Subsection 3.1.1.

Proof. We first differentiate equations (5.4) and (5.5) with respect to s, at the point $s = 0$. This yields the system (5.11) and (5.12). On the other hand, we combine (5.7) and (5.8) to obtain expression (5.13) for $W_e^{(k)}$. □

To approximate the waiting time distribution of the $M/G/1$ retrial queue, we propose to experiment with the truncated model and successively increase the threshold K until the first four decimal digits of $E[W]$ are fitted. Table 5.2 lists values of $E[W]$, calculated from formula (2.14), and K for $M/M/1$ retrial queues with $\beta_1 = 1.0$ and several choices of the pair (ρ, μ). As was expected, $E[W]$ increases with increasing values of ρ and, by contrast, is a decreasing function of μ. The above criterion yields larger values of K when ρ is closer to one and μ is smaller, and smaller values of K when ρ is smaller and μ increases.

Figure 5.4 shows the influence of ρ on the density function $f_{W^c}(x)$ for the queue with $\beta_1 = 1.0$ and $\mu = 2.5$. The density is a unimodal function with starting value $f_{W^c}(0) = 0$, irrespective of ρ. Note that heavier tails correspond to larger traffic loads ρ. In Figure 5.5, we report numerical examples to show the effect of μ on $f_{W^c}(x)$ in the queue with $\beta_1 = 1.0$ and $\rho = 0.6$. We point out that the density function exhibits a similar behavior, irrespective of the magnitude of μ. Furthermore, smaller values of μ result in heavier tails.

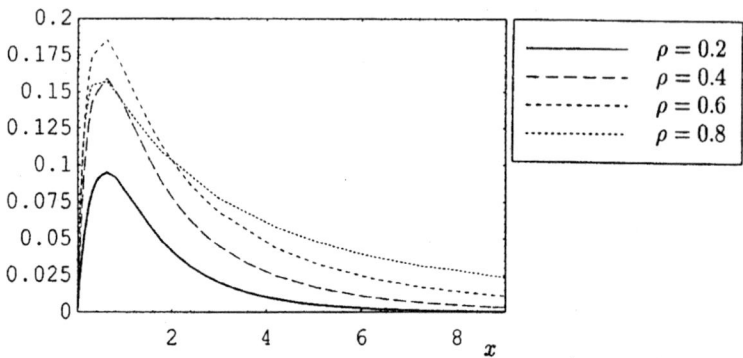

Fig. 5.4. Density functions of W^c as ρ varies

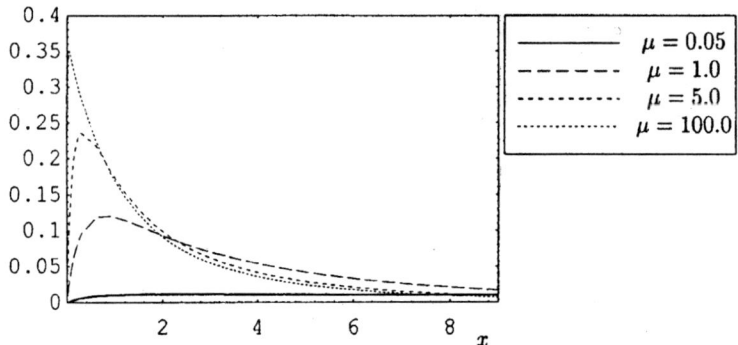

Fig. 5.5. Density functions of W^c as μ varies

In Figure 5.6, we plot $f_{W_c^c}(x)$ for various choices of the service time distribution. Specifically, we consider retrial queues with exponential, H_2 and E_4 service times, $\rho = 0.4$ and $\beta_1 = 1.0$. For the case H_2, we take the coefficient of variation $c_v = 1.25$ and choose that distribution with balanced means. Figure 5.6 shows that $f_{W_c^c}(x)$ always exhibits a bell shape in agreement with $f_{W_c^c}(0) = 0$.

Our interest in Figure 5.7 and Table 5.3 is in the convergence of the truncated counterpart towards the original waiting time distribution as $K \to \infty$.

5.1 The $M/G/1$ Retrial Queue 139

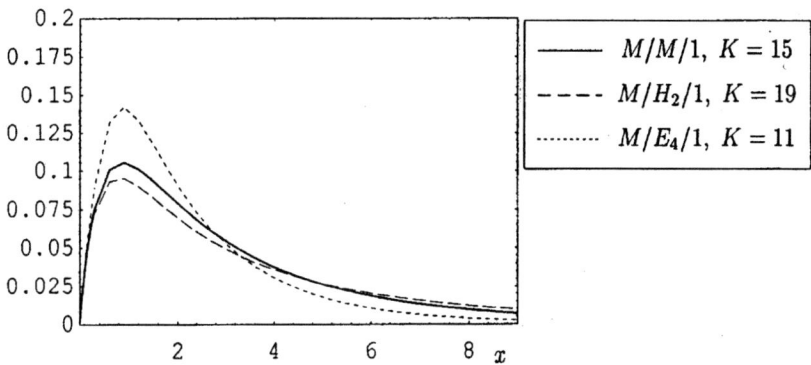

Fig. 5.6. The effect of the service time distribution on W_e^c

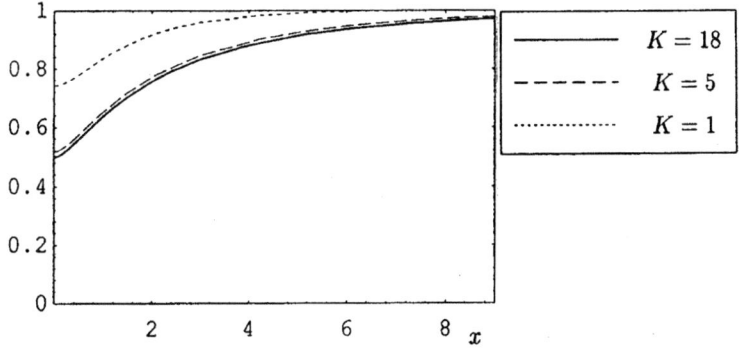

Fig. 5.7. $P(W \leq x)$ as a function of K for an $M/M/1$ queue

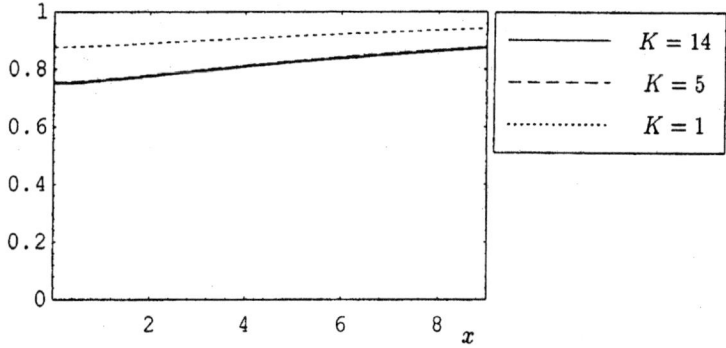

Fig. 5.8. $P(W \leq x)$ as a function of K for an $M/H_2/1$ queue

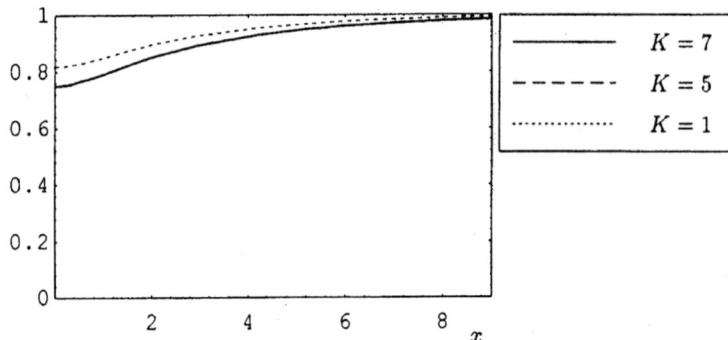

Fig. 5.9. $P(W \leq x)$ as a function of K for an $M/E_3/1$ queue

Table 5.3. $W^*(s)$ for an $M/M/1$ retrial queue as K increases

s	$K=1$	$K=5$	$K=10$	$K=18$	$K=\infty$
0.01	0.99554	0.98700	0.98557	0.98548	0.98548
0.05	0.97896	0.94137	0.93579	0.93549	0.93549
0.1	0.96063	0.89538	0.88675	0.88632	0.88632
0.25	0.91763	0.80149	0.78897	0.78842	0.78842
0.3	0.90630	0.77930	0.76618	0.76562	0.76562
0.6	0.85765	0.69286	0.67811	0.67751	0.67751
0.9	0.82853	0.64626	0.63096	0.63036	0.63035
1.2	0.80949	0.61727	0.60170	0.60109	0.60109
1.5	0.79626	0.59763	0.58191	0.58130	0.58130
10.0	0.74570	0.52320	0.50694	0.50633	0.50633
20.0	0.74302	0.51886	0.50254	0.50193	0.50193
$E[W]$	0.45161	1.33599	1.48997	1.49952	1.50000

More specifically, Figure 5.7 shows the convergence of the probability distribution function $P(W \leq x)$ as K increases for $M/M/1$ retrial queues with $\rho = 0.5$, $\beta_1 = 1.0$ and $\mu = 2.0$. The continuous curve corresponds to the threshold $K = 18$ chosen by fitting the first four decimal digits of $E[W]$. Then, we plot the curves associated with $K = 1$ (i.e., the lowest possible threshold) and the intermediate value $K = 5$. Figure 5.7 shows a rapid convergence to the continuous curve. Needless say, the jump at the origin corresponds to the probability of non-waiting. Table 5.3 lists values of the Laplace-Stieltjes transform $W^*(s)$ (given in column $K = \infty$) versus the Laplace-Stieltjes transforms of the truncated models with $K = 1, 5, 10$ and 18. As the reader may see, $K = 18$ is enough to provide an accurate approximation of the waiting time distribution in the original $M/M/1$ retrial queue.

Convergence issues are also addressed in Figures 5.8 and 5.9, and Tables 5.4 and 5.5. In particular, Figure 5.8 and Table 5.4 deal with $M/H_2/1$ retrial

5.1 The $M/G/1$ Retrial Queue   141

Table 5.4. $W^*(s)$ for an $M/H_2/1$ retrial queue as K increases

s	$K=1$	$K=5$	$K=10$	$K=14$	$K=\infty$
0.01	0.98725	0.97275	0.97198	0.97193	0.97227
0.05	0.95378	0.90379	0.90124	0.90097	0.90211
0.1	0.93132	0.85954	0.85613	0.85581	0.85706
0.25	0.90354	0.80686	0.80282	0.80262	0.80337
0.3	0.89911	0.79866	0.79455	0.79438	0.79499
0.6	0.88656	0.77552	0.77125	0.77117	0.77139
0.9	0.88205	0.76723	0.76290	0.76284	0.76294
1.2	0.87985	0.76316	0.75879	0.75874	0.75879
1.5	0.87859	0.76082	0.75642	0.75638	0.75641
10.0	0.87538	0.75478	0.75030	0.75026	0.75026
20.0	0.87528	0.75458	0.75011	0.75007	0.75007
$E[W]$	1.40729	3.01389	3.09138	3.09372	3.09375

queues with $\rho = 0.25$, $\mu = \rho/2$, $\beta_1 = 1.0$ and $c_v = 1.25$. In Figure 5.9 and Table 5.5, our interest is in $M/E_3/1$ retrial queues with $\rho = 0.25$, $\mu = 2\rho$ and $\beta_1 = 1.0$. In Figures 5.8 and 5.9, the respective thresholds $K = 14$ and 7 fit the first four decimal digits of $E[W]$ and yield values for $P(W \leq x)$ which are graphically undistinguished from those obtained for smaller thresholds ($K = 5$ in our examples). This remark shows a rapid convergence towards the probability distribution function $P(W \leq x)$ associated with such suitably chosen thresholds. Similarly to Table 5.3, entries in Tables 5.4 and 5.5 allow us to corroborate that the waiting time distributions in the truncated $M/H_2/1$ and $M/E_3/1$ retrial queues approach the corresponding distribution for the original queues with infinite retrial group.

Preceding numerical examples in Tables 5.1 and 5.3 provide a comparative analysis of the maximum entropy and truncated approaches. In the light of the columns of Table 5.1 listing values for $W^*(s)$, $W_2^*(s)$ and $W_{2,1}^*(s)$ and the column of Table 5.3 for the case $K = 18$, we may conclude that a direct truncation assumption yields more accurate estimations than those obtained from the *PME*, in agreement with Subsection 4.1.1 where such approaches were already applied to the length of a busy period.

Recently, Nobel and Tijms [552] proposed to approximate the waiting time distribution by a gamma distribution matching the first two conditional moments $E[W|W > 0]$ and $E[W^2|W > 0]$. The idea is to approximate $P(W > x|W > 0)$ by $1 - F_{gamma}(x)$, where $F_{gamma}(x)$ is the probability distribution function of a gamma distribution with shape parameter α and scale β chosen accordingly

$$\frac{\alpha}{\beta} = E[W|W > 0] = \frac{E[W]}{\rho},$$

Table 5.5. $W^*(s)$ for an $M/E_3/1$ retrial queue as K increases

s	$K=1$	$K=3$	$K=5$	$K=7$	$K=\infty$
0.01	0.99437	0.99168	0.99149	0.99149	0.99138
0.05	0.97457	0.96286	0.96217	0.96224	0.96161
0.1	0.95463	0.93452	0.93347	0.93367	0.93247
0.25	0.91450	0.87935	0.87778	0.87814	0.87621
0.3	0.90525	0.86696	0.86525	0.86561	0.86366
0.6	0.87068	0.82155	0.81929	0.81952	0.81800
0.9	0.85348	0.79941	0.79683	0.79693	0.79591
1.2	0.84353	0.78670	0.78391	0.78394	0.78325
1.5	0.83720	0.77864	0.77573	0.77571	0.77524
10.0	0.81837	0.77449	0.75125	0.75113	0.75112
20.0	0.81775	0.75367	0.75042	0.75030	0.75029
$E[W]$	0.57780	0.86595	0.88773	0.88884	0.88888

Table 5.6. Comparing $W^*_{79}(s)$ and $W^*_{gamma}(s)$ for an $M/M/1$ retrial queue

s	$W^*(s)$	$W^*_{79}(s)$	$W^*_{gamma}(s)$
0.001	0.98820	0.98820	0.98820
0.01	0.89766	0.89766	0.89789
0.05	0.67056	0.67056	0.67682
0.1	0.53785	0.53785	0.55283
0.25	0.38279	0.38279	0.41116
0.3	0.35805	0.35805	0.38841
0.6	0.28520	0.28520	0.31924
0.9	0.25668	0.25668	0.29010
1.2	0.24157	0.24157	0.27357
1.5	0.23229	0.23230	0.26277
10.0	0.20236	0.20236	0.21580
20.0	0.20073	0.20073	0.20949
50.0	0.20013	0.20013	0.20483
100.0	0.20003	0.20003	0.20290

$$\frac{\alpha(\alpha+1)}{\beta^2} = E\left[W^2 \,|\, W > 0\right] = \frac{E[W^2]}{\rho}.$$

The motivation in [552] to employ the gamma distribution is the conjecture that the tail probabilities of the waiting time distribution are asymptotically of the form $\gamma \exp\{-\delta x\}$ as $x \to \infty$. At this point, we remark that the maximum entropy estimations are computed on a finite domain $x \in (0, T)$. Thus, the maximum entropy approach cannot be employed to approximate the tail probabilities.

In Table 5.6, we take $\rho = 0.8$, $\beta_1 = 1.0$ and $\mu = 0.5$; that is, the same parameters considered in [552, Table 1] for the $M/M/1$ retrial queue. We

5.1 The $M/G/1$ Retrial Queue   143

Table 5.7. Comparing $W_{64}^*(s)$ and $W_{gamma}^*(s)$ for an $M/E_2/1$ retrial queue

s	$W^*(s)$	$W_{64}^*(s)$	$W_{gamma}^*(s)$
0.001	0.98916	0.99044	0.98916
0.01	0.90468	0.91726	0.90482
0.05	0.68412	0.72040	0.68848
0.1	0.55076	0.58913	0.56183
0.25	0.39188	0.41838	0.41410
0.3	0.36629	0.38957	0.39026
0.6	0.29049	0.30266	0.31807
0.9	0.26057	0.26799	0.28798
1.2	0.24463	0.24958	0.27108
1.5	0.23478	0.23827	0.26013
10.0	0.20251	0.20256	0.21398
20.0	0.20077	0.20077	0.20814
50.0	0.20014	0.20014	0.20398
100.0	0.20003	0.20003	0.20231

compare the Laplace-Stieltjes transforms $W_{79}^*(s)$ and $W_{gamma}^*(s)$ versus exact values of the transform $W^*(s)$ for several real choices of s. For these parameters, $K = 79$ is the first orbit threshold matching the first four decimal digits of $E[W]$. Obviously, $W_{gamma}^*(s)$ is given by

$$W_{gamma}^*(s) = 1 - \rho + \rho \left(\frac{\beta}{s+\beta}\right)^\alpha.$$

The numerical results reveal the superiority of the truncated approach for all values of s; that is, in the whole domain of the distribution.

In Table 5.7, we present the Laplace-Stieltjes transforms for a queue with E_2 service times with coefficient of variation $c_v = 1/\sqrt{2}$, $\rho = 0.8$ and $\mu = 0.5$. The entries in the table show that the gamma approximation performs better than the truncated approximation for smaller values of s. This fact is in agreement with the conjecture expressed in [552] for the tail probabilities. Other numerical experiments not reported here also give support to the conclusion that the gamma distribution gives a better estimation for the tail of the waiting time distribution. Interested readers are also referred to the numerical work in [552], where the gamma approximation is compared with simulation results.

5.1.2 The Number of Retrials Made by a Customer

Our next objective is the computation of the limiting distribution of the number of retrials made by a customer during his waiting time, which is denoted from now on by R. This characteristic, which is the discrete counterpart of W, provides a natural measure of the system performance from a customer's point of view. Indeed, R is a notable measure in itself because it determines the additional load on control devices in telephone systems.

Let us mark a primary customer arriving at the system. By observing the state of the server when this customer arrives, we have that $R = 0$, if the server is free, and $R > 0$, if the server is busy. In the former case, we derive

$$P(R = 0) = 1 - \rho. \tag{5.14}$$

In the latter case, the tagged customer will arrive during a service time, thus he is staying in orbit along the residual service in progress and during subsequent complete service times until reaching the server. We let W_1 be the residual service time and W_2 be the time from the most imminent departure until the tagged customer enters into service.

If the tagged customer produces k retrials during his waiting time, then the last one finds a free server and the first $k-1$ unsuccessful retrials are made during service times. A part of these $k-1$ vain retrials (i.e., blocked retrials) are produced in $(0, W_1)$, while the rest of them occurs over those service times belonging to $[W_1, W_1 + W_2)$. Thus, the main idea to derive the probability $P(R = k)$, for $k \geq 1$, is to consider separately those reattempts made by this customer during each one of these intervals. To this end, let $e_{lm}(\tau)$ be the probability that the tagged customer makes $l \geq 0$ retrials and $m \geq 0$ primary arrivals occur during the residual service time, given that the elapsed service time observed by him upon arrival is τ. Then, we get

$$e_{lm}(\tau) = \int_0^\infty e^{-(\lambda+\mu)x} \frac{(\mu x)^l (\lambda x)^m}{l! m!} dB_\tau(x), \quad l \geq 0, \ m \geq 0,$$

where $B_\tau(x)$ is the conditional probability distribution function of the residual service time, given that the elapsed service time is τ, and consequently $dB_\tau(x) = (1 - B(\tau))^{-1} dB(\tau + x)$, for $x \geq 0$.

An expression for $P(R = k)$, for $k \geq 1$, can be obtained by recording, at any departure epoch in $[W_1, W_1 + W_2)$, the orbit size and the accumulated number of reattempts made by the tagged customer from the departure epoch W_1 until such a departure. To be concrete, for $0 \leq r \leq k-1$ and $m \geq 1$, we denote by x_{rm}^k the conditional probability that the tagged customer makes a total of k reattempts during $[W_1, W_1 + W_2)$, given that at the previous departure epoch he accumulated r reattempts from the epoch W_1 and there were m customers in orbit. Then, we find that

$$P(R = k) = \int_0^\infty \sum_{j=0}^\infty P_{1j\tau} \sum_{l=0}^{k-1} \sum_{m=0}^\infty e_{lm}(\tau) x_{0,j+m+1}^{k-l} d\tau, \quad k \geq 1, \tag{5.15}$$

where $P_{1j\tau}$ is the limiting probability that the server is busy, the orbit size is j and the elapsed service time of the customer in service is τ.

The major drawback of (5.15) is that the probabilities $P_{1j\tau}$ have no simple analytical solution. We thus replace the distribution $B_\tau(x)$ by the distribution

$\hat{B}(x)$ of the forward renewal service time. This amounts to estimate $e_{lm}(\tau)$ by the probability $\hat{e}_{lm}$ that the marked customer makes $l \geq 0$ reattempts and $m \geq 0$ primary arrivals occur during a time distributed according to $\hat{B}(x)$.

By combining the approximating assumption $e_{lm}(\tau) \simeq \hat{e}_{lm}$, for $l \geq 0$ and $m \geq 0$, and (5.15), we have

$$P(R = k) = \sum_{j=0}^{\infty} P_{1j} \sum_{l=0}^{k-1} \sum_{m=0}^{\infty} \hat{e}_{lm} x_{0,j+m+1}^{k-l}, \quad k \geq 1. \tag{5.16}$$

It seems to be difficult to derive an exact or numerically tractable solution for the probabilities x_{rm}^k when the orbit capacity is infinite. To overcome such a difficulty, in what follows we deal with the truncated model $\{(C^K(t), N^K(t), \xi^K(t)); t \geq 0\}$. Let us keep the same definitions for $e_{lm}(\tau)$ and $\hat{e}_{lm}$ and observe that the conditional probabilities x_{rm}^k are then defined for $1 \leq m \leq K$ only.

In the truncated $M/G/1$ retrial queue, the probability mass function of the number R^K of retrials made by our tagged customer during his waiting time is given by the following formulas which replace (5.14) and (5.16):

$$P(R^K = 0) = \sum_{j=0}^{K} P_{0j}^K + P_{1K}^K, \tag{5.17}$$

$$P(R^K = k) = \sum_{j=0}^{K-1} P_{1j}^K \sum_{l=0}^{k-1} \left(\sum_{m=0}^{K-1-j} \hat{e}_{lm} x_{0,j+m+1}^{k-l} + \left(\sum_{m=K-j}^{\infty} \hat{e}_{lm} \right) x_{0K}^{k-l} \right),$$
$$k \geq 1. \tag{5.18}$$

Note that, for a fixed value of K, equation (5.17) follows by noticing that no reattempts are produced when upon arrival the tagged customer finds either a free server or a full system. To prove (5.18), we notice that, for $k \geq 1$, the event $\{R^K = k\}$ occurs when this customer finds a system in state $(1, j)$, for $0 \leq j \leq K - 1$, during the residual service he conducts at most $k - 1$ retrials and new blocked customers will be admitted in orbit whenever there were places on it; consequently, when this service ends, the tagged customer has made no retrials over an interval of length W_2 and the rest of retrials until k will be made over subsequent complete service times.

In the next result, we derive equations for the probabilities x_{rm}^k. Here, r_l is the probability that the tagged customer makes $l \geq 0$ repeated attempts during a service time and e_{lm} is the probability that he makes $l \geq 0$ reattempts and $m \geq 0$ primary arrivals occur during a service time.

Theorem 5.3. *For each fixed $k \geq 1$, the probabilities x_{rm}^k, for $0 \leq r \leq k-1$ and $1 \leq m \leq K$, can be recursively computed from the equations*

$$\mathbf{x}_{k-1} = (\mathbf{I}_K - \mathbf{A}_0)^{-1} \mathbf{g}, \tag{5.19}$$

$$\mathbf{x}_r = (\mathbf{I}_K - \mathbf{A}_0)^{-1} \sum_{n=r+1}^{k-1} \mathbf{A}_{n-r} \mathbf{x}_n, \quad 0 \leq r \leq k-2, \tag{5.20}$$

where $\mathbf{x}_r = (x_{r1}^k, ..., x_{rK}^k)'$, $\mathbf{g}$ is a column vector with mth entry defined by $\mu/(\lambda + m\mu)$, for $1 \leq m \leq K$, $\mathbf{A}_0$ is the square matrix of order K with (i,j)th entry given by

$$a_{ij}^0 = \begin{cases} \frac{\lambda e_{00} + (i-1)\mu e_{01}}{\lambda + i\mu}, & \text{if } 1 \leq i \leq K-1, \ j = i, \\ \frac{\lambda r_0 + (K-1)\mu(r_0 - e_{00})}{\lambda + K\mu}, & \text{if } i = j = K, \\ \frac{(i-1)\mu e_{00}}{\lambda + i\mu}, & \text{if } 2 \leq i \leq K, \ j = i-1, \\ \frac{\lambda e_{0,j-i} + (i-1)\mu e_{0,j-i+1}}{\lambda + i\mu}, & \text{if } 1 \leq i < j \leq K-1, \\ \frac{\lambda \left(r_0 - \sum_{n=0}^{K-i-1} e_{0n}\right)}{\lambda + i\mu} \\ \quad + \frac{(i-1)\mu \left(r_0 - \sum_{n=0}^{K-i} e_{0n}\right)}{\lambda + i\mu}, & \text{if } 1 \leq i \leq K-1, \ j = K, \\ 0, & \text{otherwise,} \end{cases}$$

and the square matrix $\mathbf{A}_{n-r}$ of order K is defined from the expression for $\mathbf{A}_0$ by replacing the quantities e_{0m} and r_0 by $e_{n-r,m}$ and r_{n-r}, respectively.

Proof. For each k, we analyze the dynamics between two successive departure epochs in order to get

$$x_{k-1,m}^k = \frac{\lambda}{\lambda + m\mu} \left(\sum_{n=0}^{K-m} e_{0n} x_{k-1,m+n}^k + \left(\sum_{n=K-m+1}^{\infty} e_{0n} \right) x_{k-1,K}^k \right)$$
$$+ \frac{(m-1)\mu}{\lambda + m\mu} \left(\sum_{n=0}^{K-m+1} e_{0n} x_{k-1,m+n-1}^k + \left(\sum_{n=K-m+2}^{\infty} e_{0n} \right) x_{k-1,K}^k \right)$$
$$+ \frac{\mu}{\lambda + m\mu}, \quad 1 \leq m \leq K. \tag{5.21}$$

The three terms contributing to the right-hand side of (5.21) are associated, respectively, with the following possibilities for the identity of the next customer entering service: (i) a new primary arrival, (ii) another customer from the orbit, and (iii) the tagged customer himself. In the cases (i) and (ii), we partition the subsequent event by depending on the number of arrivals that occur during the current service time and taking into account that the tagged customer cannot retry during such a service time.

A similar argument to the one driving to (5.21) yields

$$x_{rm}^k = \frac{\lambda}{\lambda + m\mu} \sum_{l=0}^{k-r-1} \left(\sum_{n=0}^{K-m} e_{ln} x_{r+l,m+n}^k + \left(\sum_{n=K-m+1}^{\infty} e_{ln} \right) x_{r+l,K}^k \right)$$
$$+ \frac{(m-1)\mu}{\lambda + m\mu} \sum_{l=0}^{k-r-1} \left(\sum_{n=0}^{K-m+1} e_{ln} x_{r+l,m+n-1}^k + \left(\sum_{n=K-m+2}^{\infty} e_{ln} \right) x_{r+l,K}^k \right),$$
$$0 \leq r \leq k-2, \ 1 \leq m \leq K. \tag{5.22}$$

After rearrangement, the set of equations (5.21) can be rewritten in matrix form as

$$\mathbf{x}_{k-1} = \mathbf{g} + \mathbf{A}_0 \mathbf{x}_{k-1}. \tag{5.23}$$

Now it is easy to prove that (5.23) amounts to expression (5.19).

Similarly, for $0 \leq r \leq k-2$, the set of equations (5.22) leads to the system (5.20). This completes the proof. □

As a result, equations (5.19) and (5.20) can be computed recursively, in reverse order, to get the vectors $\mathbf{x}_r$, for $0 \leq r \leq k-1$. It remains to specify expressions for the quantities $\hat{e}_{lm}$, e_{lm} and r_l. It is easy to note that these are given by

$$\hat{e}_{lm} = \frac{1}{\beta_1}\binom{l+m}{l}\frac{\lambda^m \mu^l}{(\lambda+\mu)^{l+m}}\tilde{a}_{l+m}, \quad l \geq 0, \ m \geq 0,$$

$$e_{lm} = \binom{l+m}{l}\frac{\lambda^m \mu^l}{(\lambda+\mu)^{l+m}}\bar{a}_{l+m}, \quad l \geq 0, \ m \geq 0,$$

$$r_l = \int_0^\infty e^{-\mu x}\frac{(\mu x)^l}{l!}dB(x), \quad l \geq 0,$$

where

$$\tilde{a}_j = \int_0^\infty e^{-(\lambda+\mu)x}\frac{((\lambda+\mu)x)^j}{j!}(1-B(x))dx, \quad j \geq 0,$$

$$\bar{a}_j = \int_0^\infty e^{-(\lambda+\mu)x}\frac{((\lambda+\mu)x)^j}{j!}dB(x), \quad j \geq 0.$$

By noting that $\tilde{a}_j$ and $\bar{a}_j$ are similarly defined to $\hat{a}_j$ and a_j, respectively, we may remark the following relationships:

$$\bar{a}_0 = 1 - (\lambda+\mu)\tilde{a}_0, \tag{5.24}$$

$$\bar{a}_j = (\lambda+\mu)(\tilde{a}_{j-1} - \tilde{a}_j), \quad j \geq 1. \tag{5.25}$$

From (5.24) and (5.25), the problem reduces to compute the integrals defining $\tilde{a}_j$. Since these integrals correspond to those defining $\hat{a}_j$ by replacing λ by $\lambda+\mu$, tractable expressions for $\tilde{a}_j$ can be routinely derived from Subsection 3.1.1 for the case of the most usual service time distributions.

Following [110], we recall the exact mean value of R

$$E[R] = \frac{\lambda\mu}{1-\rho}\left(\frac{\beta_1}{\mu} + \frac{\beta_2}{2}\right)$$

and then choose an enough large truncation level K as follows:

Step 1. Put $K = 1$.
Step 2. Determine the first integer $k(K) \geq 2$ such that

$$k(K)P(R^K = k(K)) < 10^{-8} \quad \text{and} \quad \sum_{n=0}^{k(K)} P(R^K = n) > 0.99.$$

Table 5.8. Performance measures related to R in an $M/H_2/1$ retrial queue

ρ	$\mu=0.05$	$\mu=0.5$	$\mu=1.0$	$\mu=2.5$	$\mu=5.0$
0.2	13,16,3	11,40,5	11,68,8	11,151,16	11,290,29
	0.26525	0.40250	0.55500	1.01250	1.77500
	0.26524	0.40249	0.55499	1.01249	1.77499
	0.36944	1.12712	2.48658	9.83395	32.97617
	0.80000	0.80000	0.80000	0.80000	0.80000
0.4	30,27,5	20,69,9	20,118,14	20,266,28	20,514,52
	0.70733	1.07333	1.48000	2.70000	4.73333
	0.70732	1.07332	1.47999	2.69998	4.73330
	1.31066	3.98876	8.82655	35.10014	118.09031
	0.60000	0.60000	0.60000	0.60000	0.60000
0.6	69,50,9	38,135,17	37,233,26	36,527,53	35,1014,97
	1.59150	2.41500	3.33000	6.07500	10.65000
	1.59148	2.41497	3.32997	6.07495	10.64990
	4.45090	13.85264	30.82960	123.00266	414.09333
	0.40000	0.40000	0.40000	0.40000	0.40000
0.8	192,122,22	92,366,42	86,634,64	82,1432,128	80,2757,234
	4.24400	6.44000	8.88000	16.20000	28.40000
	4.24395	6.43993	8.87992	16.19986	28.39973
	24.18194	79.59486	178.68350	712.90007	2393.07950
	0.20000	0.20000	0.20000	0.20000	0.20000

Step 3. Compute $\widehat{E[R^K]} = \sum_{n=1}^{k(K)} nP(R^K = n)$ and calculate the relative error

$$R(K) = \left|1 - \frac{\widehat{E[R^K]}}{E[R]}\right|.$$

Step 4. If $R(K) \geq 10^{-5}$, then put $K = K+1$ and repeat Steps 2 and 3. Otherwise, take the current threshold K as appropriate truncation level.

Tables 5.8 and 5.9 list truncation levels and several performance measures for retrial queues with H_2 and E_3 service times. In particular, the coefficient of variation of the H_2 law is $c_v = 1.2$ whereas β_1 has been normalized to be one in both cases, thus $\rho = \lambda$. For several values of ρ and μ, we display in each table the values of K, $k(K)$, the 99th percentile of R^K, $E[R]$, $\widehat{E[R^K]}$, $\widehat{Var[R^K]}$ and $P(R^K = 0)$ as ordered set values. For example, in Table 5.8, the cell associated with the pair $(\rho,\mu) = (0.2, 0.05)$, from top to bottom, lists the entry 13, 16 and 3 (first row), which means that $K = 13$, $k(K) = 16$ and the 99th percentile equals 3. Furthermore, the values 0.26525, 0.26524, 0.36944 and 0.80000 correspond to the true expectation $E[R]$, the approximate values $\widehat{E[R^K]}$ and $\widehat{Var(R^K)}$, and $P(R^K = 0)$, respectively.

From Tables 5.8 and 5.9, we stress that the above mentioned characteristics for the $M/H_2/1$ retrial queue exhibit higher variability than those for

Table 5.9. Performance measures related to R in an $M/E_3/1$ retrial queue

ρ	$\mu=0.05$	$\mu=0.5$	$\mu=1.0$	$\mu=2.5$	$\mu=5.0$
0.2	11,15,3	7,25,4	7,36,5	7,69,9	7,124,15
	0.25833	0.33333	0.41666	0.66666	1.08333
	0.25833	0.33333	0.41666	0.66666	1.08332
	0.34272	0.66870	1.14478	3.29210	9.26803
	0.80000	0.80000	0.80000	0.80000	0.80000
0.4	25,25,5	13,44,7	12,66,9	12,132,16	12,243,27
	0.68888	0.88888	1.11111	1.77777	2.88888
	0.68888	0.88888	1.11110	1.77777	2.88887
	1.21708	2.37647	4.09858	12.00578	34.31995
	0.60000	0.60000	0.60000	0.60000	0.60000
0.6	55,46,9	24,87,13	22,135,18	21,277,31	20,513,53
	1.55000	2.00000	2.50000	4.00000	6.50000
	1.54998	1.99998	2.49998	3.99997	6.49994
	4.12281	8.25352	14.47935	43.48985	126.28021
	0.40000	0.40000	0.40000	0.40000	0.40000
0.8	153,110,21	58,238,32	51,378,44	47,792,77	45,1475,133
	4.13333	5.33333	6.66666	10.66666	17.33333
	4.13329	5.33328	6.66660	10.66658	17.33317
	22.23097	47.15749	85.07995	263.43659	775.97845
	0.20000	0.20000	0.20000	0.20000	0.20000

the $M/E_3/1$ retrial queue, which may be explained by taking into account the coefficients of variation of both service time laws. As was expected, the truncation level K increases with increasing values of ρ and, by contrast, decreases as a function of μ. The measures $E[R]$, $\widehat{E[R^K]}$, $\widehat{Var(R^K)}$, the 99th percentile of R^K and $k(K)$ are increasing functions of ρ and μ. Indeed, a rapid reattempt for service has a significant chance of being blocked, so that the expectation and the quantities measuring dispersion (i.e., $\widehat{Var(R^K)}$, the 99th percentile and $k(K)$) increase with increasing values of μ. Finally, we also observe that the probability $P(R^K=0)$ gives an excellent approximation of the true probability $P(R=0) = 1-\rho$.

5.2 The $M/M/c$ Retrial Queue

5.2.1 Waiting Time Distribution

In Sections 3.4 and 4.2, we stressed the necessity of replacing the analysis of the performance characteristics of the $M/M/c$ retrial queue with $c>2$ by the corresponding approximation in terms of several truncated models. In what follows, we focus on the most natural and traditional approach consisting in assuming a certain truncation limit K in the orbit capacity. Then, the system state can be represented by the bidimensional process $\mathcal{X}^W$ defined in Subsection 3.4.1.

In Theorem 5.4, we provide two alternative expressions for the Laplace-Stieltjes transform of the unconditional waiting time. Previously, once more, we use a first-step analysis to derive the dynamics of the Laplace-Stieltjes transforms of the residual waiting times of the tagged customer, given that the system state is (i,j); that is, we have

$$W_{ij}^*(s) = \frac{\lambda}{s+\lambda+i\nu+j\mu} W_{i+1,j}^*(s) + \frac{i\nu}{s+\lambda+i\nu+j\mu} W_{i-1,j}^*(s)$$
$$+ \frac{(j-1)\mu}{s+\lambda+i\nu+j\mu} W_{i+1,j-1}^*(s) + \frac{\mu}{s+\lambda+i\nu+j\mu},$$
$$0 \le i \le c-1,\ 1 \le j \le K, \quad (5.26)$$

$$W_{cj}^*(s) = \frac{(1-\delta_{jK})\lambda}{s+(1-\delta_{jK})\lambda+c\nu} W_{c,j+1}^*(s) + \frac{c\nu}{s+(1-\delta_{jK})\lambda+c\nu} W_{c-1,j}^*(s),$$
$$1 \le j \le K. \quad (5.27)$$

Theorem 5.4. *The Laplace-Stieltjes transform of the unconditional waiting time is given by any of the following alternative expressions:*

$$W^*(s) = 1 - \sum_{j=0}^{K-1} P_{cj}^W + \sum_{j=0}^{K-1} P_{cj}^W W_{c,j+1}^*(s) \quad (5.28)$$

$$= 1 - \frac{s}{\lambda} \sum_{i=0}^{c} \sum_{j=1}^{K} j P_{ij}^W W_{ij}^*(s). \quad (5.29)$$

Proof. Clearly, by the *PASTA* property, we have that $W^*(s)$ is given by formula (5.28).

To obtain (5.29), we multiply equation (5.26) by $j(s+\lambda+i\nu+j\mu)P_{ij}^W$ and equation (5.27) by $j(s+(1-\delta_{jK})\lambda+c\nu)P_{cj}^W$; then, we replace $(\lambda+i\nu+j\mu)P_{ij}^W$ and $((1-\delta_{jK})\lambda+c\nu)P_{cj}^W$ on the resulting equations with the help of formulas (3.65)-(3.68) for the limiting probabilities. By summing in j, we get after some algebra that

$$\lambda \sum_{j=0}^{K-1} P_{cj}^W = \lambda \sum_{j=0}^{K-1} P_{cj}^W W_{c,j+1}^*(s) + s \sum_{i=0}^{c} \sum_{j=1}^{K} j P_{ij}^W W_{ij}^*(s). \quad (5.30)$$

Combining (5.28) and (5.30) yields expression (5.29) for $W^*(s)$. □

It should be pointed out that (5.28) includes both the discrete contribution $P(W=0) = 1 - \sum_{j=0}^{K-1} P_{cj}^W$ and the transform of the continuous contribution with density $f_{W^c}(x)$ on the interval $(0,\infty)$. Note that, since the underlying retrial queue for $\mathcal{X}^W$ has a finite retrial group, the probability of the event $\{W=0\}$ corresponds to the probability of non-waiting $\sum_{i=0}^{c-1} \sum_{j=0}^{K} P_{ij}^W$ and the loss probability P_{cK}^W.

5.2 The $M/M/c$ Retrial Queue

After solving the system of linear equations (5.26) and (5.27) for $W_{ij}^*(s)$, the transform $W^*(s)$ can be computed for any given s. This amounts to the application of the EULER and POST-WIDDER algorithms in [2] for the numerical inversion of $f_{W^c}(x)$. Alternatively, we may obtain the residual waiting time distribution function $P(W > x)$ by noticing that the relationship between $W^*(s)$ and $R^*(s) = \int_0^\infty e^{-sx} P(W > x) dx$ is $W^*(s) = 1 - sR^*(s)$.

In the next result, we give expressions for the recursive computation of the moments $W^{(k)}$, for $k \geq 0$.

Proposition 5.5. *(i) For each fixed $k \geq 1$, the conditional moments $W_{ij}^{(k)}$, for $0 \leq i \leq c$ and $1 \leq j \leq K$, can be obtained as the solution to the set of equations*

$$(\lambda + i\nu + j\mu) W_{ij}^{(k)} = k W_{ij}^{(k-1)} + \lambda W_{i+1,j}^{(k)} + i\nu W_{i-1,j}^{(k)}$$
$$+ (j-1)\mu W_{i+1,j-1}^{(k)}, \quad 0 \leq i \leq c-1, \; 1 \leq j \leq K, \quad (5.31)$$

$$((1-\delta_{jK})\lambda + c\nu) W_{cj}^{(k)} = k W_{cj}^{(k-1)} + (1-\delta_{jK})\lambda W_{c,j+1}^{(k)} + c\nu W_{c-1,j}^{(k)},$$
$$1 \leq j \leq K. \quad (5.32)$$

(ii) The moments of the unconditional waiting time can be computed from any of the following alternative formulas:

$$W^{(k)} = \delta_{0k} + (1-\delta_{0k}) \sum_{j=0}^{K-1} P_{cj}^W W_{c,j+1}^{(k)} \quad (5.33)$$

$$= \delta_{0k} + \frac{k}{\lambda} \sum_{i=0}^{c} \sum_{j=1}^{K} j P_{ij}^W W_{ij}^{(k-1)}, \quad k \geq 0. \quad (5.34)$$

Proof. To derive (5.31) and (5.32) we differentiate equations (5.26) and (5.27), respectively. Expectations for $k = 0$ are trivially equal to one. Then, the above set of equations allows us to compute the $(c+1)K$ unknowns $W_{ij}^{(k)}$ in terms of the moments of one order less.

By differentiating (5.28) with respect to s, at the point $s = 0$, we obtain expression (5.33), which gives the kth moment of W in terms of K conditional moments of the same order. On the other hand, we may differentiate the alternative formula (5.29). This leads to a second expression involving more terms, but concerning moments of one order less. □

In the rest of this subsection, for ease of notation, we simply denote the mean and the second order moment by $E[W]$ and $E[W^2]$.

In particular, formula (5.34) shows that the expectation agrees with Little's formula

$$E[N] = \lambda E[W],$$

and the second order moment depends only on the expectations of the conditional waiting times W_{ij}; that is,

Table 5.10. Moments of W

ρ		$\mu = 0.05$	$\mu = 0.5$	$\mu = 1.0$	$\mu = 2.5$	$\mu = 5.0$
0.2	$E[W]$	0.06340	0.00726	0.00415	0.00226	0.00162
	$\sigma(W)$	1.59403	0.17256	0.09467	0.04894	0.03456
0.4	$E[W]$	0.89655	0.11135	0.06719	0.03997	0.03043
	$\sigma(W)$	6.06095	0.69572	0.40205	0.23044	0.17621
0.6	$E[W]$	3.97467	0.52138	0.32732	0.20726	0.16501
	$\sigma(W)$	13.26655	1.62377	0.99183	0.62237	0.50420
0.8	$E[W]$	14.81297	2.03227	1.31428	0.87440	0.72198
	$\sigma(W)$	28.80943	3.90971	2.55305	1.75731	1.50076

$$E\left[W^2\right] = \frac{2}{\lambda} \sum_{i=0}^{c} \sum_{j=1}^{K} j P_{ij}^{W} W_{ij}^{(1)}.$$

We may also note how the Tauberian theorems are again helpful to determine the value of the density function at the point $x = 0$. In this sense, by taking limits in (5.26) and (5.27), we observe that

$$f_{W_{ij}}(0) = \lim_{s \to \infty} s W_{ij}^*(s) = (1 - \delta_{ic})\mu. \tag{5.35}$$

Finally, from (5.28) and (5.35), we easily find

$$f_{W^c}(0) = 0. \tag{5.36}$$

We present numerical examples involving the first two moments of W and the numerical inversion of the continuous contribution of W and the conditional waiting times W_{ij}. Since we desire to use the truncated process $\mathcal{X}^W$ instead of $\mathcal{X}$, we consider Criterion I of Subsection 3.4.2. According to this criterion, we will assume, in what follows, that our numerical results can be considered as the true characteristics of the model with infinite retrial group. Appropriate truncation thresholds for our numerical examples were reported in Table 3.9 for a model with $c = 5$ and $\nu = 1.0$.

We first concentrate on the moments of W. After solving the set of equations (5.31) and (5.32) for $k = 1$, we display in Table 5.10 the mean and the standard deviation for different values of the traffic intensity ρ and the retrial rate μ. We observe that both descriptors decrease with increasing values of μ and increase as functions of ρ.

We invert numerically formula (5.28) and get the density $f_{W^c}(x)$. In Figure 5.10, we fix $c = 5$, $\nu = 1.0$ and $\mu = 2.5$ and display four curves corresponding to $\rho = 0.2, 0.4, 0.6$ and 0.8. We notice that all curves are unimodal which could be expected in the light of the behavior at the origin. The total mass accumulated by $f_{W^c}(x)$ accounts for the probability of waiting $P(W > 0) = \sum_{j=0}^{K-1} P_{cj}^W$. This fact explains the shape of the densities and, in particular, it helps to understand why the curve associated with $\rho = 0.2$ is very close to the horizontal axis. Concretely, the probabilities $P(W > 0)$ are given, in increasing order of ρ, by 0.00340, 0.05034, 0.19721 and 0.48232.

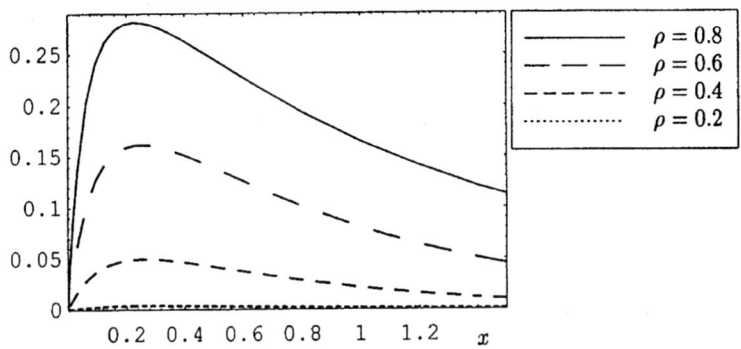

Fig. 5.10. Density functions of W^c as ρ varies

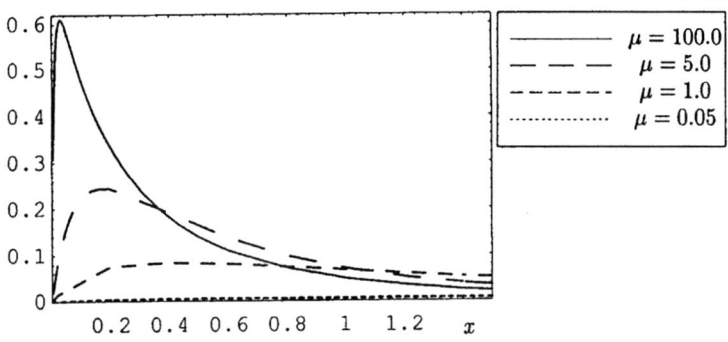

Fig. 5.11. Density functions of W^c as μ varies

The influence of the retrial rate is analyzed in Figure 5.11. We keep the values of c and ν fixed and take $\rho = 0.6$. Then, we plot the density $f_{W^c}(x)$ for $\mu = 0.05$, 1.0, 5.0 and 100.0. The case $\mu = 100.0$ can be considered close to the standard $M/M/5$ queue operating under random order. Now the probability of waiting varies from 0.16255 (case $\mu = 0.05$) to 0.23356 (case $\mu = 100.0$). It should be noticed that as long as μ decreases the distribution of the waiting time becomes sparser. This fact is corroborated in the figure in the light of the rate of convergence of the different densities to the horizontal axis. Obviously, the curve associated with $\mu = 0.05$ exhibits a heavier tail.

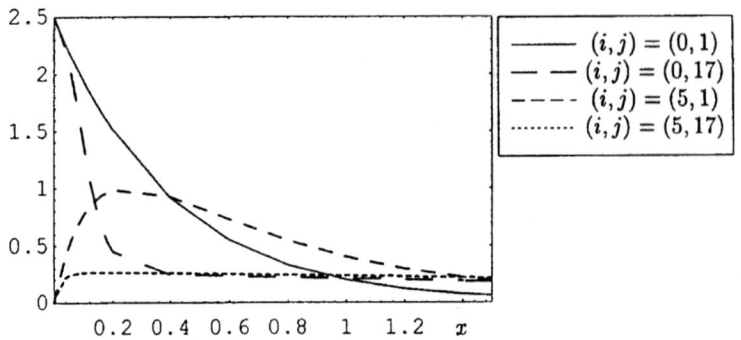

Fig. 5.12. The density $f_{W_{ij}}(x)$ versus (i,j)

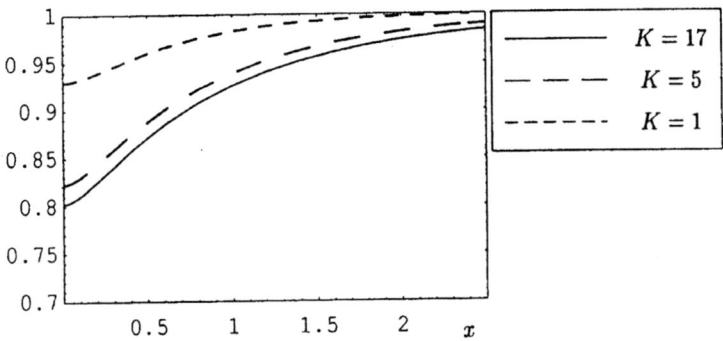

Fig. 5.13. $P(W \leq x)$ as a function of K

For $\rho = 0.6$ and $\mu = 2.5$, in Figure 5.12 we turn to the conditional waiting times W_{ij}. We display the densities $f_{W_{ij}}(x)$ for the initial states $(i,j) = (0,1)$, $(5,1)$, $(0,17)$ and $(5,17)$. We take $c = 5$ and so, according to Table 3.9, we have $K = 17$. Thus, these four cases represent extreme initial configurations. All densities $f_{W_{ij}}(x)$ accumulate a mass equal to one. By solving (5.31) and (5.32), we find that $W_{01}^{(1)} = 0.40877$, $W_{51}^{(1)} = 0.82686$, $W_{0,17}^{(1)} = 1.70635$ and $W_{5,17}^{(1)} = 2.73322$. These mean values agree with the graphics and, in particular, with the rate of decrease to the horizontal axis. It should be pointed out that $f_{W_{01}}(x)$ and $f_{W_{0,17}}(x)$ are decreasing functions whereas the shapes of $f_{W_{51}}(x)$ and $f_{W_{5,17}}(x)$ are unimodal.

In a last set of experiments, we focus on the convergence of the waiting time of the truncated process $\mathcal{X}^W$ as $K \to \infty$. According to the intuitive expectations and the existing asymptotic results for the limiting distribution of the system state [288, 604], we expect that the waiting time distribution of the truncated process should approach the corresponding distribution in the $M/M/c$ queue with infinite retrial group. Figure 5.13 and Table 5.3 give numerical support to our claim.

Once more, in Figure 5.13, we deal with the parameters $c = 5$, $\nu = 1.0$, $\mu = 2.5$ and $\rho = 0.6$. We use numerical inversion to obtain the probability distribution function $P(W \leq x)$. The continuous curve is associated with the threshold $K = 17$ proposed in Table 3.9. Then, we plot the curve corresponding to $K = 1$ (i.e., the lowest possible threshold) and an intermediate value, say $K = 5$. The graph shows a rapid convergence to the continuous curve. In fact, no more distribution functions are plotted because they tend to be graphically undistinguished. The jump at the origin corresponds to the probability $P(W = 0)$ which is 0.80278 (case $K = 17$), 0.82300 (case $K = 5$) and 0.92981 (case $K = 1$).

5.2.2 The Number of Retrials Made by a Customer

While the analysis of the number of retrials R made by a customer in the $M/G/1$ retrial queue was performed in Subsection 5.1.2 over the sequence of departure epochs, we may readily update the state of R at any transition of the process $\mathcal{X}$ in the case of an $M/M/c$ retrial queue, since $\mathcal{X}$ is a continuous-time Markov chain.

Indeed, we may replace the computation of the probabilities x_{rm}^k by parallel quantities $z_{l,(i,j)}^k$ defined as the conditional probabilities that the tagged customer makes k retrials before entering into service, given that he has accumulated l retrials and the current state of $\mathcal{X}$ is (i,j).

Note that

$$P(R = 0) = 1 - \sum_{j=0}^{\infty} P_{cj}, \qquad (5.37)$$

$$P(R = k) = \sum_{j=0}^{\infty} P_{cj} z_{0,(c,j+1)}^k, \quad k \geq 1. \qquad (5.38)$$

To derive the mass function of R from (5.37) and (5.38), we should face two major difficulties. The first one is related to the recursive computation of $\{P_{ij}; (i,j) \in \mathcal{S}\}$, which cannot be performed when $c > 2$. The second difficulty concerns the investigation of the infinite non-homogeneous set of equations governing the unknowns $z_{l,(i,j)}^k$, which indeed seems to be a more intricate problem. Therefore, we next concentrate on the estimation of R through the corresponding number of retrials R^K made by a tagged customer in the truncated process $\mathcal{X}^W$ investigated in Subsection 3.4.1. This implies that, for a suitably chosen value of K, we approximate the probability mass function of R in (5.37) and (5.38) by means of

$$P(R^K = 0) = 1 - \sum_{j=0}^{K-1} P_{cj}^W,$$

$$P(R^K = k) = \sum_{j=0}^{K-1} P_{cj}^W z_{0,(c,j+1)}^k, \quad k \geq 1.$$

We describe the dynamics of the unknowns $z_{l,(i,j)}^k$ in the following.

Proposition 5.6. *For each fixed $k \geq 1$, the probabilities $z_{l,(i,j)}^k$, for $0 \leq l \leq k-1$, $0 \leq i \leq c$ and $1 \leq j \leq K$, satisfy that*

$$z_{k-1,(i,j)}^k = \frac{\lambda}{\lambda + i\nu + j\mu} z_{k-1,(i+1,j)}^k + \frac{i\nu}{\lambda + i\nu + j\mu} z_{k-1,(i-1,j)}^k$$
$$+ \frac{(j-1)\mu}{\lambda + i\nu + j\mu} z_{k-1,(i+1,j-1)}^k + \frac{\mu}{\lambda + i\nu + j\mu},$$
$$0 \leq i \leq c-1, \ 1 \leq j \leq K, \quad (5.39)$$

$$z_{k-1,(c,j)}^k = \frac{(1-\delta_{jK})\lambda}{(1-\delta_{jK})\lambda + c\nu + \mu} z_{k-1,(c,j+1)}^k$$
$$+ \frac{c\nu}{(1-\delta_{jK})\lambda + c\nu + \mu} z_{k-1,(c-1,j)}^k, \quad 1 \leq j \leq K, \quad (5.40)$$

$$z_{l,(i,j)}^k = \frac{\lambda}{\lambda + i\nu + j\mu} z_{l,(i+1,j)}^k + \frac{i\nu}{\lambda + i\nu + j\mu} z_{l,(i-1,j)}^k$$
$$+ \frac{(j-1)\mu}{\lambda + i\nu + j\mu} z_{l,(i+1,j-1)}^k,$$
$$0 \leq l \leq k-2, \ 0 \leq i \leq c-1, \ 1 \leq j \leq K, \quad (5.41)$$

$$z_{l,(c,j)}^k = \frac{(1-\delta_{jK})\lambda}{(1-\delta_{jK})\lambda + c\nu + \mu} z_{l,(c,j+1)}^k + \frac{c\nu}{(1-\delta_{jK})\lambda + c\nu + \mu} z_{l,(c-1,j)}^k$$
$$+ \frac{\mu}{(1-\delta_{jK})\lambda + c\nu + \mu} z_{l+1,(c,j)}^k, \quad 0 \leq l \leq k-2, \ 1 \leq j \leq K.$$
$$(5.42)$$

Proof. For any fixed $k \geq 1$, first-step analysis leads to the above finite set of equations (5.39)-(5.42). We notice that in order to prove these equations, we avoid the consideration of vain retrials made by the non-tagged customers, which neither affect to the event under study nor modify the current state of the system. □

For any fixed value of k, we need firstly to solve (5.39) and (5.40) in order to get $z_{k-1,(i,j)}^k$, for $0 \leq i \leq c$ and $1 \leq j \leq K$. We then iterate the solution of (5.41) and (5.42) backward, from $l = k-2$ to 0, so that at each iterative step we need only those values for $z_{l+1,(c,j)}^k$ evaluated at the preceding step.

To choose a suitable truncation level K, we first remark that the exact value of $E[R]$ is now unknown. Thus, by appealing to Little's formula and Wald's identity, we find that

$$E[R] = \frac{\mu}{\lambda} E[N],$$

5.2 The $M/M/c$ Retrial Queue   157

Table 5.11. Performance measures related to R in an $M/M/5$ retrial queue

ρ	$\mu=0.05$	$\mu=0.5$	$\mu=1.0$	$\mu=2.5$	$\mu=5.0$
0.2	5,8,5,0	3,8,9,0	3,9,12,0	3,9,21,0	3,9,36,0
	0.00317	0.00367	0.00420	0.00575	0.00825
	0.00317	0.00367	0.00420	0.00575	0.00825
	0.00326	0.00469	0.00653	0.01370	0.03103
	0.99688	0.99677	0.99670	0.99656	0.99644
0.4	16,20,8,1	11,16,14,1	10,15,20,2	9,15,40,3	9,16,72,4
	0.04483	0.05569	0.06723	0.10006	0.15237
	0.04483	0.05569	0.06723	0.10006	0.15237
	0.04850	0.07993	0.12322	0.30350	0.76876
	0.95783	0.95511	0.95309	0.94963	0.94690
0.6	47,53,13,2	22,30,25,3	28,29,39,4	17,28,82,7	16,28,152,12
	0.19874	0.26077	0.32733	0.51887	0.82672
	0.19874	0.26077	0.32733	0.51886	0.82672
	0.24822	0.46607	0.78801	2.23678	6.22700
	0.83743	0.82578	0.81724	0.80273	0.79136
0.8	163,167,28,5	57,77,65,8	47,70,107,11	42,66,230,20	47,65,436,35
	0.74066	1.01649	1.31534	2.18849	3.61114
	0.74065	1.01648	1.31533	2.18846	3.61111
	1.35411	3.00018	5.59792	18.13375	54.61594
	0.58193	0.56051	0.54465	0.51761	0.49652

from which we approximate $E[R]$ by the value $E[R^K]$ obtained as

$$E[R^K] = \frac{\mu}{\lambda} E[N^W].$$

To speed up the calculation of the threshold K, we take an initial value K_0 defined as the first positive integer such that the four decimal digits of $E[N]$ are fitted by $E[N^W]$. Then, by starting from K_0, we implement Steps 2-4 in Subsection 5.1.2.

To illustrate the behavior of R for $M/M/5$ retrial queues, Table 5.11 lists values of K_0, K, $k(K)$ and the 99th percentile of R^K (first row), and $E[R^K]$, $\widehat{E[R^K]}$, $Var(\widehat{R^K})$ and $P(R^K=0)$ (subsequent rows). The arrival rate λ has been normalized to have traffic loads ρ similarly to Tables 5.8 and 5.9. The behavior of the performance measures related to R, as a function of the system parameters, is similar to that observed in Subsection 5.1.2. We remark here that the orbit threshold K is moderately larger than the initial value K_0. Furthermore, the probability $P(R^K=0)$ decreases with increasing values of ρ and μ, which corroborates our intuitive expectations since increasing values of ρ imply higher congestion and an increase in μ gives support to the existence of vain reattempts.

5.3 Bibliographical Notes

The paper by Falin and Fricker [283] is a basic reference for the waiting time of the $M/G/1$ retrial queue. The main result of the paper is the analysis of the distribution of W in terms of its Laplace-Stieltjes transform. For the computation of the second order moment and further results related to the maximum entropy estimation, see [83, 94]. For the study of the waiting time distribution in the $M/M/c$ retrial queue, an alternative approach based on an auxiliary absorbing Markov chain can be found in [93].

More material on the waiting time analysis includes extensions to advanced models with priorities [182], constant retrial policy [184] and finite population [289, 290], among others. Although most applications of retrial queues are based on models operating under the random order discipline, the case with customers in orbit served on a first-come-first-served (*FCFS*) basis has been considered by several authors (see, for instance, [183, 184, 315]).

Although our arguments in Subsections 5.1.2 and 5.2.2 are based on the paper [110], the reader is also referred to [267] for expressions for $E[R]$, $Var(R)$ and the asymptotic behavior of the mass function of R as $\mu \to 0$ in the particular case of the $M/M/1$ retrial queue. Early literature dealing with R provides its expected value in terms of the mean waiting time W; see [380]. Falin and Fricker [283] obtain the generating function of R in terms of an integral expression involving the Laplace-Stieltjes transforms of the service time distribution and of the length of a busy period in the standard $M/G/1$ queue.

Throughout this chapter, we have found several finite sets of linear equations governing the Laplace-Stieltjes transforms and the moments of conditional waiting times, as well as auxiliary probabilities related to the number of retrials made by a customer. A comparison of different methods for solving systems of linear equations is not our aim here, but it is clear that a number of numerical methods, such as Gaussian elimination, LU algorithms and aggregate-disaggregate methods, can be used to solve efficiently the sets of equations under consideration; see e.g. [160, 573, 644].

6
Other Descriptors

This chapter ends up the discussion on performance descriptors in retrial queues through various measures of intrinsic interest, but less studied in the literature. In Section 6.1, we give an alternative Markovian description of the $M/G/1$ retrial queue that replaces the state of the server by the total number of arrivals trying to enter into service since the last service completion. Our interest in Section 6.2 is to distinguish between the different events occurring along a busy period of the $M/G/1$ retrial queue. In particular, we study four characteristics allowing us to record the numbers of successful and blocked retrials, as well as the numbers of successful and blocked primary arrivals. Sections 6.3 and 6.4 concern the multiserver case and analyze the distribution of the server idle periods and the time to reach a certain orbit level. A brief section of bibliographical notes concludes the chapter.

6.1 Attempts Since the Last Service Completion

Our aim in this section is to show, in the case of the $M/G/1$ retrial queue, that the state of the server $C(t)$ at time t can be described in more detail by introducing the total number $M(t)$ of attempts to get service since the last departure epoch. We may see this by noting that the server is free at time t when $M(t) = 0$. Similarly, $M(t) = k \geq 1$ implies that the server is busy at time t and, consequently, since the last departure epoch, exactly k customers arrive at the system, including both primary arrivals and retrials, as well as the customer being served at time t. The resulting Markovian process $\{(M(t), N(t), \xi(t)); t \geq 0\}$ can be used instead of the commonly studied process $\{(C(t), N(t), \xi(t)); t \geq 0\}$, where $\xi(t)$ denotes the elapsed service time.

We begin by considering the limiting probabilities $\{P_{0j}; j \geq 0\}$ and defining

$$P_{kj\tau} d\tau = \lim_{t \to \infty} P(M(t) = k, N(t) = j, \tau \leq \xi(t) < \tau + d\tau), \quad k \geq 1, \ j \geq 0.$$

Assuming steady state is achievable, it is found that these probabilities satisfy the set of equations

$$(\lambda + j\mu)P_{0j} = \sum_{k=1}^{\infty} \int_0^{\infty} P_{kj\tau} b(\tau) d\tau, \quad j \geq 0, \tag{6.1}$$

$$\frac{dP_{kj\tau}}{d\tau} = (1 - \delta_{1k})\left((1 - \delta_{0j})\lambda P_{k-1,j-1,\tau} + j\mu P_{k-1,j,\tau}\right)$$
$$- (\lambda + j\mu + b(\tau))P_{kj\tau}, \quad k \geq 1, \; j \geq 0, \tag{6.2}$$

where $b(\tau) = (1 - B(\tau))^{-1} dB(\tau)$ represents the instantaneous service rate, given that the elapsed time of the service in process equals τ. Boundary conditions for the equations (6.1) and (6.2) are given by

$$P_{kj0} = \begin{cases} \lambda P_{0j} + (j+1)\mu P_{0,j+1}, & \text{if } k = 1, \; j \geq 0, \\ 0, & \text{if } k \geq 2, \; j \geq 0. \end{cases} \tag{6.3}$$

By introducing the generating function

$$H(y, z; \tau) = \sum_{k=1}^{\infty} y^k \sum_{j=0}^{\infty} z^j P_{kj\tau},$$

it follows, from (6.2) and (6.3), that $H(y, z; \tau)$ is the solution to

$$\frac{\partial H(y, z; \tau)}{\partial \tau} = -(\lambda - \lambda yz + b(\tau))H(y, z; \tau) + \mu(y - 1)z\frac{\partial H(y, z; \tau)}{\partial z}, \tag{6.4}$$
$$H(y, z; 0) = \lambda y(P_0(z) + P_1(z)). \tag{6.5}$$

In order to compute the factorial moments defined by

$$F(m, n) = \frac{\partial^m}{\partial y^m} \frac{\partial^n}{\partial z^n} H(y, z)\bigg|_{y=z=1},$$

with $H(y, z) = \int_0^{\infty} H(y, z; \tau) d\tau$, differentiating (6.4) and evaluating the result at $y = z = 1$ yields

$$\frac{\partial F(m, n; \tau)}{\partial \tau} = -b(\tau)F(m, n; \tau) + mn\lambda F(m-1, n-1; \tau)$$
$$+ m(\lambda + n\mu)F(m-1, n; \tau) + n\lambda F(m, n-1; \tau)$$
$$+ m\mu F(m-1, n+1; \tau), \quad m \geq 0, \; n \geq 0, \tag{6.6}$$

where $F(m, n; \tau) = \frac{\partial^m}{\partial y^m} \frac{\partial^n}{\partial z^n} H(y, z; \tau)\big|_{y=z=1}$. Similarly, the boundary equation (6.5) leads to

$$F(m, n; 0) = \begin{cases} \lambda M_n, & \text{if } m \in \{0, 1\}, \; n \geq 0, \\ 0, & \text{if } m \geq 2, \; n \geq 0, \end{cases} \tag{6.7}$$

where we recall that M_n denotes the nth factorial moment of the orbit size. Since $F(0, n; \tau) = \frac{\partial^n}{\partial z^n} H(1, z; \tau)\big|_{z=1}$, we find that

$$F(0, n; \tau) = \lambda(1 - B(\tau)) \sum_{l=0}^{n} \binom{n}{l} (\lambda \tau)^l M_{n-l}, \quad n \geq 0. \qquad (6.8)$$

To prove (6.8), we solve the differential equation

$$\frac{\partial H(1, z; \tau)}{\partial \tau} = -(\lambda - \lambda z + b(\tau)) H(1, z; \tau),$$

which is obtained from (6.4) by taking $y = 1$.

For convenience, we now introduce the function $G(m, n; \tau)$ defined from

$$F(m, n; \tau) = (1 - B(\tau)) G(m, n; \tau), \quad m \geq 0, \ n \geq 0, \ \tau \geq 0.$$

Then, the equations (6.6)-(6.8) can be rewritten as follows:

$$\frac{\partial G(m, n; \tau)}{\partial \tau} = mn\lambda G(m - 1, n - 1; \tau) + m(\lambda + n\mu) G(m - 1, n; \tau)$$
$$+ n\lambda G(m, n - 1; \tau) + m\mu G(m - 1, n + 1; \tau), \ m \geq 0, \ n \geq 0, \quad (6.9)$$

$$G(m, n; 0) = \begin{cases} \lambda M_n, & \text{if } m \in \{0, 1\}, \ n \geq 0, \\ 0, & \text{if } m \geq 2, \ n \geq 0, \end{cases} \qquad (6.10)$$

$$G(0, n; \tau) = \lambda \sum_{l=0}^{n} \binom{n}{l} (\lambda \tau)^l M_{n-l}, \quad n \geq 0. \qquad (6.11)$$

It should be noted that $G(0, n; \tau)$, as a function of τ, is a polynomial of degree n with coefficients

$$G_l(0, n) = \lambda^{l+1} \binom{n}{l} M_{n-l}, \quad 0 \leq l \leq n. \qquad (6.12)$$

More generally, the structural form of $G(m, n; \tau)$ is given in the next theorem.

Theorem 6.1. *The function $G(m, n; \tau)$ can be expressed as*

$$G(m, n; \tau) = \sum_{l=0}^{2m+n} G_l(m, n) \tau^l, \quad m \geq 0, \ n \geq 0, \qquad (6.13)$$

where the coefficients $G_l(m, n)$ verify

$$G_0(m, n) = G(m, n; 0), \quad m \geq 0, \ n \geq 0, \qquad (6.14)$$

$$lG_l(m, n) = mn\lambda G_{l-1}(m - 1, n - 1) + m(\lambda + n\mu) G_{l-1}(m - 1, n)$$
$$+ n\lambda G_{l-1}(m, n - 1) + m\mu G_{l-1}(m - 1, n + 1),$$
$$1 \leq l \leq 2m + n, \ m \geq 0, \ n \geq 0, \quad (6.15)$$

$$G_l(m, n) = 0, \quad l > 2m + n, \ m \geq 0, \ n \geq 0. \qquad (6.16)$$

162   6 Other Descriptors

Proof. The validity of (6.13)-(6.16) for $m = 0$ is readily derived from (6.11) and (6.12).

To prove (6.13) for a fixed pair (m,n) with $m \geq 1$, we first assume the validity of (6.13)-(6.16) for the functions $G(0,n;\tau), ..., G(m-1,n;\tau)$, for $n \geq 0$, and $G(m,0;\tau), ..., G(m,n-1;\tau)$. Then, formula (6.9) implies

$$G(m,n;\tau) = G(m,n;0) + mn\lambda \int_0^\tau G(m-1,n-1;u)du$$
$$+ m(\lambda + n\mu) \int_0^\tau G(m-1,n;u)du + n\lambda \int_0^\tau G(m,n-1;u)du$$
$$+ m\mu \int_0^\tau G(m-1,n+1;u)du, \quad m \geq 0, \ n \geq 0. \quad (6.17)$$

By inserting expressions (6.13) for $G(m-1,n-1;u)$, $G(m-1,n;u)$, $G(m,n-1;u)$ and $G(m-1,n+1;u)$ into (6.17) and using (6.14)-(6.16), we get the desired expression (6.13). □

This has the following immediate consequence.

Corollary 6.2. *The factorial moments $F(m,n)$, for $m \geq 0$ and $m \geq 0$, are given by*

$$F(m,n) = \sum_{l=0}^{2m+n} G_l(m,n) \frac{\beta_{l+1}}{l+1},$$

where the coefficients $G_l(m,n)$ can be recursively computed from (6.14)-(6.16).

In particular, the steady state mean and the variance of $M(t)$, and the covariance of $M(t)$ and $N(t)$ are given by

$$E[M] = \rho + \lambda \left((\lambda + \mu M_1) \frac{\beta_2}{2} + \lambda\mu \frac{\beta_3}{6} \right),$$
$$Var(M) = \lambda(\lambda + \mu M_1)\beta_2$$
$$+ \lambda \left(\lambda(\lambda + 2\mu) + \mu(2\lambda + \mu)M_1 + \mu^2 M_2 \right) \frac{\beta_3}{3}$$
$$+ \lambda^2 \mu(3\lambda + \mu + 3\mu M_1) \frac{\beta_4}{12} + \lambda^3 \mu^2 \frac{\beta_5}{20} + E[M] - (E[M])^2,$$
$$Cov(M,N) = \rho M_1 + \lambda \left(2\lambda + (\lambda + \mu)M_1 + \mu M_2 \right) \frac{\beta_2}{2}$$
$$+ \lambda^2 (2\lambda + \mu + 3\mu M_1) \frac{\beta_3}{6} + \lambda^3 \mu \frac{\beta_4}{8} - E[M]E[N].$$

For the sake of completeness, we next give a recursive scheme for computing the factorial moments of the orbit size. This scheme is indeed an alternative to the expression for M_n given in Subsection 2.2.2.

6.1 Attempts Since the Last Service Completion

Table 6.1. $E[M]$, $Var(M)$ and $Cov(M, N)$ versus ρ and μ

ρ		$\mu = 0.05$	$\mu = 0.5$	$\mu = 1.0$	$\mu = 2.5$	$\mu = 5.0$
0.2	$E[M]$	0.25250	0.27500	0.30000	0.37500	0.50000
	$Var(M)$	0.32791	0.50437	0.78000	2.10937	6.00000
	$Cov(M,N)$	0.11662	0.15375	0.19500	0.31875	0.52500
0.4	$E[M]$	0.68000	0.80000	0.93333	1.33333	2.00000
	$Var(M)$	1.21680	2.48000	4.54222	14.88888	46.00000
	$Cov(M,N)$	0.74400	1.04000	1.36888	2.35555	4.00000
0.6	$E[M]$	1.54500	1.95000	2.40000	3.75000	6.00000
	$Var(M)$	4.26367	10.56750	21.12000	75.18750	240.00000
	$Cov(M,N)$	3.31650	4.81500	6.48000	11.47500	20.80000
0.8	$E[M]$	4.16000	5.60000	7.20000	12.00000	20.00000
	$Var(M)$	23.49120	66.72000	139.68000	516.00000	1668.00000
	$Cov(M,N)$	20.25600	29.76000	40.32000	72.00000	124.80000

Proposition 6.3. *The factorial moment M_k can be evaluated as $M_k = M_k^0 + M_k^1$, where*

$$M_0^0 = 1 - \rho, \quad M_0^1 = \rho, \quad M_1^0 = \frac{\lambda \rho}{\mu}, \quad M_k^1 = \frac{\mu}{\lambda} M_{k+1}^0, \; k \geq 1,$$

and M_k^0, for $k \geq 2$, satisfies

$$k\mu(1-\rho)M_k^0 = (1-\rho)\lambda^{k+1}\beta_k$$
$$+ \sum_{n=1}^{k-1}\left(\binom{k}{n}\beta_{k-n} + \mu\binom{k}{n-1}\beta_{k+1-n}\right)\lambda^{k+1-n}M_n^0.$$

Proof. The proof is based on the well-known relation $\lambda P_{1j} = (j+1)\mu P_{0,j+1}$, for $j \geq 0$, and the differentiation of the equality $\mu(\beta(\lambda - \lambda z) - z)P_0'(z) = \lambda(1 - \beta(\lambda - \lambda z))P_0(z)$. □

To illustrate the preceding analysis, Table 6.1 lists values of the descriptors $E[M]$, $Var(M)$ and $Cov(M, N)$ for an $M/M/1$ retrial queue with $\beta_1 = 1.0$. The descriptors are seen to be increasing functions of ρ and μ. Moreover, positive values of $Cov(M, N)$ are always obtained, which is in agreement with our expectations.

In Table 6.2, we study the effect of the service time distribution. In particular, we consider E_4, exponential and H_2 laws with $\beta_1 = 1.0$. For the H_2 case, the coefficient of variation is assumed to be $c_v = 1.25$. The retrial rate is chosen as $\mu = 0.5$. The entries of the table show that the three descriptors are increasing functions of ρ. We also observe a monotone behavior with respect to the variability of the service time distribution. Note that the differences in magnitude become more apparent as ρ increases.

Table 6.2. $E[M]$, $Var(M)$ and $Cov(M,N)$ versus ρ and the service law

ρ		$E[M]$	$Var(M)$	$Cov(M,N)$
0.2	E_4	0.23945	0.29490	0.07416
	Exp	0.27500	0.50437	0.15375
	H_2	0.31230	0.89557	0.26034
0.4	E_4	0.61250	0.98121	0.44020
	Exp	0.80000	2.48000	1.04000
	H_2	0.98937	5.67672	1.92187
0.6	E_4	1.32421	3.06775	1.83331
	Exp	1.95000	10.56750	4.81500
	H_2	2.55670	27.47812	9.21323
0.8	E_4	3.40000	14.88825	10.50500
	Exp	5.60000	66.72000	29.76000
	H_2	7.64187	184.32681	56.52234

6.2 Successful versus Blocked Events

In this section, we refer to a busy period in the $M/G/1$ retrial queue and present a detailed computational analysis of four measures related to the numbers of successful retrials, blocked retrials, successful primary arrivals and blocked primary arrivals.

First of all, we remember that the performance of telephone systems modelled as retrial queues differs of standard waiting lines because typically the retrial group is an invisible queue which cannot be observed. Furthermore, the original flow of primary arrivals and the flow of repeated attempts become undistinguished. These facts provide an initial motivation to investigate the distribution of the number of successful and blocked events made by primary customers and retrial customers. We propose direct methods for the computation of the probability mass functions instead of using an approach based on generating functions. The main advantage of the direct approach is in avoiding the numerical inversion of the generating functions.

Let us define the new performance measures

R^s : the number of successful retrials during a busy period (i.e., those repeated attempts which find the server free),
R^b : the number of blocked retrials during a busy period,
Λ^s : the number of successful primary arrivals during a busy period,
Λ^b : the number of blocked primary arrivals during a busy period.

By employing a level cross argument, we notice that the number of times per busy period that the process crosses up from $j-1$ to j customers in orbit (i.e., a blocked arrival occurs) equals the number of down crosses from j to $j-1$ (i.e., a successful retrial takes place). Thus, we have

$$R^s = \Lambda^b. \tag{6.18}$$

Since both descriptors are coincident, we only need to study one of them.

We also remark the following relationships:

$$I = 1 + A^s + A^b, \tag{6.19}$$

$$\sum_{i=1}^{I} R_i = R^s + R^b, \tag{6.20}$$

where R_i denotes the number of repeated attempts made by the ith arriving customer until he reaches a free server.

Taking expectations on equations (6.18)-(6.20), it can be proved that

$$E[R^s] = \rho E[I], \tag{6.21}$$
$$E[A^s] = (1-\rho)E[I] - 1, \tag{6.22}$$
$$E[R^b] = E[I]\left(E[R] - \rho\right), \tag{6.23}$$

where the expressions for $E[I]$ and $E[R]$ can be found in Subsections 4.1.2 and 5.1.2, respectively. The above formulas (6.21)-(6.23) can be easily computed and they give exact mean values of the descriptors under study.

First, we derive recursive equations for the exact computation of the probability mass function $P(R^s = r)$, for $r \geq 0$. Let $x_j^s(r)$ be the probability of having exactly $r \geq 0$ successful retrials during the remaining busy period, given that a service time has just been completed leaving behind j customers in orbit, for $0 \leq j \leq r$. We notice that $x_0^s(r) = \delta_{0r}$, for $r \geq 0$.

Then, the probability distribution of R^s satisfies that

$$P(R^s = r) = \sum_{j=0}^{r} a_j x_j^s(r), \quad r \geq 0. \tag{6.24}$$

In the next theorem, we derive equations for the probabilities $x_j^s(r)$.

Theorem 6.4. *For each fixed $r \geq 1$, the probabilities $x_j^s(r)$ can be written in matrix form as*

$$\mathbf{M}_r \mathbf{x}_r^s = \mathbf{B}_r \tilde{\mathbf{x}}_r, \tag{6.25}$$

where $\mathbf{x}_r^s = (x_1^s(r), ..., x_r^s(r))'$, $\tilde{\mathbf{x}}_1 = 1$, $\tilde{\mathbf{x}}_r = \left(0, \mathbf{x}_{r-1}^s\right)'$, for $r \geq 2$, and $\mathbf{M}_r = (m_{ij})$ and $\mathbf{B}_r = (b_{ij})$ are square matrices of order r with elements

$$m_{ij} = \begin{cases} \lambda(1-a_0) + i\mu, & \text{if } 1 \leq i \leq r,\ j = i, \\ -\lambda a_{j-i}, & \text{if } 1 \leq i < j \leq r, \\ 0, & \text{otherwise,} \end{cases}$$

$$b_{ij} = \begin{cases} i\mu a_{j-i}, & \text{if } 1 \leq i \leq j \leq r, \\ 0, & \text{otherwise.} \end{cases}$$

166   6 Other Descriptors

Proof. We first analyze the dynamics between two successive service completion epochs in order to derive equations for the probabilities $x_j^s(r)$. This gives

$$x_j^s(r) = \frac{\lambda}{\lambda + j\mu} \sum_{k=j}^{r} a_{k-j} x_k^s(r) + \frac{j\mu}{\lambda + j\mu} \sum_{k=j-1}^{r-1} a_{k-j+1} x_k^s(r-1), \quad 1 \leq j \leq r.$$

(6.26)

Equation (6.26) is derived by noting that the previous departure left j customers in orbit. Then, the next customer getting service may be a primary arrival (with probability $\lambda/(\lambda + j\mu)$) or may come from the orbit (with probability $j\mu/(\lambda + j\mu)$). In either case, we must record the number of arrivals occurring during the subsequent service time so that, at its completion, the accumulated number of customers who visited the orbit never reaches $r + 1$.

For each fixed r, we put (6.26) in matrix form, which leads to (6.25). □

The triangular system (6.25) can be solved recursively to get the probabilities $x_j^s(r)$ and consequently $P(R^s = r)$ is determined from (6.24).

We now turn our attention to the study of the number of successful primary arrivals. To this end, we approximate the $M/G/1$ retrial queue with infinite orbit capacity by the truncated model with orbit capacity $K \geq 1$.

Let $y_j^s(a)$ be the probability of having $a \geq 0$ successful arrivals during the remaining busy period, given that a service time has been completed leaving behind j customers in orbit, for $0 \leq j \leq K$. We notice that $y_0^s(a) = \delta_{0a}$, for $a \geq 0$. Now, we have

$$P(A^s = a) = \sum_{j=0}^{K-1} a_j y_j^s(a) + \left(1 - \sum_{j=0}^{K-1} a_j\right) y_K^s(a), \quad a \geq 0. \quad (6.27)$$

Theorem 6.5. *For each fixed $a \geq 0$, the probabilities $y_j^s(a)$ satisfy the following system:*

$$\mathbf{M} \mathbf{y}_0^s = \mathbf{b}_0^s, \quad (6.28)$$

$$\mathbf{M} \mathbf{y}_a^s = \mathbf{B} \mathbf{y}_{a-1}^s, \quad a \geq 1, \quad (6.29)$$

where $\mathbf{y}_a^s = (y_1^s(a), ..., y_K^s(a))'$, for $a \geq 0$, $\mathbf{b}_0^s = (\mu a_0, 0, ..., 0)'$, and $\mathbf{M} = (m_{ij})$ and $\mathbf{B} = (b_{ij})$ are square matrices of order K with elements

$$m_{ij} = \begin{cases} \lambda + i\mu - i\mu a_1, & \text{if } 1 \leq i \leq K-1, \ j = i, \\ \lambda + K\mu - K\mu(1 - a_0), & \text{if } i = j = K, \\ -i\mu a_0, & \text{if } 2 \leq i \leq K, \ j = i-1, \\ -i\mu a_{j-i+1}, & \text{if } 1 \leq i < j \leq K-1, \\ -i\mu \left(1 - \sum_{k=0}^{K-i} a_k\right), & \text{if } 1 \leq i \leq K-1, \ j = K, \\ 0, & \text{otherwise,} \end{cases}$$

$$b_{ij} = \begin{cases} \lambda a_{j-i}, & if\ 1 \leq i \leq j \leq K-1, \\ \lambda \left(1 - (1-\delta_{iK}) \sum_{k=0}^{K-i-1} a_k \right), & if\ 1 \leq i \leq K,\ j=K, \\ 0, & otherwise. \end{cases}$$

Proof. We notice that

$$y_j^s(a) = (1-\delta_{0a}) \frac{\lambda}{\lambda+j\mu} \left((1-\delta_{jK}) \sum_{k=j}^{K-1} a_{k-j} y_k^s(a-1) \right.$$
$$\left. + \left(1 - (1-\delta_{jK}) \sum_{k=0}^{K-j-1} a_k \right) y_K^s(a-1) \right)$$
$$+ \frac{j\mu}{\lambda+j\mu} \left(\sum_{k=j-1}^{K-1} a_{k-j+1} y_k^s(a) + \left(1 - \sum_{k=0}^{K-j} a_k \right) y_K^s(a) \right),$$
$$1 \leq j \leq K,\ a \geq 0. \quad (6.30)$$

With the help of Kronecker's function δ_{ab}, we have written the above compact formula, but it is convenient to derive the expression by thinking in three different cases: (i) $a = 0$, (ii) $a \geq 1$ and $1 \leq j < K$, and (iii) $a \geq 1$ and $j = K$. If the primary arrival gains the competition to occupy the free server, then the index a decreases one unit. Once more, we must update the orbit state by counting the number of arrivals during the service time in progress.

After rearrangement, the system (6.30) can be written in matrix form as claimed in (6.28) and (6.29). □

In the light of formulas (6.27) and (6.30), we understand the need of considering a truncated model. Otherwise, the first sums would turn into infinite series. As a result, for each fixed a, we would have an infinite system of equations with a coefficient matrix of level dependent $M/G/1$-type. Unfortunately, such a system has no known solution. The discussion regarding the choice of K is postponed to the numerical examples.

The number of blocked retrials taking place during a given service time depends on the arbitrary number of blocked primary arrivals occurring during the service time in progress. Thus, it seems difficult, or even impossible, to obtain the exact distribution of R^b. We next propose two methods for the computation of the probability mass function of R^b for the model with finite capacity $K \geq 1$.

Firstly, we condition on the number of arrivals and retrials occurring during the first service time of the busy period. Then, for $r \geq 0$, we have

$$P(R^b = r) = \delta_{0r} c_0^{0,0} + (1-\delta_{1K}) \sum_{j=1}^{K-1} \sum_{k=0}^{r} c_0^{j,k} x_j^b(r-k) + \sum_{k=0}^{r} d_0^{K,k} x_K^b(r-k),$$
$$(6.31)$$

168   6 Other Descriptors

where $c_i^{j,k}$ is the probability that j primary arrivals and k retrials occur during a service time, given that immediately after the beginning of the service time the orbit size is i, for $0 \leq i \leq K$, $j \geq 0$ and $k \geq 0$. Then, the accumulative probabilities $d_i^{K-i,k}$ are defined by

$$d_i^{K-i,k} = \sum_{j=K-i}^{\infty} c_i^{j,k}, \quad 0 \leq i \leq K, \ k \geq 0.$$

The quantities $x_j^b(r)$ denote the probability of having r blocked retrials during the remaining busy period, given that a service time has been completed leaving j customers in orbit, for $0 \leq j \leq K$ and $r \geq 0$. For $j = 0$, we have $x_0^b(r) = \delta_{0r}$, for $r \geq 0$.

The dynamics of the probabilities $x_j^b(r)$ is summarized in the following result.

Theorem 6.6. *For each fixed $r \geq 0$, the probabilities $x_j^b(r)$ satisfy the following block tridiagonal system:*

$$\mathbf{N}_0 \mathbf{x}_0^b = \mathbf{b}_0^b, \tag{6.32}$$

$$\mathbf{N}_0 \mathbf{x}_r^b = \sum_{m=1}^{r} \mathbf{N}_m \mathbf{x}_{r-m}^b, \quad r \geq 1, \tag{6.33}$$

where $\mathbf{x}_r^b = (x_1^b(r), ..., x_K^b(r))'$, for $r \geq 0$, $\mathbf{b}_0^b = \left(\mu \widehat{c}_0^{0,0}, 0, ..., 0\right)'$ and $\mathbf{N}_m = \left(n_{ij}^m\right)$ is the square matrix of order K with elements

$$n_{ij}^0 = \begin{cases} \lambda + i\mu - \lambda \widehat{c}_i^{0,0} - i\mu \widehat{c}_{i-1}^{1,0}, & \text{if } 1 \leq i \leq K-1, \ j = i, \\ \lambda + K\mu - \lambda \widehat{d}_K^{0,0} - K\mu \widehat{d}_{K-1}^{1,0}, & \text{if } i = j = K, \\ -i\mu \widehat{c}_{i-1}^{0,0}, & \text{if } 2 \leq i \leq K, \ j = i-1, \\ -\lambda \widehat{c}_i^{j-i,0} - i\mu \widehat{c}_{i-1}^{j-i+1,0}, & \text{if } 1 \leq i < j \leq K-1, \\ -\lambda \widehat{d}_i^{K-i,0} - i\mu \widehat{d}_{i-1}^{K-i+1,0}, & \text{if } 1 \leq i \leq K-1, \ j = K, \\ 0, & \text{otherwise,} \end{cases}$$

and, for $1 \leq m \leq r$,

$$n_{ij}^m = \begin{cases} \lambda \widehat{c}_i^{0,m} + i\mu \widehat{c}_{i-1}^{1,m}, & \text{if } 1 \leq i \leq K-1, \ j = i, \\ \lambda \widehat{d}_K^{0,m} + K\mu \widehat{d}_{K-1}^{1,m}, & \text{if } i = j = K, \\ i\mu \widehat{c}_{i-1}^{0,m}, & \text{if } 2 \leq i \leq K, \ j = i-1, \\ \lambda \widehat{c}_i^{j-i,m} + i\mu \widehat{c}_{i-1}^{j-i+1,m}, & \text{if } 1 \leq i < j \leq K-1, \\ \lambda \widehat{d}_i^{K-i,m} + i\mu \widehat{d}_{i-1}^{K-i+1,m}, & \text{if } 1 \leq i \leq K-1, \ j = K, \\ 0, & \text{otherwise.} \end{cases}$$

The estimations $\widehat{c}_i^{j,k}$ and $\widehat{d}_i^{K-i,k}$ will be specified in the sequel.

6.2 Successful versus Blocked Events

Proof. Conditioning on the identity of the customer who occupies the server and counting the number of primary arrivals and retrials taking place during the service in progress, we find that

$$x_j^b(r) = \frac{\lambda}{\lambda + j\mu} \sum_{k=0}^{r} \left((1-\delta_{jK}) \sum_{i=j}^{K-1} c_j^{i-j,k} x_i^b(r-k) + d_j^{K-j,k} x_K^b(r-k) \right)$$

$$+ \frac{j\mu}{\lambda + j\mu} \sum_{k=0}^{r} \left(\sum_{i=j-1}^{K-1} c_{j-1}^{i-j+1,k} x_i^b(r-k) + d_{j-1}^{K-j+1,k} x_K^b(r-k) \right),$$

$$1 \leq j \leq K, \ r \geq 0. \quad (6.34)$$

To specify the auxiliary quantities $c_i^{j,k}$ and $d_i^{K-i,k}$, suppose that the length of the service time is x and j primary customers arrive at epochs $x_1, ..., x_j$, with $0 < x_1 < ... < x_j < x$. It implies that they may perform retrials during the remaining service time of length $x - x_1, ..., x - x_j$, respectively. It seems intricate to manage these multidimensional constraints. Alternatively, we assume that the j customers arrive at the mean point $x/2$. This simple approximating assumption affects in average only $\rho < 1$ customers per service. As a result of this reallocation, the constants $c_i^{j,k}$ are approximated by

$$\widetilde{c}_i^{j,k} = \int_0^\infty e^{-\lambda x} \frac{(\lambda x)^j}{j!} \sum_{m=0}^{k} e^{-i\mu x} \frac{(i\mu x)^m}{m!} e^{-\frac{j\mu x}{2}} \frac{(\frac{j\mu x}{2})^{k-m}}{(k-m)!} dB(x).$$

Now, it is easy to derive that

$$\widetilde{c}_i^{j,k} = \frac{\lambda^j \left(i\mu + \frac{j\mu}{2} \right)^k}{j!k!} \int_0^\infty e^{-\left(\lambda + i\mu + \frac{j\mu}{2}\right)x} x^{j+k} dB(x), \quad (6.35)$$

$$\widetilde{d}_i^{K-i,k} = \frac{\left(\frac{(K+i)\mu}{2} \right)^k}{k!} \left(\int_0^\infty e^{-\frac{(K+i)\mu x}{2}} x^k dB(x) \right.$$

$$\left. - (1-\delta_{iK}) \sum_{j=0}^{K-i-1} \frac{\lambda^j}{j!} \int_0^\infty e^{-\left(\lambda + \frac{(K+i)\mu}{2}\right)x} x^{j+k} dB(x) \right). \quad (6.36)$$

Then, by expressing (6.34) in matrix form, we obtain (6.32) and (6.33). □

Explicit expressions for approximated quantities (6.35) and (6.36) can be given for the most usual service time distributions. We next give expressions in the cases of hyperexponential and Erlang service times.

(i) *Hyperexponential service times*

$$\widetilde{c}_i^{j,k} = \binom{j+k}{j} \sum_{l=1}^{m} p_l \frac{\lambda^j \nu_l \left(i\mu + \frac{j\mu}{2} \right)^k}{\left(\lambda + \nu_l + i\mu + \frac{j\mu}{2} \right)^{j+k+1}},$$

170   6 Other Descriptors

$$\widehat{d}_i^{K-i,k} = \left(\frac{(K+i)\mu}{2}\right)^k \sum_{l=1}^m p_l \left(\frac{\nu_l}{\left(\nu_l + \frac{(K+i)\mu}{2}\right)^{k+1}}\right.$$

$$\left. -(1-\delta_{iK})\nu_l \sum_{j=0}^{K-i-1} \binom{j+k}{k} \frac{\lambda^j}{\left(\lambda + \nu_l + \frac{(K+i)\mu}{2}\right)^{j+k+1}}\right).$$

(ii) *Erlang service times*

$$\widetilde{c}_i^{j,k} = \frac{(j+k+m-1)!}{j!k!(m-1)!} \frac{\lambda^j \nu^m \left(i\mu + \frac{i\mu}{2}\right)^k}{\left(\lambda + \nu + i\mu + \frac{i\mu}{2}\right)^{j+k+m}},$$

$$\widehat{d}_i^{K-i,k} = \left(\frac{(K+i)\mu}{2}\right)^k \left(\binom{k+m-1}{k}\frac{\nu^m}{\left(\nu + \frac{(K+i)\mu}{2}\right)^{k+m}}\right.$$

$$\left. -(1-\delta_{iK})\frac{\nu^m}{k!(m-1)!} \sum_{j=0}^{K-i-1} \frac{(j+k+m-1)!\lambda^j}{j!\left(\lambda + \nu + \frac{(K+i)\mu}{2}\right)^{j+k+m}}\right).$$

The combination of (6.31) and the iterative solution of the systems (6.32) and (6.33) gives a first possibility for the computation of the probabilities $P(R^b = r)$. The key point of this first approximation is the reallocation of those primary customers arriving during a service time.

In what follows, we consider a second alternative approach yielding a tractable model for accurately representing the service times. To this end, we assume a phase-type representation; see Subsection 7.1.4. It might be possible to derive the system of equations governing the unknowns $x_i^b(r)$ in the truncated $M/PH/1$ retrial queue. However, for the sake of easiness, we present only the particular cases where service times follow hyperexponential and Erlang laws.

In the case of hyperexponential service times, we define $x_{ij}^b(r)$ as the probability of having r blocked retrials during the remaining busy period, given that the current system state is (i,j), for $0 \leq i \leq m$ and $0 \leq j \leq K$. The index i denotes the phase of the service time in progress. The case $i = 0$ means that the server is idle.

The probabilities $P(R^b = r)$, for $r \geq 0$, can be calculated as

$$P(R^b = r) = \sum_{i=1}^m p_i x_{i0}^b(r), \quad r \geq 0, \tag{6.37}$$

where the unknowns $x_{ij}^b(r)$ satisfy the following system of equations:

$$x_{00}^b(r) = \delta_{0r}, \tag{6.38}$$

$$x_{0j}^b(r) = \frac{\lambda}{\lambda+j\mu}\sum_{i=1}^m p_i x_{ij}^b(r) + \frac{j\mu}{\lambda+j\mu}\sum_{i=1}^m p_i x_{i,j-1}^b(r), \quad 1\le j\le K, \quad (6.39)$$

$$x_{ij}^b(r) = \frac{\lambda}{\lambda+\nu_i+j\mu} x_{i,j+1}^b(r) + \frac{\nu_i}{\lambda+\nu_i+j\mu} x_{0j}^b(r)$$
$$+ (1-\delta_{0r})\frac{j\mu}{\lambda+\nu_i+j\mu} x_{ij}^b(r-1), \quad 1\le i\le m,\ 0\le j\le K-1,$$
$$(6.40)$$

$$x_{iK}^b(r) = \frac{\nu_i}{\nu_i+K\mu} x_{0K}^b(r) + (1-\delta_{0r})\frac{K\mu}{\nu_i+K\mu} x_{iK}^b(r-1), \quad 1\le i\le m.$$
$$(6.41)$$

In the case of Erlang service times, we have

$$P(R^b = r) = x_{10}^b(r), \quad r\ge 0, \quad (6.42)$$

and

$$x_{00}^b(r) = \delta_{0r}, \quad (6.43)$$

$$x_{0j}^b(r) = \frac{\lambda}{\lambda+j\mu} x_{1j}^b(r) + \frac{j\mu}{\lambda+j\mu} x_{1,j-1}^b(r), \quad 1\le j\le K, \quad (6.44)$$

$$x_{ij}^b(r) = \frac{\lambda}{\lambda+\nu+j\mu} x_{i,j+1}^b(r) + \frac{\nu}{\lambda+\nu+j\mu} x_{i+1,j}^b(r)$$
$$+ (1-\delta_{0r})\frac{j\mu}{\lambda+\nu+j\mu} x_{ij}^b(r-1), \quad 1\le i\le m-1,\ 0\le j\le K-1,$$
$$(6.45)$$

$$x_{iK}^b(r) = \frac{\nu}{\nu+K\mu} x_{i+1,K}^b(r) + (1-\delta_{0r})\frac{K\mu}{\nu+K\mu} x_{iK}^b(r-1), \quad 1\le i\le m-1,$$
$$(6.46)$$

$$x_{mj}^b(r) = \frac{\lambda}{\lambda+\nu+j\mu} x_{m,j+1}^b(r) + \frac{\nu}{\lambda+\nu+j\mu} x_{0j}^b(r)$$
$$+ (1-\delta_{0r})\frac{j\mu}{\lambda+\nu+j\mu} x_{mj}^b(r-1), \quad 0\le j\le K-1, \quad (6.47)$$

$$x_{mK}^b(r) = \frac{\nu}{\nu+K\mu} x_{0K}^b(r) + (1-\delta_{0r})\frac{K\mu}{\nu+K\mu} x_{mK}^b(r-1). \quad (6.48)$$

In order to illustrate the performance of the descriptors under study, we next present some numerical experiments.

In Table 6.3, we display the expected value $E[R^s]$ for different choices of the service time distribution. To this end, we consider E_3, exponential and H_2 service times. The coefficient of variation of the hyperexponential law is 1.25 whereas β_1 has been normalized to be 1.0 in all cases. The traffic intensity ρ and the retrial rate μ take values 0.2, 0.4, 0.6 and 0.8, and 0.05, 0.5, 2.5, 25.0 and 100.0, respectively. Each cell is associated with a pair (ρ,μ) and gives

Table 6.3. The value of $E[R^s]$ as a function of the service time law

ρ		$\mu = 0.05$	$\mu = 0.5$	$\mu = 2.5$	$\mu = 25.0$	$\mu = 100.0$
0.2	E_3	0.62980	0.27419	0.25466	0.25046	0.25011
	Exp	0.61035	0.27334	0.25450	0.25044	0.25011
	H_2	0.59722	0.27274	0.25439	0.25043	0.25010
0.4	E_3	53.96221	1.03449	0.72790	0.67255	0.66813
	Exp	39.69161	1.00320	0.72344	0.67213	0.66803
	H_2	33.09761	0.98513	0.72081	0.67189	0.66796
0.6	E_3	346968.20544	5.15835	1.92032	1.53751	1.50929
	Exp	89406.96716	4.50421	1.86894	1.53335	1.50826
	H_2	44224.80293	4.19806	1.84281	1.53119	1.50773
0.8	E_3	7.9274×10^{13}	85.46082	7.37906	4.25259	4.06170
	Exp	6.1035×10^{11}	52.53055	6.69468	4.21140	4.05183
	H_2	6.3259×10^{10}	41.87613	6.39795	4.19235	4.04724

Table 6.4. The main characteristics of A^s

ρ		$\mu = 0.05$	$\mu = 0.5$	$\mu = 2.5$	$\mu = 25.0$	$\mu = 100.0$
0.2	$P(A^s = 0)$	0.86312	0.94290	0.98422	0.99827	0.99956
	$P(A^s \leq 100)$	0.99998	0.99999	0.99999	0.99999	0.99999
	$E[A^s]$	1.38890	0.09098	0.01756	0.00174	0.00043
0.4	$P(A^s = 0)$	0.74505	0.83760	0.93888	0.99240	0.99806
	$P(A^s \leq 100)$	0.86801	0.99999	0.99999	0.99999	0.99999
	$E[A^s]$	48.64642	0.47770	0.08122	0.00784	0.00195
0.6	$P(A^s = 0)$	0.65750	0.73599	0.86983	0.98055	0.99492
	$P(A^s \leq 100)$	0.68505	0.99999	0.99999	0.99999	0.99999
	$E[A^s]$	29482.20195	1.79870	0.22854	0.02079	0.00515
0.8	$P(A^s = 0)$	0.58965	0.65154	0.78725	0.95836	0.98862
	$P(A^s \leq 100)$	0.60272	0.98334	0.99999	0.99999	0.99999
	$E[A^s]$	1.5814×10^{10}	9.46903	0.59948	0.04808	0.01181

the expectation for the three service time laws. An examination of the table reveals that $E[R^s]$ is an increasing function of ρ, but it decreases as a function of μ. We also notice that $E[R^s(H_2)] < E[R^s(Exp)] < E[R^s(E_3)]$.

In Table 6.4, we summarize some of the main characteristics of A^s for the model with H_2 service times and truncation threshold $K = 150$. Since the approach in this subsection is oriented to the direct computation of the probability mass function, it seems consistent to determine the value of K by employing a point criterion. For $K = 150$, the maximum norm defined by $\mathcal{P}(150) = \max_{0 \leq a \leq 100} |P(A^s(149) = a) - P(A^s(150) = a)| < 10^{-14}$, where $P(A^s(K) = a)$ indicates that the corresponding probability is calculated from (6.27) after solving the system (6.28) and (6.29) with orbit capacity K. Similar accuracy levels are kept in all experiments throughout this section. Firstly, we observe that the initial probability $P(A^s = 0)$ decreases

6.2 Successful versus Blocked Events 173

Table 6.5. Comparing the approximations of R^b

r	$\overline{F}_{R^b}^{E_3}(r)$	$\widehat{F}_{R^b}^{E_3}(r)$	$\overline{F}_{R^b}^{Exp}(r)$	$\widehat{F}_{R^b}^{Exp}(r)$	$\overline{F}_{R^b}^{H_2}(r)$	$\widehat{F}_{R^b}^{H_2}(r)$
0	0.49398	0.49205	0.55797	0.55571	0.58403	0.58160
1	0.49591	0.49207	0.56035	0.55603	0.58663	0.58199
2	0.49786	0.49210	0.56271	0.55649	0.58919	0.58257
3	0.49981	0.49216	0.56503	0.55708	0.59170	0.58330
4	0.50176	0.49226	0.56732	0.55780	0.59418	0.58417
5	0.50373	0.49241	0.56958	0.55863	0.59662	0.58518
6	0.50570	0.49261	0.57181	0.55956	0.59902	0.58631
7	0.50767	0.49288	0.57402	0.56059	0.60138	0.58755
8	0.50965	0.49323	0.57619	0.56172	0.60371	0.58889
9	0.51163	0.49366	0.57834	0.56292	0.60600	0.59032
10	0.51361	0.49418	0.58046	0.56420	0.60825	0.59183
⋮	⋮	⋮	⋮	⋮	⋮	⋮
100	0.65630	0.66124	0.69733	0.69691	0.72417	0.72386

with ρ and increases with μ. The value of $P(A^s \leq 100)$ shows that the tail of the distribution becomes heavier as far as ρ increases or μ decreases. With respect to the influence of the service time distribution, it can be observed [36] that $P(A^s(E_3) = 0) < P(A^s(Exp) = 0) < P(A^s(H_2) = 0)$, and $P(A^s(E_3) \leq 100) < P(A^s(Exp) \leq 100) < P(A^s(H_2) \leq 100)$.

It should be noticed that, as long as the system congestion increases, the distribution of the descriptors under study becomes sparser. This fact is corroborated in Tables 6.3 and 6.4 for the entry $(\rho, \mu) = (0.8, 0.05)$. The dispersion can be explained because the descriptors are referred to a busy period which becomes stochastically larger when the congestion increases.

We now focus on the distribution function of R^b. The approximation based on a reallocation of the primary customers is denoted by $\widehat{F}_{R^b}(r)$, whereas $\overline{F}_{R^b}(r)$ denotes the second approximation based on the direct truncated equations (6.37)-(6.48). In both cases, we take $K = 150$. The advantage of $\overline{F}_{R^b}(r)$ is that it employs only exact equations for the truncated model. However, we may wish to deal with any arbitrary service time, not necessarily of phase-type. This can be done with the help of the first approximation $\widehat{F}_{R^b}(r)$.

For small values of μ, the service time tends to expire before a repeated attempt takes place. As a result, the reallocation of customers has no perturbing effect on the system dynamics. Thus, we may expect high accuracy for small retrial rates. In Table 6.5, we show that $\widehat{F}_{R^b}(r)$ also gives a good approximation for higher values of μ. More concretely, we choose $(\rho, \mu) = (0.8, 100.0)$. Then, the entries in the table show that $\widehat{F}_{R^b}(r)$ is close enough to $\overline{F}_{R^b}(r)$ (i.e., the exact distribution of the truncated model) for E_3, exponential and H_2 service times.

The efficiency of the approximations can be measured in terms of the convergence of the approximate expectations $\overline{E}[R^b]$ and $\widehat{E}[R^b]$ to the true value $E[R^b]$. For the numerical experiment in Table 6.5 and $K = 150$, both approximate expected values fit the five first decimal digits of $E[R^b]$ regardless of the service time pattern.

6.3 Server Idle Periods

We saw in Section 2.1 that a period during which the server is idle always occurs when a service time is completed. The analysis in that section showed that the distribution of the server idle period is simple when dealing with a single server retrial queue. In this section, we obtain a complete picture by analyzing the server idle periods for the $M/M/c$ retrial queue.

Let us assume that the system state at time $t = 0$ is (i, j), for any state with $i \leq c - 1$. Now mark one of the $c - i$ free servers and define X_{ij}, for $0 \leq i \leq c-1$ and $j \geq 0$, as the length of time during which the marked server remains free. We need to specify rules for the assignment of incoming primary and retrial customers to the free servers. Two possibilities are considered:

(i) *Random order (RO) assignment.* All incoming (primary or retrial) customers have identical chance to occupy any of the $c-i$ free servers. Hence, the marked server becomes busy with probability $1/(c-i)$.
(ii) *First-free-first-busy (FFFB) assignment.* The system may record and order the identity of the servers at the service completion epochs. A sequential assignment mechanism is then employed to assign the free server to the incoming customers.

Let us first consider the RO assignment. We denote, for $0 \leq i \leq c-1$ and $j \geq 0$, the Laplace-Stieltjes transform of X_{ij} by $X^*_{ij}(s) = E[\exp\{-sX_{ij}\}]$. By using first-step analysis to condition on the most imminent event, we obtain

$$X^*_{ij}(s) = \frac{1}{c-i}\frac{\lambda+j\mu}{s+\lambda+i\nu+j\mu} + \frac{i\nu}{s+\lambda+i\nu+j\mu}X^*_{i-1,j}(s)$$
$$+ \frac{c-i-1}{c-i}\left(\frac{\lambda}{s+\lambda+i\nu+j\mu}X^*_{i+1,j}(s) + \frac{j\mu}{s+\lambda+i\nu+j\mu}X^*_{i+1,j-1}(s)\right).$$

After some algebraic manipulations which are omitted here, we get the following algorithmic solution.

Theorem 6.7. *The Laplace-Stieltjes transforms $X^*_{ij}(s)$, for $0 \leq i \leq c-1$ and $j \geq 0$, can be computed at a given value s as follows:*

Step 1. Put $j = 0$.
Step 2. For $i = 0, ..., c-1$, calculate sequentially the following coefficients:

$$B_{ij} = \frac{1}{c-i}\frac{\lambda+j\mu}{s+\lambda+i\nu+j\mu} + \frac{c-i-1}{c-i}\frac{j\mu}{s+\lambda+i\nu+j\mu}X^*_{i+1,j-1}(s),$$
$$A_{i+1,j} = \frac{c-i-1}{c-i}\frac{\lambda}{s+\lambda+i\nu+j\mu},$$

$$D_{i-1,j} = \frac{i\nu}{s + \lambda + i\nu + j\mu}.$$

Step 3. Set $a_{0j} = A_{1j}$ and $b_{0j} = B_{0j}$. For $i = 1, ..., c-1$, calculate

$$a_{ij} = \frac{A_{i+1,j}}{1 - D_{i-1,j}a_{i-1,j}},$$

$$b_{ij} = \frac{B_{ij} + D_{i-1,j}b_{i-1,j}}{1 - D_{i-1,j}a_{i-1,j}}.$$

Step 4. Set $X^*_{c-1,j}(s) = b_{c-1,j}$. For $i = c-2, ..., 0$, compute recursively in reverse order $X^*_{ij}(s) = b_{ij} + a_{ij}X^*_{i+1,j}(s)$.

Step 5. Put $j = j + 1$ and return to Step 2.

If we are interested in the computation of the kth moment of $X^*_{ij}(s)$, which is denoted by

$$X^{(k)}_{ij} = E[X^k_{ij}],$$

then, an appropriate recursive scheme is proposed in the following result.

Theorem 6.8. *The moments $X^{(k)}_{ij}$, for $k \geq 0$, $0 \leq i \leq c-1$ and $0 \leq j \leq J$, can be computed by following the steps:*

Step 1. Put $k = 0$. Then, we have $X^{(0)}_{ij} = 1$, for $0 \leq i \leq c-1$ and $0 \leq j \leq J$.
Step 2. Put $k = k + 1$.
Step 3. Put $j = 0$.
Step 4. For $i = 0, ..., c-1$, calculate the coefficients

$$B^{(k)}_{ij} = \frac{k!}{c-i}\frac{\lambda + j\mu}{(\lambda + i\nu + j\mu)^{k+1}} + \frac{c-i-1}{c-i}\frac{j\mu}{\lambda + i\nu + j\mu}X^{(k)}_{i+1,j-1}$$

$$+ \frac{1}{\lambda + i\nu + j\mu}\sum_{n=0}^{k-1}\binom{k}{n}\frac{(k-n)!}{(\lambda + i\nu + j\mu)^{k-n}}$$

$$\times \left(i\nu X^{(n)}_{i-1,j} + \frac{c-i-1}{c-i}\left(\lambda X^{(n)}_{i+1,j} + j\mu X^{(n)}_{i+1,j-1}\right)\right),$$

$$A_{i+1,j} = \frac{c-i-1}{c-i}\frac{\lambda}{\lambda + i\nu + j\mu},$$

$$D_{i-1,j} = \frac{i\nu}{\lambda + i\nu + j\mu}.$$

Step 5. Set $a_{0j} = A_{1j}$ and $b^{(k)}_{0j} = B^{(k)}_{0j}$. For $i = 1, ..., c-1$, compute

$$a_{ij} = \frac{A_{i+1,j}}{1 - D_{i-1,j}a_{i-1,j}},$$

$$b^{(k)}_{ij} = \frac{B^{(k)}_{ij} + D_{i-1,j}b^{(k)}_{i-1,j}}{1 - D_{i-1,j}a_{i-1,j}}.$$

176   6 Other Descriptors

Step 6. Set $X_{c-1,j}^{(k)} = b_{c-1,j}^{(k)}$. For $i = c-2, ..., 0$, calculate recursively in reverse order $X_{ij}^{(k)} = b_{ij}^{(k)} + a_{ij} X_{i+1,j}^{(k)}$.

Step 7. Put $j = j+1$. If $j \leq J$, then return to Step 4. Otherwise, if $j = J+1$, then go to Step 2.

Now we focus on the *FFFB* assignment. Assume that, at time $t = 0$, a service time ends and leaves the system at the state (i,j). Then the server becoming free is marked with the level $c-i$ meaning that there are other $c-i-1$ servers that will be occupied by the first $c-i-1$ incoming customers. To make emphasis on the order assigned to the marked server, we rewrite the random variable X_{ij} as $Y_{c-i,j}$. It is quite obvious that $Y_{c-i,j}$ depends on the next $c-i$ incoming customers, but it does not depend on future service times.

Let $Y_{c-i,j}^*(s)$ be the Laplace-Stieltjes transform of $Y_{c-i,j}$. By conditioning on the next state visited, we get

$$Y_{1j}^*(s) = \frac{\lambda + j\mu}{s + \lambda + j\mu},$$

$$Y_{c-i,j}^*(s) = \frac{\lambda}{s+\lambda+j\mu} Y_{c-i-1,j}^*(s) + \frac{j\mu}{s+\lambda+j\mu} Y_{c-i-1,j-1}^*(s),$$

$$0 \leq i \leq c-2,\ j \geq 0.$$

If $j = 0$, then the $c-i$ subsequent incoming arrivals are primary customers and thus $Y_{c-i,0}$ has an Erlang distribution with parameters $(c-i, \lambda)$.

The following result provides an algorithmic scheme for the computation of $Y_{c-i,j}^*(s)$.

Theorem 6.9. *The Laplace-Stieltjes transforms $Y_{c-i,j}^*(s)$, for $0 \leq i \leq c-1$ and $j \geq 0$, can be computed by following the steps:*

Step 1. For $i = 0, ..., c-1$, calculate

$$Y_{c-i,0}^*(s) = \left(\frac{\lambda}{s+\lambda}\right)^{c-i}.$$

Step 2. Put $j = 1$.
Step 3. For $i = 0, ..., c-2$, compute

$$R_{c-i,j} = \frac{j\mu}{s+\lambda+j\mu} Y_{c-i-1,j-1}^*(s).$$

Step 4. Set $Y_{1j}^*(s) = (\lambda + j\mu)/(s+\lambda+j\mu)$. For $i = c-2, ..., 0$, calculate recursively in reverse order

$$Y_{c-i,j}^*(s) = \frac{\lambda}{s+\lambda+j\mu} Y_{c-i-1,j}^*(s) + R_{c-i,j}.$$

Step 5. Put $j = j+1$ and return to Step 3.

We can also compute the moments $Y_{c-i,j}^{(k)} = E[Y_{c-i,j}^k]$ by using the recursive scheme proposed in the following theorem.

Theorem 6.10. *The moments $Y_{c-i,j}^{(k)}$, for $k \geq 0$, $0 \leq i \leq c-1$ and $0 \leq j \leq J$, can be computed as follows:*

Step 1. Put $k = 0$. Then, we have $Y_{c-i,j}^{(0)} = 1$, for $0 \leq i \leq c-1$ and $0 \leq j \leq J$.
Step 2. Put $k = k + 1$.
Step 3. Put $j = 0$. For $i = 0, ..., c-1$, compute

$$Y_{c-i,0}^{(k)} = \frac{1}{\lambda^k} \frac{(k+c-i-1)!}{(c-i-1)!}.$$

Step 4. Put $j = 1$.
Step 5. For $i = 0, ..., c-2$, calculate

$$S_{c-i,j}^{(k)} = \frac{j\mu}{\lambda + j\mu} Y_{c-i-1,j-1}^{(k)}$$
$$+ \sum_{n=1}^{k} \frac{k!}{(k-n)!} \left(\frac{1}{\lambda + j\mu}\right)^{n+1} \left(\lambda Y_{c-i-1,j}^{(k-n)} + j\mu Y_{c-i-1,j-1}^{(k-n)}\right).$$

Step 6. Set $Y_{1j}^{(k)} = k!/(\lambda + j\mu)^k$. For $i = c-2, ..., 0$, compute in reverse order

$$Y_{c-i,j}^{(k)} = S_{c-i,j}^{(k)} + \frac{\lambda}{\lambda + j\mu} Y_{c-i-1,j}^{(k)}.$$

Step 7. Put $j = j + 1$. If $j \leq J$, then return to Step 5. When $j = J+1$, go to Step 2.

In the remaining of this section, we compare the performance of the two proposed assignment rules. With the help of the algorithmic recursions described in Theorems 6.7 and 6.9, we carry out the numerical inversion of the Laplace-Stieltjes transforms $X_{ij}^*(s)$ and $Y_{c-i,j}^*(s)$. In Figure 6.1, we deal with the RO assignment rule and assume that $c = 4$, $\lambda = 2.0$, $\nu = 1.0$ and $\mu = 0.5$. For the case $j = 2$, we display four curves corresponding to $i = 0, 1, 2$ and 3. The same system parameters are kept in Figure 6.2, where we display the corresponding density function for the *FFFB* assignment. Both figures exhibit dramatically different shapes as a result of the incidence of the assignment rule on the distribution of the idle server periods.

In Figure 6.3 we turn to the influence of the service rate ν. To this end, we consider $c = 5$, $\lambda = 1.0$ and $(i,j) = (2,1)$. We present three curves corresponding to $\mu = 0.5, 2.5$ and 12.5, for each expectation $E[X_{ij}]$ and $E[Y_{c-i,j}]$. We notice that the mean value is very sensitive with respect to ν and the choice of the assignment rule. In agreement with its definition, the *FFFB* assignment rule is absolutely insensitive with respect to ν. Indeed, the mean value $E[Y_{c-i,j}]$ remains constant. However, in the case of the RO assignment,

178   6 Other Descriptors

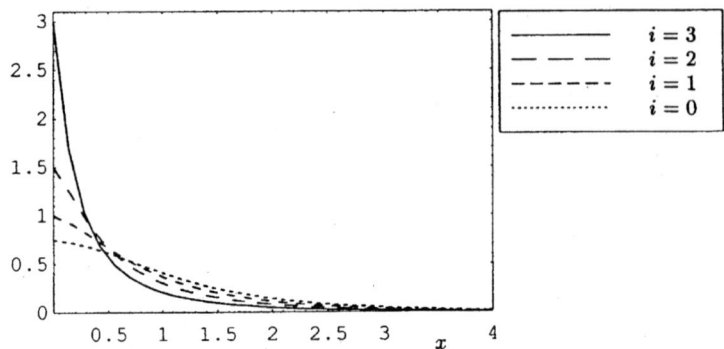

Fig. 6.1. Densities corresponding to the *RO* assignment rule

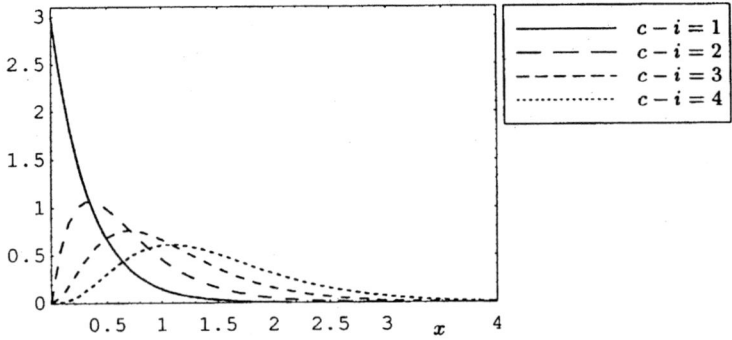

Fig. 6.2. Densities corresponding to the *FFFB* assignment rule

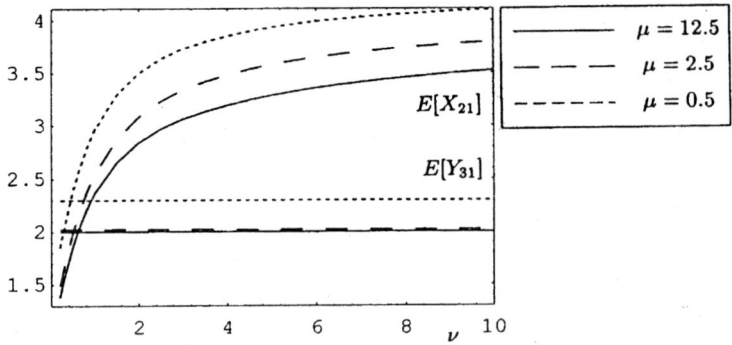

Fig. 6.3. The effect of ν on the expected value

6.4 Time to Reach a Certain Orbit Level   179

Table 6.6. $Var(X_{ij})$ versus $Var(Y_{c-i,j})$

ρ	i	μ	$j=0$	$j=1$	$j=2$	$j=3$	$j=4$
0.5	0	0.1	8.47572	6.77906	5.45484	4.42725	3.62986
			5.33333	**4.29705**	**3.50809**	**2.90376**	**2.43553**
		1.0	8.47572	5.70556	2.87406	1.09212	0.40254
			5.33333	**3.34462**	**1.52383**	**0.45255**	**0.22685**
		10.0	8.47572	6.52711	3.01978	0.20832	0.01030
			5.33333	**3.55339**	**1.75202**	**0.01177**	**0.00393**
	1	0.1	7.65642	5.98958	4.72678	3.77213	3.04784
			3.55555	**2.81703**	**2.27673**	**1.87353**	**1.56632**
		1.0	7.65642	4.15057	1.51998	0.49772	0.19047
			3.55555	**1.62450**	**0.41433**	**0.19262**	**0.11162**
		10.0	7.65642	4.25174	0.31062	0.01357	0.00176
			3.55555	**1.76284**	**0.01081**	**0.00335**	**0.00165**
	2	0.1	6.45791	4.89597	3.76307	2.93701	2.32873
			1.77777	**1.38408**	**1.10803**	**0.90702**	**0.75614**
		1.0	6.45791	2.17169	0.61369	0.19559	0.08438
			1.77777	**0.32653**	**0.13223**	**0.07111**	**0.04432**
		10.0	6.45791	0.46489	0.01738	0.00152	0.00065
			1.77777	**0.00865**	**0.00232**	**0.00105**	**0.00060**

we observe an initial rapid growth and then $E[X_{ij}]$ continues increasing but moderately.

Finally, Table 6.6 shows the behavior of the variances $Var(X_{ij})$ and $Var(Y_{c-i,j})$ (in bold) as functions of i and j, for $c = 3$, $\lambda = 0.75$, $\nu = 0.5$ and $\mu = 0.1, 1.0$ and 10.0. The results show that the variance is a decreasing function of i and j. The behavior inferred with respect to the retrial rate is more complicate. For example, the entry $(c - i, j) = (3, 1)$ shows a non-monotone behavior of $Var(Y_{c-i,j})$. As is to be expected, a comparison of both variances shows that the variability is greater for the RO assignment rule.

6.4 Time to Reach a Certain Orbit Level

The purpose of the present section is to analyze the first-passage time to a certain orbit level $K \geq 1$ in the $M/M/c$ retrial queue. This descriptor gives some insight on the orbit behavior providing a natural measure of system congestion. Visiting frequently high orbit levels degrades the quality of service. If this occurs, then the system manager should take special actions either to control the admission of fresh customers or to switch on any removable server.

We carry out an extensive analysis of the time to reach the orbit level K by including the algorithmic analysis of the length of this first-passage time, the number of customers served during this period and the computation of their moments.

We first introduce some definitions and notation:

L_{ij}^K: the first-passage time to reach the critical level K, given that the initial state is (i,j),

I_{ij}^K: the number of customers served during L_{ij}^K.

We consider $L_{ij}^{*K}(s) = E\left[\exp\{-sL_{ij}^K\}\right]$ and $I_{ij}^K(z) = E\left[z^{I_{ij}^K}\right]$. The corresponding moments are denoted by

$$L_{ij}^{K,(k)} = E\left[(L_{ij}^K)^k\right], \quad k \geq 1,$$

$$I_{ij}^{K,(k)} = E\left[I_{ij}^K(I_{ij}^K - 1)...(I_{ij}^K - k + 1)\right], \quad k \geq 1.$$

These transforms and moments can be gathered according to the orbit levels as follows:

$$\mathbf{l}_j^{*K}(s) = \left(L_{0j}^{*K}(s), ..., L_{cj}^{*K}(s)\right)', \quad 0 \leq j \leq K-1,$$

$$\mathbf{l}^{*K}(s) = \left(\mathbf{l}_0^{*K}(s), ..., \mathbf{l}_{K-1}^{*K}(s)\right)',$$

$$\mathbf{i}_j^K(z) = \left(I_{0j}^K(z), ..., I_{cj}^K(z)\right)', \quad 0 \leq j \leq K-1,$$

$$\mathbf{i}^K(z) = \left(\mathbf{i}_0^K(z), ..., \mathbf{i}_{K-1}^K(z)\right)',$$

$$\mathbf{l}_j^{K,(k)} = \left(L_{0j}^{K,(k)}, ..., L_{cj}^{K,(k)}\right)', \quad 0 \leq j \leq K-1,$$

$$\mathbf{l}^{K,(k)} = \left(\mathbf{l}_0^{K,(k)}, ..., \mathbf{l}_{K-1}^{K,(k)}\right)',$$

$$\mathbf{i}_j^{K,(k)} = \left(I_{0j}^{K,(k)}, ..., I_{cj}^{K,(k)}\right)', \quad 0 \leq j \leq K-1,$$

$$\mathbf{i}^{K,(k)} = \left(\mathbf{i}_0^{K,(k)}, ..., \mathbf{i}_{K-1}^{K,(k)}\right)'.$$

We finally denote the matrix obtained by truncation of $\mathbf{Q}$ at the level K by

$$\overline{\mathbf{Q}}(K) = \begin{pmatrix} \mathbf{A}_{00} & \mathbf{A}_{01} & & & \\ \mathbf{A}_{10} & \mathbf{A}_{11} & \mathbf{A}_{12} & & \\ & \ddots & \ddots & \ddots & \\ & & \mathbf{A}_{K-2,K-3} & \mathbf{A}_{K-2,K-2} & \mathbf{A}_{K-2,K-1} \\ & & & \mathbf{A}_{K-1,K-2} & \mathbf{A}_{K-1,K-1} \end{pmatrix}.$$

For the matrix notation in the rest of this section, we refer the reader to Subsection 7.1.1.

Theorem 6.11. *The Laplace-Stieltjes transforms $L_{ij}^{*K}(s)$, for $0 \leq i \leq c$ and $0 \leq j \leq K-1$, satisfy the block tridiagonal system*

$$\mathbf{Q}^K(s)\mathbf{l}^{*K}(s) = \mathbf{f}^K, \tag{6.49}$$

where $\mathbf{Q}^K(s) = s\mathbf{I}_{(c+1)K} - \overline{\mathbf{Q}}(K)$ *and* $\mathbf{f}^K = \lambda\mathbf{e}_{(c+1)K}((c+1)K)$.

6.4 Time to Reach a Certain Orbit Level

Proof. We assume that the initial state is (i,j) and use first-step analysis to get

$$L_{ij}^{*K}(s) = \frac{\lambda}{s+\lambda+i\nu+j\mu}L_{i+1,j}^{*K}(s) + \frac{i\nu}{s+\lambda+i\nu+j\mu}L_{i-1,j}^{*K}(s)$$
$$+ \frac{j\mu}{s+\lambda+i\nu+j\mu}L_{i+1,j-1}^{*K}(s), \quad 0 \le i \le c-1, \ 0 \le j \le K-1,$$
(6.50)

$$L_{cj}^{*K}(s) = \frac{\lambda}{s+\lambda+c\nu}L_{c,j+1}^{*K}(s) + \frac{c\nu}{s+\lambda+c\nu}L_{c-1,j}^{*K}(s), \quad 0 \le j \le K-1.$$
(6.51)

Since $L_{cK}^{K} = 0$, we have the boundary condition

$$L_{cK}^{*K}(s) = 1.$$
(6.52)

For each orbit level j, equations (6.50) and (6.51) can be expressed as

$$-\mathbf{A}_{j-1,j-2}\mathbf{l}_{j-1}^{*K}(s) + \mathbf{A}_{jj}(s)\mathbf{l}_{j}^{*K}(s) - \mathbf{A}_{j+1,j+2}\mathbf{l}_{j+1}^{*K}(s) = \mathbf{0}_{c+1}, \quad (6.53)$$

for $0 \le j \le K-1$, where $\mathbf{A}_{jj}(s) = s\mathbf{I}_{c+1} - \mathbf{A}_{jj}$. Then, by combining (6.52) and (6.53), we derive the desired matrix expression (6.49). □

The next corollary gives a recursive scheme for computing the vector $\mathbf{l}^{K,(k)}$ in terms of moments of one less order. The proof follows trivially after appropriate differentiation.

Corollary 6.12. *For each $k \ge 0$, the moments $L_{ij}^{K,(k)}$, for $0 \le i \le c$ and $0 \le j \le K-1$, can be computed as the solution of the following tridiagonal system:*

$$\mathbf{l}^{K,(0)} = \mathbf{e}_{(c+1)K},$$
$$\overline{\mathbf{Q}}(K)\mathbf{l}^{K,(k)} = -k\mathbf{l}^{K,(k-1)}, \quad k \ge 1.$$

The analysis of the number of customers served during the time to reach the level K can be derived as a variant of the scheme proposed for the continuous descriptor L_{ij}^{K}. The following results summarize the corresponding systems for $\mathbf{i}^{K}(z)$ and its moments $\mathbf{i}^{K,(k)}$.

Theorem 6.13. *The generating functions $I_{ij}^{K}(z)$, for $0 \le i \le c$ and $0 \le j \le K-1$, verify the block tridiagonal system*

$$\mathbf{Q}^{K}(z)\mathbf{i}^{K}(z) = \mathbf{f}^{K},$$

where $\mathbf{Q}^{K}(z) = (1-z)\nu\mathbf{I}_{K} \otimes \mathbf{W} - \overline{\mathbf{Q}}(K)$ and $\mathbf{W} = (w_{ij})$ is a square matrix of order $c+1$ with entries given by

$$w_{ij} = \begin{cases} i-1, & \text{if } 2 \le i \le c+1, \ j = i-1, \\ 0, & \text{otherwise}. \end{cases}$$

Table 6.7. The moments $L_{10}^{K,(k)}$, $k = 1, 2$

K	$\mu = 0.05$	$\mu = 0.375$	$\mu = 0.75$	$\mu = 1.5$	$\mu = 50.0$
1	3.85096	3.85096	3.85096	3.85096	3.85096
	24.62988	24.62988	24.62988	24.62988	24.62988
2	5.41171	5.61307	5.72634	5.83745	6.10808
	46.25999	51.58987	54.44844	57.16480	63.52798
3	7.02576	7.67372	8.05484	8.43682	9.36489
	74.46935	95.31947	107.91620	120.74662	153.01437
4	8.69564	10.08860	10.94565	11.82272	13.95436
	109.94259	163.77821	200.31210	240.09245	347.05293
5	10.42406	12.92497	14.53547	16.21774	20.31895
	153.43657	268.52731	356.09077	457.80693	749.27843
10	20.05211	36.99417	50.83212	66.85343	107.73710
	523.93554	2281.08399	4612.41743	8280.15942	22296.89580
20	45.90468	249.60736	525.79402	920.38758	2086.41311
	2644.57363	116160.88119	537658.26188	1670542.84583	8660127.81690

Corollary 6.14. *The factorial moments $I_{ij}^{K,(k)}$, for $0 \leq i \leq c$ and $0 \leq j \leq K - 1$, can be obtained as the solution of the following tridiagonal system:*

$$\mathbf{i}^{K,(0)} = \mathbf{e}_{(c+1)K},$$
$$\overline{\mathbf{Q}}(K)\mathbf{i}^{K,(k)} = -k\nu\left(\mathbf{I}_K \otimes \mathbf{W}\right)\mathbf{i}^{K,(k-1)}, \quad k \geq 1.$$

To shed light on the performance of the time to reach the orbit level K, we next present some numerical results for the moments of L_{10}^K and I_{10}^K. The initial state $(1,0)$ indicates that a busy period just starts. More concretely, in Tables 6.7 and 6.8, we display the moments $L_{10}^{K,(k)}$ and $I_{10}^{K,(k)}$, for $k = 1$ and 2, in a queue with $c = 5$, $\lambda = 3.75$ and $\nu = 1.0$, so the traffic load is $\rho = 0.75$. We vary the retrial rate and the orbit level and, for each choice of (μ, K), the entry in the table gives the first moment (upper position) and the second moment (lower position).

As was expected, the obtained values show that the characteristics of the time to reach the critical level and the number of customers served increase with increasing values of μ and K. It is not reported on these tables, but our numerical experience also showed that both descriptors increase when ρ decreases.

6.5 Bibliographical Notes

Falin [286] introduced the characterization of the $M/M/1$ retrial queue based on the total number of attempts to get service since the last service completion. Our analysis in Section 6.1 is inspired in the paper by Rodrigo et al. [584], where the characterization is extended to the $M/G/1$ retrial queue. In this

Table 6.8. The moments $I_{10}^{K,(k)}$, $k = 1, 2$

K	$\mu = 0.05$	$\mu = 0.375$	$\mu = 0.75$	$\mu = 1.5$	$\mu = 50.0$
1	9.44110	9.44110	9.44110	9.44110	9.44110
	166.00309	166.00309	166.00309	166.00309	166.00309
2	14.29393	15.04902	15.47380	15.89046	16.90530
	347.40797	404.06228	434.84978	464.34619	534.05711
3	19.34661	21.77645	23.20567	24.63808	28.11835
	594.22843	822.20830	963.55549	1109.78290	1485.00110
4	24.60866	29.83225	33.04620	36.33523	44.32887
	914.49813	1517.49470	1941.29322	2412.13446	3710.61531
5	30.09025	39.46867	45.50802	51.81655	67.19607
	1317.15637	2633.75291	3677.93603	4918.78159	8572.33302
10	61.19542	124.72815	176.62047	236.70037	390.01413
	4957.78259	26722.63358	57074.60044	105856.63733	295917.41517
20	148.14255	912.02761	1947.72757	3427.45343	7800.04918
	27929.23500	1564025.47944	7405897.50486	2.32141×10^7	1.21141×10^8

paper, the authors also show the usefulness of the descriptor in the estimation of the retrial rate.

The analysis of successful and blocked events is based on the paper by Amador and Artalejo [36]. As related work, we also mention the paper [35] where the descriptors were introduced for the $M/M/c$ retrial queue.

The paper by Artalejo and Gómez-Corral [86] is the basic reference for the analysis of the server idle periods for the $M/M/c$ retrial queue. Section 4 of the paper shows the applicability of the server idle periods to optimal design problems based mainly on the optimization of a given cost function or on the consideration of probabilistic criteria (independently of costs).

The study of the time to reach a certain orbit level follows the lines given by Apaolaza and Artalejo [45]. In a previous work, Lebedev [466] investigated some asymptotic properties of this first-passage time, as far as the orbit level increases. Remiche [578] also studies times to congestion, but in the context of level independent QBD models in discrete-time.

Part III

Retrial Queueing Systems Analyzed Through the Matrix-Analytic Formalism

7

The Matrix-Analytic Formalism

This third part of the book deals with retrial queues analyzed by matrix-analytic methods. We begin with the present chapter on some generalities of the matrix-analytic formalism for two reasons. First and foremost, this chapter summarizes basic results of the matrix-analytic methods, presents the main tools employed in Chapters 8 and 9, and fixes some notation. In addition, it helps the reader to acquire some feeling of the constant interplay between algebraic manipulations and reasoning on the probabilistic interpretation of the underlying expressions.

Although it is assumed that the reader will have some knowledge of the basic matrix algebra, we give in Subsection 7.1.1 a short glossary of the conventions we use in Part III of this book and of some notation used earlier. In the rest of Section 7.1, we introduce the QBD processes and the Markov chains of $GI/M/1$- and $M/G/1$-types. Then, we describe a point process, namely the batch Markovian arrival process, which generalizes and unifies several Markovian processes commonly used in queueing theory. We next examine the phase-type distribution and the semi-Markov service process. In Sections 7.2 and 7.3, we study in some detail concrete results on the QBD, $GI/M/1$ and $M/G/1$ structures. Our intention is only to facilitate the understanding of the material in later chapters, hence we omit proofs. Bibliographical notes in Section 7.4 are then useful for finding the corresponding sources.

7.1 A General Overview

The matrix-analytic formalism is popular as a modelling tool where scalar quantities are replaced by matrices. Such a procedure is closely related to structured Markov chains and the method of stages, and it gives one the ability to construct and study, in a unified manner and in an algorithmically tractable way, a wide range of queueing models.

7.1.1 Notation

To begin with, matrices and vectors are in boldface $\mathbf{A}$, $\mathbf{a}$, $\mathbf{\Pi}$, $\mathbf{\pi}$, etc. In particular, matrices have uppercase letters and vectors lowercase letters. The transpose of $\mathbf{A}$ is written $\mathbf{A}'$. The matrix $\text{diag}(a_1, ..., a_p)$ is the square matrix having elements $a_1, ..., a_p$ along its diagonal and zeros elsewhere.

We denote by $\mathbf{I}_p$ and $\mathbf{0}_{p \times q}$ the identity matrix of order p and the null matrix of dimension $p \times q$, respectively. We let $\mathbf{e}_p$ and $\mathbf{0}_p$ be column vectors of order p of 1s and 0s, respectively. These are simply denoted by $\mathbf{e}$ and $\mathbf{0}$ if they have an infinite number of entries. The vector $\mathbf{e}_i(j)$ is a column vector of dimension i such that all entries are equal to 0, except for the jth one which is equal to 1.

For a square matrix $\mathbf{A}$, the matrix exponential, denoted by $\exp\{\mathbf{A}\}$ and $e^{\mathbf{A}}$, is defined by

$$\exp\{\mathbf{A}\} = \sum_{k=0}^{\infty} \frac{1}{k!} \mathbf{A}^k.$$

Consider a matrix $\mathbf{A} = (a_{ij})$ of dimension $p \times q$ and a matrix $\mathbf{B}$ of dimension $r \times s$. The Kronecker product of these matrices, denoted by $\mathbf{A} \otimes \mathbf{B}$, is defined as the partitioned matrix of dimension $pr \times qs$

$$\mathbf{A} \otimes \mathbf{B} = \begin{pmatrix} a_{11}\mathbf{B} & a_{12}\mathbf{B} & \cdots & a_{1q}\mathbf{B} \\ a_{21}\mathbf{B} & a_{22}\mathbf{B} & \cdots & a_{2q}\mathbf{B} \\ \vdots & \vdots & & \vdots \\ a_{p1}\mathbf{B} & a_{p2}\mathbf{B} & \cdots & a_{pq}\mathbf{B} \end{pmatrix}.$$

Given two square matrices $\mathbf{A}$ and $\mathbf{B}$ of orders p and q, respectively, their Kronecker sum, denoted by $\mathbf{A} \oplus \mathbf{B}$, is defined as the expression $\mathbf{A} \oplus \mathbf{B} = \mathbf{A} \otimes \mathbf{I}_q + \mathbf{I}_p \otimes \mathbf{B}$.

7.1.2 The Main Structured Markov Chains

We gave in Section 2.1 an example of a structured Markov chain. Specifically, the continuous-time Markov chain $\mathcal{X}$ describing the system state in the $M/M/c$ retrial queue has the infinitesimal generator

$$\mathbf{Q} = \begin{pmatrix} \mathbf{A}_{00} & \mathbf{A}_{01} & & & \\ \mathbf{A}_{10} & \mathbf{A}_{11} & \mathbf{A}_{12} & & \\ & \mathbf{A}_{21} & \mathbf{A}_{22} & \mathbf{A}_{23} & \\ & & \ddots & \ddots & \ddots \end{pmatrix}. \qquad (7.1)$$

The state space is two-dimensional and may be partitioned into levels as $\mathcal{S} = \cup_{j=0}^{\infty} l(j)$, where $l(j) = \{(0, j), ..., (c, j)\}$, for $j \geq 0$. Transitions are allowed between adjacent levels only and infinitesimal rates out of the state (i,j) depend on the level j.

Infinitesimal generators of the form (7.1) characterize level dependent quasi-birth-and-death (QBD) processes, which are treated in their full generality in [169] and [464, Chapter 12]. Structural properties for these processes are essentially derived by extending results for level independent QBD processes. Computational issues are, however, much more complex in the level dependent case. Therefore, their resolution must necessarily exploit further properties of the specific model under study or replace the structured infinitesimal generator (7.1) by another simpler matrix.

In Section 3.4, we discussed various approximations for the $M/M/c$ retrial queue based on finite QBD processes and level independent QBD processes. Specifically, the truncated model $\mathcal{X}^W$ is an example of finite QBD process, whereas the models $\mathcal{X}^F$ and $\mathcal{X}^{NR}$ can be formulated as level independent QBD processes with a large number of boundary states.

In a finite QBD process, we assume that the state space is restricted to $\cup_{j=0}^{K} l(j)$ where K is finite, and that the infinitesimal generator has the form

$$Q = \begin{pmatrix} \mathbf{A}_{00} & \mathbf{A}_{01} & & & & & \\ \mathbf{A}_{10} & \mathbf{A}_{11} & \mathbf{A}_{12} & & & & \\ & \mathbf{A}_{21} & \mathbf{A}_{22} & \mathbf{A}_{23} & & & \\ & & \ddots & \ddots & \ddots & & \\ & & & & \mathbf{A}_{K-1,K-2} & \mathbf{A}_{K-1,K-1} & \mathbf{A}_{K-1,K} \\ & & & & & \mathbf{A}_{K,K-1} & \mathbf{A}_{KK} \end{pmatrix}. \quad (7.2)$$

To determine the stationary distribution of a finite QBD process, many approaches may be followed, which are frequently algorithmic. The most commonly used solution is given by Gaver et al. [307]. The reader is referred to Subsection 7.2.1 for further details. In later chapters, examples of this structure can be found in Subsection 8.2.1, when we study the $MAP/M/c$ retrial queue, and in Section 8.3, when we focus on a queue with finite population and phase-type service and retrial times.

A Markov chain defined on the state space $\cup_{j=0}^{\infty} l(j)$ is called a level independent QBD process if one-step transitions from a state in the level j are restricted to states in the same level or in the two adjacent levels $j-1$ and $j+1$, provided that $j \geq 1$, but with infinitesimal rates that do not depend on the initial level j. Thus, its infinitesimal generator is a variant of (7.1) with $\mathbf{A}_{j,j-1} = \mathbf{A}_2$, for $j \geq 2$, and $\mathbf{A}_{jj} = \mathbf{A}_1$ and $\mathbf{A}_{j,j+1} = \mathbf{A}_0$, for $j \geq 1$. We describe in Subsection 7.2.2 the matrix-geometric form of the stationary probability distribution and the conditions for its existence. Then, we examine in Subsection 7.2.3 the question of its numerical evaluation by limiting ourselves to the successive substitution algorithm and the logarithmic reduction algorithm [463]. This material provides notable ease in deriving the stationary distribution of the system state in the $MAP/PH/1$ retrial queue, as it is shown in Subsection 8.1.1.

190   7 The Matrix-Analytic Formalism

In the literature on retrial queues, matrix-analytic methods are also applied to more elaborate structures than QBD processes and, in particular, to level dependent Markov chains for which one-step transitions are allowed across several levels in one direction. Examples of them are related to the $Geo/Geo/c$ and the $BMAP/SM/1$ retrial queues, which are studied in Sections 9.1 and 9.2, respectively. As the reader will notice, the corresponding approximating approaches lead to level independent $GI/M/1$ and $M/G/1$ structures.

A discrete-time Markov chain of $GI/M/1$-type has one-step transition probability matrix of the form

$$\mathbf{P} = \begin{pmatrix} \mathbf{B}_0 & \mathbf{C}_0 & & & \\ \mathbf{B}_1 & \mathbf{A}_1 & \mathbf{A}_0 & & \\ \mathbf{B}_2 & \mathbf{A}_2 & \mathbf{A}_1 & \mathbf{A}_0 & \\ \vdots & \vdots & \ddots & \ddots & \ddots \end{pmatrix}, \qquad (7.3)$$

thus resembling the structure of the Markov chain embedded at arrival epochs for the standard $GI/M/1$ queue. It is also called skip-free to the right to indicate that it can move up by one level only, although it may move in the downward direction several levels in one transition. Section 7.3.1 shows how a matrix-geometric form characterizes its stationary probability distribution.

Similarly, a Markov chain of $M/G/1$-type, also called skip-free to the left, has one-step transition probability matrix

$$\mathbf{P} = \begin{pmatrix} \mathbf{B}_0 & \mathbf{B}_1 & \mathbf{B}_2 & \mathbf{B}_3 & \cdots \\ \mathbf{A}_0 & \mathbf{A}_1 & \mathbf{A}_2 & \mathbf{A}_3 & \cdots \\ & \mathbf{A}_0 & \mathbf{A}_1 & \mathbf{A}_2 & \cdots \\ & & \mathbf{A}_0 & \mathbf{A}_1 & \cdots \\ & & & \ddots & \ddots \end{pmatrix}. \qquad (7.4)$$

In this case, we remark that the Markov chain embedded at service completion epochs for the standard $M/G/1$ queue has a transition matrix of this form, but with scalar entries instead of block matrices. A key ingredient to the solution of the stationary probability vector of (7.4) is the matrix $\mathbf{G}$ that is related to the fundamental period of the model. This matrix is briefly discussed in Subsection 7.3.2.

Needless to say, QBD processes can be seen as particular cases of the $GI/M/1$ and $M/G/1$ structures.

7.1.3 The Batch Markovian Arrival Process

The Poisson process has served as the main arrival flow for many years and generalizations have frequently concentrated on renewal processes. Their simplifying feature is the independence and equidistribution of successive interrenewal intervals. Thus, in queueing applications, the class of renewal processes is not flexible enough and, in particular, arrivals that tend to occur in bursts cannot be modelled in this way.

We present here the batch Markovian arrival process $(BMAP)$, which is thought to be a fairly general point process where the correlation aspect is not ignored. It has the feature of making many analytic properties explicit or at least computationally tractable.

We start with a constructive description of this process. Assume that a background Markov chain $\{J(t); t \geq 0\}$ with $m < \infty$ states, called phases, is in some state i. Let $\mathbf{D}$ be its infinitesimal generator, which is assumed to be irreducible. At the end of a sojourn time in i, which is exponentially distributed with parameter λ_i, there occurs a transition to another or (possibly) the same state. That transition may or may not correspond to an arrival epoch. With probability $P_{ij}(k)$, it corresponds to a transition to state j with a batch arrival of size k, for $k \geq 1$, and similarly, with probability $P_{ij}(0)$, the transition corresponds to no arrival and state of the Markov chain is j, for $j \neq i$. Therefore, $J(t)$ can go from state i to state i only through an arrival and

$$\sum_{j=1,\ j\neq i}^{m} P_{ij}(0) + \sum_{j=1}^{m}\sum_{k=1}^{\infty} P_{ij}(k) = 1, \quad 1 \leq i \leq m.$$

Define the matrices $\mathbf{D}_k = (d_{ij}(k))$ with entries $d_{ii}(0) = -\lambda_i$, $d_{ij}(0) = \lambda_i P_{ij}(0)$, for $j \neq i$, and $d_{ij}(k) = \lambda_i P_{ij}(k)$, for $k \geq 1$, from which it is clear that $\mathbf{D} = \sum_{k=0}^{\infty} \mathbf{D}_k$. The particular choice $\mathbf{D}_0 \neq \mathbf{D}$ and $\mathbf{D}_k = \mathbf{0}_{m \times m}$, for $k \geq 2$, means single arrivals and yields the Markovian arrival process (MAP).

Define the counting process $\{M(t); t \geq 0\}$, where $M(t)$ represents the number of arrivals up to time t. Our preceding construction shows that $\{(M(t), J(t)); t \geq 0\}$ is a Markov chain with the structured infinitesimal generator

$$\mathbf{Q} = \begin{pmatrix} \mathbf{D}_0 & \mathbf{D}_1 & \mathbf{D}_2 & \mathbf{D}_3 & \cdots \\ & \mathbf{D}_0 & \mathbf{D}_1 & \mathbf{D}_2 & \cdots \\ & & \mathbf{D}_0 & \mathbf{D}_1 & \cdots \\ & & & \ddots & \ddots \end{pmatrix}.$$

The sequence $\{\mathbf{D}_k; k \geq 0\}$ contains all information for $\mathbf{Q}$ and thus is usually called the characteristic sequence of a $BMAP$. A complete specification also requires specification of the distribution of $J(0)$. We do this in terms of a row vector $\boldsymbol{\alpha}$ with ith entry given by $P(J(0) = i)$, for $1 \leq i \leq m$.

By assuming $\mathbf{D}_0$ to be non-singular, the inter-arrival times are finite, with probability one. Our last assumption is that the vector $\mathbf{d} = \sum_{k=1}^{\infty} k\mathbf{D}_k \mathbf{e}_m$ is finite. This condition is equivalent to require that $E[M(t)] < \infty$ over finite intervals. The fundamental arrival rate is then defined by $\lambda = \boldsymbol{\theta}\mathbf{d}$, where $\boldsymbol{\theta}$ is the stationary probability vector of $\mathbf{D}$, and gives the expected number of arrivals per unit of time in the stationary version of a $BMAP$ [464, Section 3.6].

In [173], we find an expression for the coefficient of correlation between two successive arrivals in terms of

$$c_c = \frac{\lambda_g^{-1}\boldsymbol{\theta}(-\mathbf{D}_0^{-1})(\mathbf{D}-\mathbf{D}_0)(-\mathbf{D}_0^{-1})\mathbf{e}_m - \lambda_g^{-2}}{2\lambda_g^{-1}\boldsymbol{\theta}(-\mathbf{D}_0^{-1})\mathbf{e}_m - \lambda_g^{-2}},$$

where $\lambda_g = \boldsymbol{\theta}(-\mathbf{D}_0)\mathbf{e}_m$ gives the arrival rate of groups. For a MAP, we have that $\lambda_g = \lambda$.

Here is a selected sample of point processes subsumed under a $BMAP$ as special cases.

(i) *Poisson process.* The Poisson process of rate λ is a MAP of order $m = 1$ with $\mathbf{D}_0 = -\lambda$ and $\mathbf{D}_1 = \lambda$. For a batch Poisson process, the characteristic sequence can be defined as $\mathbf{D}_0 = -\lambda$, $\mathbf{D}_k = \lambda p_k$, for $k \geq 1$, where $\{p_k; k \geq 1\}$ is the probability mass function of the batch size.

(ii) *Markov modulated Poisson process (MMPP).* This is a Poisson process in which the instantaneous arrival rate depends on a finite Markov chain serving as a random environment. Let $\mathbf{Q}_a$ be its infinitesimal generator. Assume that arrivals occur at rate λ_i when the environment is in state i. The resulting process, called Markov modulated Poisson process, is a MAP with $\mathbf{D}_0 = \mathbf{Q}_a - \boldsymbol{\Delta}(\boldsymbol{\lambda})$ and $\mathbf{D}_1 = \boldsymbol{\Delta}(\boldsymbol{\lambda})$, where $\boldsymbol{\Delta}(\boldsymbol{\lambda}) = \text{diag}(\lambda_1, ..., \lambda_m)$.

(iii) *Phase-type renewal process.* Suppose that inter-renewal times in a renewal process follow a phase-type distribution with representation $(\boldsymbol{\tau}, \mathbf{T})$ of order m. Then, we deal with a MAP defined by $\mathbf{D}_0 = \mathbf{T}$ and $\mathbf{D}_1 = \mathbf{t}\boldsymbol{\tau}$, where $\mathbf{t} = -\mathbf{T}\mathbf{e}_m$.

Among other special cases, we mention the batch versions of processes described in (ii) and (iii), as well as point processes obtained from superpositions and thinning of Markovian arrival processes or as departure processes in Markovian queues.

To finish, we remark that the $BMAP$ results in a class of arrival processes, which is dense in a suitable sense in the space of point processes defined over $[0, \infty)$; see [115].

7.1.4 The Phase-Type Distribution and the SM Service Process

Our interest here is in the phase-type distribution and a special SM process, which are used to govern service and retrial times in later chapters.

The class of probability distributions of phase-type (PH) provides a simple framework to demonstrate how one may extend many results on exponential distributions to more complex models, but without losing computational tractability. We can find two parallel discussions in the literature, one for distributions in $[0, \infty)$, and the other for discrete distributions on $\mathbb{Z}_+$. For the sake of brevity, we shall only discuss the former.

To define a PH distribution we consider an absorbing Markov chain on the state space $\{0, 1, ..., r\}$ with initial probability vector $(\tau_0, \boldsymbol{\tau})$ and infinitesimal generator

7.1 A General Overview

$$Q = \begin{pmatrix} 0 & 0'_r \\ \mathbf{t} & \mathbf{T} \end{pmatrix},$$

with $\mathbf{t} = -\mathbf{T}e_r$. Then, a PH distribution corresponds to the distribution of the time until absorption into the state 0.

An important question to be examined is when the absorption occurs in a finite interval almost surely. By using the expression $F(x) = 1 - \tau \exp\{\mathbf{T}x\}e_r$ for the distribution function, it is readily verified that $F(\infty) = 1$ if and only if the matrix $\mathbf{T}$ is non-singular. Furthermore, this is certain if and only if states in $\{1, ..., r\}$ are all transient. We shall henceforth only consider the case $F(\infty) = 1$ and take $\tau_0 = 0$ to avoid instantaneous PH service and retrial times.

To show the versatility of the phase-type distribution, we give a few examples below.

(i) *Exponential distribution.* It is obtained by $r = 1$, $\tau = 1$ and $\mathbf{T} = -\nu$.
(ii) *Erlang distribution.* The Erlang law with r phases and rate ν can be described as a PH distribution of order r with $\tau = (1, 0, ..., 0)$ and

$$\mathbf{T} = \begin{pmatrix} -\nu & \nu & & \\ & \ddots & \ddots & \\ & & -\nu & \nu \\ & & & -\nu \end{pmatrix}.$$

A slight variant is obtained if we admit the exponential stages to have different rates.

(iii) *Hyperexponential distribution.* This leads to a PH representation by $\tau = (p_1, ..., p_r)$ and $\mathbf{T} = \mathrm{diag}(-\nu_1, ..., -\nu_r)$.

For practical use, the class of PH distributions provides ease in conditioning arguments, results in a Markovian structure of models involving exponential assumptions and leads to significant simplifications in various integral and differential equations arising in their analysis. An excellent summary of closure properties can be found in [116] and [464, Section 2.6]. Among these, we emphasize two properties. First, this class is dense, in the sense of weak convergence, in the class of all distributions on $(0, \infty)$. Second, sums and mixtures of a finite number of independent phase-type random variables are phase-type random variables.

Next, we focus on the semi-Markov (SM) service process. Suppose that, for each $n \in \mathbb{N}$, we define a sequence $\{X_n; n \geq 0\}$ of random variables on a countable set $\mathcal{S}$ and a point process $\{T_n; n \geq 0\}$, with $T_0 = 0$, such that

$$P(X_{n+1} = j, T_{n+1} - T_n \leq x \,|\, X_0, ..., X_n; T_0, ..., T_n)$$
$$= P(X_{n+1} = j, T_{n+1} - T_n \leq x \,|\, X_n),$$

for all $n \geq 0$, $j \in \mathcal{S}$ and $x \geq 0$. In other words, let $\{(X_n, T_n); n \geq 0\}$ be a Markov renewal process with state space $\mathcal{S}$, which is also assumed time-homogeneous; that is, for any $i, j \in \mathcal{S}$ and $x \geq 0$, the probabilities

$$P(X_{n+1} = j, T_{n+1} - T_n \leq x | X_n = i) = b_{ij}(x) \tag{7.5}$$

do not depend on n.

The family of matrices $\{\mathbf{B}(x) = (b_{ij}(x)); x \geq 0\}$ is called a semi-Markov kernel over $\mathcal{S}$. For each pair (i,j), the function $x \to b_{ij}(x)$ has all properties of a distribution function except that $b_{ij}(\infty)$ is not necessarily one, but we have that $\sum_{j \in \mathcal{S}} b_{ij}(\infty) = 1$, for $i \in \mathcal{S}$.

We assume that our Markov renewal processes are regular; that is, they have a finite number of transitions in a finite amount of time. Then, the SM service process can be thought of as the semi-Markov process [197, Section 10.5] with embedded Markov renewal process $\{(X_n, T_n)); n \geq 0\}$ defined by

$$K(t) = X_n, \quad \text{if } T_n \leq t < T_{n+1}.$$

We notice that if $\mathcal{S}$ is finite or $\{X_n; n \geq 0\}$ is recurrent, then the process is regular.

By definition $\{K(t); t \geq 0\}$ is a pure jump process, which is determined by its embedded Markov renewal process. The sequence $\{T_{n+1} - T_n; n \geq 0\}$ defines successive service times and $\{X_n; n \geq 0\}$ describes states visited immediately after the service completion epochs. This means that, though the server becomes free after each service completion, the state visited immediately after a service completion necessarily corresponds to the starting state for the next service time.

In view of (7.5), the length $T_{n+1} - T_n$ of a service time is a random variable whose distribution depends on both the state X_n being visited and the state X_{n+1} to be visited next. The successive states visited $\{X_n; n \geq 0\}$ form a Markov chain with state space $\mathcal{S}$ and one-step transition probability matrix $\mathbf{B}(\infty)$. Given this Markov chain, the successive service times are conditionally independent.

7.2 Some General Tools for QBD Structures

7.2.1 The Finite Case

Let $\mathbf{p}$, partitioned as $(\mathbf{p}(0), ..., \mathbf{p}(K))$, be the stationary probability vector of the finite QBD process defined by (7.2). Denote by $d(j)$ the cardinality of the jth level.

To evaluate $\mathbf{p}$, Gaver et al. [307] propose an efficient method, which is inspired from block-Gaussian elimination. The approach proceeds in two steps. In the first step, we progressively reduce the state space by removing one level at each iteration, until we are left with a Markov chain on the level K only. Once this Markov chain is solved, we construct the vector $\mathbf{p}$ in the second step by adding back one level at each iteration.

If $\mathcal{S}_0$ denotes the state space $\cup_{j=0}^{K} l(j)$ in (7.2), then we should restrict the range of the QBD process successively to the sets $\mathcal{S}_1, \mathcal{S}_2, ..., \mathcal{S}_K$, where

$\mathcal{S}_j = \cup_{i=j}^{K} l(i)$. Then, the infinitesimal generator of the resulting censored process on $\mathcal{S}_j$ is given by

$$\mathbf{Q}(j) = \begin{pmatrix} \mathbf{B}_j & \mathbf{A}_{j,j+1} & & & & \\ \mathbf{A}_{j+1,j} & \mathbf{A}_{j+1,j+1} & \mathbf{A}_{j+1,j+2} & & & \\ & \ddots & \ddots & \ddots & & \\ & & & \mathbf{A}_{K-1,K-2} & \mathbf{A}_{K-1,K-1} & \mathbf{A}_{K-1,K} \\ & & & & \mathbf{A}_{K,K-1} & \mathbf{A}_{KK} \end{pmatrix}, \quad 0 \leq j \leq K-1,$$

and $\mathbf{Q}(K) = \mathbf{B}_K$, where the matrices $\mathbf{B}_j$ are recursively defined

$$\mathbf{B}_0 = \mathbf{A}_{00}, \tag{7.6}$$

$$\mathbf{B}_j = \mathbf{A}_{jj} + \mathbf{A}_{j,j-1}\left(-\mathbf{B}_{j-1}^{-1}\right)\mathbf{A}_{j-1,j}, \quad 1 \leq j \leq K. \tag{7.7}$$

We observe that $\mathbf{B}_j$ is the infinitesimal generator of the process defined on $l(j)$ as the restriction of $\mathbf{Q}$, observed during those intervals of time spent at level j, before the original process enters level $j+1$ for the first time.

Thus, the stationary probability vector of $\mathbf{B}_K$ is proportional to $\mathbf{p}(K)$, so that $\mathbf{p}(K)\mathbf{B}_K = \mathbf{0}'_{d(K)}$. Similarly, for $0 \leq j \leq K-1$, the stationary probability vector of the censored process on $\mathcal{S}_j$ is proportional to $(\mathbf{p}(j), \mathbf{p}(j+1), ..., \mathbf{p}(K))$. Then, by recurrence, we find that

$$\mathbf{p}(j) = \mathbf{p}(j+1)\mathbf{A}_{j+1,j}\left(-\mathbf{B}_j^{-1}\right), \quad 0 \leq j \leq K-1. \tag{7.8}$$

This yields the procedure described in the next theorem.

Theorem 7.1. *The stationary probability vector $\mathbf{p}$ of a finite QBD process can be computed from the following steps:*

Step 1. For $0 \leq j \leq K$, compute the matrices $\mathbf{B}_j$ from (7.6) and (7.7).
Step 2. Solve $\mathbf{p}(K)\mathbf{B}_K = \mathbf{0}'_{d(K)}$ and $\mathbf{p}(K)\mathbf{e}_{d(K)} = 1$, and put $\delta = 1$.
Step 3. For $j = K-1, ..., 0$, compute recursively $\mathbf{p}(j)$ from (7.8) and evaluate
  $\delta = \delta + \mathbf{p}(j)\mathbf{e}_{d(j)}$.
Step 4. For $0 \leq j \leq K$, compute $\mathbf{p}(j) = \delta^{-1}\mathbf{p}(j)$.

There exits a risk of finding overflow problems when the mass of the Kth level is very small. In this case, one might re-normalize the vector $\mathbf{p}$ each time a new subvector in Step 3 is computed.

A variant of the above solution is readily derived if, instead of cutting the levels off, starting from the level 0 and moving up to the level K, we proceed in the reverse direction.

7.2.2 The Matrix-Geometric Distribution

Assume that the level independent QBD process introduced in Subsection 7.1.2 is irreducible and positive recurrent. Its stationary probability vector

$\mathbf{p} = (\mathbf{p}(0), \mathbf{p}(1), ...)$, where $\mathbf{p}(0)$ has $d(0)$ entries and $\mathbf{p}(j)$ has d entries, for $j \geq 1$, is given by the modified matrix-geometric form

$$\mathbf{p}(j) = \mathbf{p}(1)\mathbf{R}^{j-1}, \quad j \geq 1, \tag{7.9}$$

where the matrix $\mathbf{R}$, called rate matrix, records rates of sojourn in states of $l(j+1)$ per unit of the local time of $l(j)$. This can be re-expressed as $\mathbf{R} = \mathbf{A}_0 \mathbf{N}$, where the matrix $\mathbf{N}$ records expected sojourn times in states of $l(j)$, starting from $l(j)$, before the first visit to $l(j-1)$ [464, Section 6.4].

In order to completely specify the stationary distribution, we consider the censored process on $l(0) \cup l(1)$. If the QBD process is positive recurrent, then so is the censored process on $l(0) \cup l(1)$ and consequently the subvectors $\mathbf{p}(0)$ and $\mathbf{p}(1)$ satisfy

$$(\mathbf{p}(0), \mathbf{p}(1)) \begin{pmatrix} \mathbf{A}_{00} & \mathbf{A}_{01} \\ \mathbf{A}_{10} & \mathbf{A}_1 + \mathbf{R}\mathbf{A}_2 \end{pmatrix} = \mathbf{0}'_{d(0)+d}. \tag{7.10}$$

The normalization factor is then determined by $\mathbf{pe} = 1$; that is,

$$\mathbf{p}(0)\mathbf{e}_{d(0)} + \mathbf{p}(1)\left(\mathbf{I}_d - \mathbf{R}\right)^{-1}\mathbf{e}_d = 1. \tag{7.11}$$

If the infinitesimal generator $\mathbf{A} = \mathbf{A}_2 + \mathbf{A}_1 + \mathbf{A}_0$ is irreducible, then the QBD process is positive recurrent if and only if

$$\boldsymbol{\beta}\mathbf{A}_0\mathbf{e}_d < \boldsymbol{\beta}\mathbf{A}_2\mathbf{e}_d, \tag{7.12}$$

where $\boldsymbol{\beta}$ is the unique solution of the set of equations $\boldsymbol{\beta}\mathbf{A} = \mathbf{0}'_d$ and $\boldsymbol{\beta}\mathbf{e}_d = 1$. It is null recurrent if $\boldsymbol{\beta}\mathbf{A}_0\mathbf{e}_d = \boldsymbol{\beta}\mathbf{A}_2\mathbf{e}_d$, and it is transient if $\boldsymbol{\beta}\mathbf{A}_0\mathbf{e}_d > \boldsymbol{\beta}\mathbf{A}_2\mathbf{e}_d$. For a reducible matrix $\mathbf{A}$, see [464, Section 7.3] and [544, Section 1.4].

Condition (7.12) has an appealing interpretation. By defining the average transition rates to the right and to the left from the level j, for $j > 1$, by the left and right sides of (7.12), respectively, this condition expresses that the rate to the left should exceed that to the right. The entries of $\boldsymbol{\beta}$ provide correct stationary weights for an appropriate definition of the average transition rates.

We may remark that the positive recurrence condition (7.12) holds if and only if the spectral radius of $\mathbf{R}$, denoted by η, is such that $\eta < 1$. Recall that this value η is defined as the eigenvalue of $\mathbf{R}$ with largest modulus, which is real and positive.

When the spectral radius η needs to be explicitly computed, there are two cases depending on whether the rate matrix is explicitly available or not. In the former case, we should assume that the matrix $\mathbf{R}$ is known and irreducible. Then, the maximal eigenvalue η and its corresponding left eigenvector may be readily computed by Elsner's algorithm; see [544, Section 1.9]. In the second case, for which it suffices that $\mathbf{A}$ is irreducible, we evaluate η by solving the equation $z = \chi(z)$ on $(0, 1]$, where $\chi(z)$ is the maximal eigenvalue of the matrix polynomial $\mathbf{A}(z) = \mathbf{A}_2 z^2 + \mathbf{A}_1 z + \mathbf{A}_0$. To do this, we may apply an elementary procedure, such as the bisection method or the secant method.

7.2.3 The Computation of the Rate Matrix

In the positive recurrent case, the rate matrix $\mathbf{R}$ is characterized as the minimal solution to the matrix-quadratic equation

$$\mathbf{R}^2 \mathbf{A}_2 + \mathbf{R}\mathbf{A}_1 + \mathbf{A}_0 = \mathbf{0}_{d \times d}, \qquad (7.13)$$

in the set of non-negative matrices.

To solve (7.13), we next present the successive substitution (*SS*) algorithm and the logarithmic reduction (*LR*) algorithm. The first algorithm is linearly convergent, with convergence rate equal to the spectral radius η. The second algorithm is more involved but has the advantage of being quadratically convergent.

In the *SS* algorithm, we start somewhat arbitrarily with $\mathbf{R}_0 = \mathbf{0}_{d \times d}$ and then iteratively evaluate the sequence of matrices $\{\mathbf{R}_k; k \geq 1\}$ as follows:

$$\mathbf{R}_{k+1} = \left(\mathbf{R}_k^2 \mathbf{A}_2 + \mathbf{A}_0\right)\left(-\mathbf{A}_1^{-1}\right), \quad k \geq 0.$$

Neuts [544] has shown that $\{\mathbf{R}_k; k \geq 1\}$ is a monotonically increasing sequence that converges to the minimal non-negative solution of (7.13) and that this is the solution that uniquely provides the matrix $\mathbf{R}$ satisfying this equation. To define a stopping criterion, the standard practice is to keep track of two successive matrices in each iteration and stop iterations when $\|\mathbf{R}_k - \mathbf{R}_{k-1}\|_\infty < \varepsilon$, for a small value $\varepsilon > 0$, where we denote $\|\mathbf{A}\|_\infty = \max_{1 \leq i \leq p, 1 \leq j \leq q} |a_{ij}|$ for a matrix $\mathbf{A} = (a_{ij})$ of dimension $p \times q$.

The *LR* algorithm, which is summarized in Theorem 7.2, involves a number of iterations that is equal to the logarithm of the number of iterations in the *SS* algorithm. In conjunction with its numerical stability, the efficiency of the *LR* approach has made it the algorithm of choice for the solution of *QBD* models for a long time.

Theorem 7.2. *The rate matrix* $\mathbf{R}$ *of a level independent QBD process can be computed iteratively from the following steps:*

Step 1. Compute the matrices
$\Pi_0 = \mathbf{A}_0(-\mathbf{A}_1^{-1})$,
$\Pi_1 = \mathbf{A}_2(-\mathbf{A}_1^{-1})$,
$\Pi_2 = \Pi_0$,
$\Pi_3 = \mathbf{I}_d$.

Step 2. Repeat the following computations until $\|\Pi_0 \Pi_3\|_\infty < \varepsilon$:
$\Pi_3 = \Pi_1 \Pi_3$,
$\Sigma_0 = \Pi_0^2$,
$\Sigma_1 = \Pi_0 \Pi_1 + \Pi_1 \Pi_0$,
$\Sigma_2 = \Pi_1^2$,
$\Pi_0 = \Sigma_0 (\mathbf{I}_d - \Sigma_1)^{-1}$,
$\Pi_1 = \Sigma_2 (\mathbf{I}_d - \Sigma_1)^{-1}$,
$\Pi_2 = \Pi_2 + \Pi_0 \Pi_3$.

Step 3. Put $\mathbf{R} = \Pi_2$.

In the original approach [463], the qualities of the algorithm are proved for the related matrix $\mathbf{G}$, which is solution of the following dual equation to (7.13):

$$\mathbf{A}_0 \mathbf{G}^2 + \mathbf{A}_1 \mathbf{G} + \mathbf{A}_2 = \mathbf{0}_{d \times d}. \tag{7.14}$$

This matrix records probabilities of moving down by one level, from $l(j)$ to $l(j-1)$, for $j > 1$, in a finite first-passage time, and is related to $\mathbf{R}$ by

$$\mathbf{R} = \mathbf{A}_0 \left(-\mathbf{A}_1 - \mathbf{A}_0 \mathbf{G} \right)^{-1}. \tag{7.15}$$

By (7.15), the advantage of considering $\mathbf{G}$ instead of $\mathbf{R}$ is that its entries have a very intuitive probabilistic interpretation, which is the basis for deriving the *LR* algorithm. To be concrete, by writing (7.14) as

$$\mathbf{G} = (-\mathbf{A}_1^{-1}) \mathbf{A}_0 \mathbf{G}^2 + (-\mathbf{A}_1^{-1}) \mathbf{A}_2, \tag{7.16}$$

we observe that the term $(-\mathbf{A}_1^{-1})\mathbf{A}_2$ records the probability that, starting from a state in $l(j)$, the process eventually leaves $l(j)$ and moves down to $l(j-1)$, thereby terminating the first-passage time. In a similar manner, with probabilities recorded in $(-\mathbf{A}_1^{-1})\mathbf{A}_0$, the process eventually leaves $l(j)$ and moves up to $l(j+1)$. In this case, the matrix $\mathbf{G}^2$ records probabilities of moving down by two levels (i.e., from $l(j+1)$ to $l(j-1)$) [464, Section 6.4].

Equation (7.16) motivates the analysis of the censored process restricted to the levels $l(2^k)$, for $k \geq 0$. Thus, the resulting algorithm doubles the number of considered levels in each iteration.

7.3 Some General Tools for $GI/M/1$ and $M/G/1$ Structures

7.3.1 The Matrix-Geometric Distribution

Let us now focus on the solution to the $GI/M/1$ structure defined by (7.3).

In what follows, we assume that $\mathbf{P}$ is aperiodic and irreducible, and that the matrix $\mathbf{A} = \sum_{j=0}^{\infty} \mathbf{A}_j$ is stochastic and irreducible. Then, the Markov chain in (7.3) is positive recurrent if and only if $\boldsymbol{\beta} \mathbf{a} > 1$ and the stochastic matrix

$$\mathbf{B}(\mathbf{R}) = \begin{pmatrix} \mathbf{B}_0 & \mathbf{C}_0 \\ \sum_{j=1}^{\infty} \mathbf{R}^{j-1} \mathbf{B}_j & \sum_{j=1}^{\infty} \mathbf{R}^{j-1} \mathbf{A}_j \end{pmatrix}$$

has a strictly positive left invariant vector. Here, the row vector $\boldsymbol{\beta}$ is the unique solution to $\boldsymbol{\beta} = \boldsymbol{\beta} \mathbf{A}$ and $\boldsymbol{\beta} \mathbf{e}_d = 1$, and the column vector $\mathbf{a}$ is defined as $\mathbf{a} = \sum_{j=1}^{\infty} j \mathbf{A}_j \mathbf{e}_d$. The matrix $\mathbf{R}$ is the minimal non-negative solution to the matrix equation

$$\mathbf{R} = \sum_{j=0}^{\infty} \mathbf{R}^j \mathbf{A}_j. \qquad (7.17)$$

In practice, $\beta \mathbf{a} > 1$ is the only condition requiring detailed verification. It is usually obvious that $\mathbf{B}(\mathbf{R})$ is irreducible, so its invariant vector is positive [544].

We may interpret the matrix $\mathbf{R}$ as recording the expected numbers of visits to states of $l(j+1)$ before the first return to $\cup_{i=0}^{j} l(i)$, given that the Markov chain starts in $l(j)$, for $j \geq 1$ [544, Section 1.2].

The spectral radius η of $\mathbf{R}$ is strictly less than one, if the Markov chain is positive recurrent. The value of η can be computed in a similar manner to the case of a QBD process, but with the matrix function $\mathbf{A}(z)$ replaced by $\sum_{j=0}^{\infty} \mathbf{A}_j z^j$. Then, the partitioned stationary probability vector $\mathbf{p}$ satisfies the matrix-geometric property

$$\mathbf{p}(j) = \mathbf{p}(1) \mathbf{R}^{j-1}, \quad j \geq 1,$$

and the subvectors $\mathbf{p}(0)$ and $\mathbf{p}(1)$ are the solution to

$$(\mathbf{p}(0), \mathbf{p}(1)) = (\mathbf{p}(0), \mathbf{p}(1)) \mathbf{B}(\mathbf{R}),$$
$$\mathbf{p}(0) \mathbf{e}_{d(0)} + \mathbf{p}(1) (\mathbf{I}_d - \mathbf{R})^{-1} \mathbf{e}_d = 1.$$

Various algorithms have been designed to solve (7.17). In [544], two iterative schemes are suggested, by starting with $\mathbf{R}_0 = \mathbf{0}_{d \times d}$, from

$$\mathbf{R}_{k+1} = \sum_{j=0}^{\infty} \mathbf{R}_k^j \mathbf{A}_j, \quad k \geq 0, \qquad (7.18)$$

$$\mathbf{R}_{k+1} = \sum_{j=0, j \neq 1}^{\infty} \mathbf{R}_k^j \mathbf{A}_j (\mathbf{I}_d - \mathbf{A}_1)^{-1}, \quad k \geq 0. \qquad (7.19)$$

Equations (7.18) and (7.19) define monotonically increasing sequences that converge to the minimal non-negative solution of (7.17). In general, both schemes suffer from slow convergence when η is close to one. However, the number of iterations notably decreases when the above series become finite sums, which is the case in Section 9.1. It is also pointed out that the iteration (7.19) converges faster than (7.18).

7.3.2 The Computation of the Matrix G

We now turn our attention to solving Markov chains whose transition matrix is given by (7.4). We assume that $\mathbf{P}$ is aperiodic and irreducible.

Neuts [546] gives the theory for this class of structured Markov chains. The key role in his approach is played by the minimal non-negative solution of the matrix equation

$$\mathbf{G} = \sum_{j=0}^{\infty} \mathbf{A}_j \mathbf{G}^j. \tag{7.20}$$

The matrix $\mathbf{G}$ now records the probabilities, starting from $l(j)$, of visiting states of $l(j-1)$, for $j \geq 1$, in a finite first-passage time [464, Section 13.1].

With regard to the stationary probability vector $\mathbf{p} = (\mathbf{p}(0), \mathbf{p}(1), \ldots)$, we do not obtain a recursion as simple as the matrix-geometric result. Instead, we have $\mathbf{p}(0)\overline{\mathbf{B}}_0 = \mathbf{p}(0)$ and

$$\mathbf{p}(j) = \left(\mathbf{p}(0)\overline{\mathbf{B}}_j + (1 - \delta_{1j}) \sum_{i=1}^{j-1} \mathbf{p}(i) \overline{\mathbf{A}}_{j-i+1} \right) \left(\mathbf{I}_d - \overline{\mathbf{A}}_1 \right)^{-1}, \quad j \geq 1,$$

where $\overline{\mathbf{A}}_j = \sum_{i=0}^{\infty} \mathbf{A}_{j+i} \mathbf{G}^i$ and $\overline{\mathbf{B}}_j = \sum_{i=0}^{\infty} \mathbf{B}_{j+i} \mathbf{G}^i$. In the case of an irreducible and stochastic matrix $\mathbf{A} = \sum_{j=0}^{\infty} \mathbf{A}_j$, the above equations provide us with the stationary probability vector if the Markov chain is positive recurrent. This holds if and only if $\boldsymbol{\beta}\mathbf{a} < 1$ and the matrix $\sum_{j=1}^{\infty} j \mathbf{B}_j$ has finite entries, where $\boldsymbol{\beta}$ is the invariant probability vector of $\mathbf{A}$ and $\mathbf{a} = \sum_{j=1}^{\infty} j \mathbf{A}_j \mathbf{e}_d$ [146].

A common algorithm for computing the solution of (7.20) is the fixed point iteration

$$\mathbf{G}_{k+1} = \sum_{j=0}^{\infty} \mathbf{A}_j \mathbf{G}_k^j,$$

with $\mathbf{G}_0 = \mathbf{0}_{d \times d}$. An excellent treatment that applies cyclic reduction to (7.20) can be found in [146, Sections 7.4 and 7.5].

7.3.3 Asymptotically Quasi-Toeplitz Markov Chains

We next introduce a special class of level dependent $M/G/1$ structure, called asymptotically quasi-Toeplitz Markov chain ($AQTMC$), which applies to the $BMAP/SM/1$ retrial queue in Section 9.2.

We consider an aperiodic and irreducible Markov chain on the state space $\mathcal{S} = \{(j,n); j \geq 0, 1 \leq n \leq m\}$, which we partition as $\cup_{j=0}^{\infty} l(j)$, with the level j defined as $l(j) = \{(j,n); 1 \leq n \leq m\}$, for $j \geq 0$. Its one-step transition probability matrix is given by

$$\mathbf{P} = \begin{pmatrix} \mathbf{A}_{00} & \mathbf{A}_{01} & \mathbf{A}_{02} & \mathbf{A}_{03} & \cdots \\ \mathbf{A}_{10} & \mathbf{A}_{11} & \mathbf{A}_{12} & \mathbf{A}_{13} & \cdots \\ & \mathbf{A}_{21} & \mathbf{A}_{22} & \mathbf{A}_{23} & \cdots \\ & & \ddots & \ddots & \ddots \end{pmatrix}.$$

The Markov chain is called an $AQTMC$ if and only if its block matrices $\mathbf{A}_{ij}$ can be expressed as

7.3 Some General Tools for $GI/M/1$ and $M/G/1$ Structures

$$\mathbf{A}_{0j} = \mathbf{V}_j, \quad j \geq 0, \tag{7.21}$$

$$\mathbf{A}_{ij} = \sum_{k=1}^{k_0} \mathbf{Q}_k^{(i)} \mathbf{Y}_{j-i+1}^{(k)}, \quad i \geq 1, \; j \geq i-1, \tag{7.22}$$

where the matrices $\mathbf{Q}_k^{(i)}$ are such that the limits

$$\mathbf{Q}_k = \lim_{i \to \infty} \mathbf{Q}_k^{(i)}, \quad 1 \leq k \leq k_0, \tag{7.23}$$

define matrices with finite entries, the matrices $\mathbf{V}_j$, $\mathbf{Y}_j^{(k)}$ and $\mathbf{Q}_k$ are sub-stochastic, for $j \geq 0$ and $1 \leq k \leq k_0$, and the matrices $\sum_{j=0}^{\infty} \mathbf{V}_j$, $\sum_{j=0}^{\infty} \mathbf{Y}_j^{(k)}$, for $1 \leq k \leq k_0$, and $\sum_{k=1}^{k_0} \mathbf{Q}_k$ are stochastic.

Equations (7.21)-(7.23) try to reflect how the dependence of the matrices $\mathbf{A}_{ij}$ on the level i vanishes as i tends to ∞. Indeed, we may associate a limit Markov chain of $M/G/1$-type to facilitate the study of the $AQTMC$. More precisely, a Markov chain of $M/G/1$-type is referred as the limit chain of the $AQTMC$ if and only if its block matrices, denoted by $\widetilde{\mathbf{A}}_{ij}$, are given by

$$\widetilde{\mathbf{A}}_{0j} = \mathbf{A}_{0j}, \quad j \geq 0,$$

$$\widetilde{\mathbf{A}}_{ij} = \sum_{k=1}^{k_0} \mathbf{Q}_k \mathbf{Y}_{j-i+1}^{(k)}, \quad i \geq 1, \; j \geq i-1.$$

Let us assume that the $AQTMC$ is positive recurrent; for details, we refer to [229]. Partition its stationary probability vector $\boldsymbol{\pi}$ by levels into sub-vectors $\boldsymbol{\pi}(j)$, for $j \geq 0$, and define the vector generating function $\boldsymbol{\Pi}(z) = \sum_{j=0}^{\infty} \boldsymbol{\pi}(j) z^j$, as well as the matrix functions $\mathbf{V}(z) = \sum_{j=0}^{\infty} \mathbf{V}_j z^j$ and $\mathbf{Y}^{(k)}(z) = \sum_{j=0}^{\infty} \mathbf{Y}_j^{(k)} z^j$, for $1 \leq k \leq k_0$.

The equilibrium equations are seen to be given by

$$\boldsymbol{\pi}(j) = \boldsymbol{\pi}(0) \mathbf{V}_j + \sum_{i=1}^{j+1} \boldsymbol{\pi}(i) \sum_{k=1}^{k_0} \mathbf{Q}_k^{(i)} \mathbf{Y}_{j-i+1}^{(k)}, \quad j \geq 0. \tag{7.24}$$

By assuming the matrices $\sum_{k=1}^{k_0} \mathbf{Q}_k^{(j)} \mathbf{Y}_0^{(k)}$, for $j \geq 1$, to be non-singular, equation (7.24) can be rewritten as follows:

$$\boldsymbol{\pi}(j) = \boldsymbol{\pi}(0) \mathbf{F}_j, \quad j \geq 0, \tag{7.25}$$

where $\mathbf{F}_0 = \mathbf{I}_m$ and

$$\mathbf{F}_j = \left(\mathbf{F}_{j-1} - \mathbf{V}_{j-1} - \sum_{i=1}^{j-1} \mathbf{F}_i \sum_{k=1}^{k_0} \mathbf{Q}_k^{(i)} \mathbf{Y}_{j-i}^{(k)} \right) \left(\sum_{k=1}^{k_0} \mathbf{Q}_k^{(j)} \mathbf{Y}_0^{(k)} \right)^{-1}, \quad j \geq 1.$$

To avoid differences in the above iterative procedure, Klimenok and Dudin suggest the use of expressions like (29) and (30) in [411].

From (7.24), one also obtains

$$\Pi(z) = \pi(0)\mathbf{V}(z) + \sum_{i=1}^{\infty} \pi(i) z^{i-1} \sum_{k=1}^{k_0} \mathbf{Q}_k^{(i)} \mathbf{Y}^{(k)}(z). \tag{7.26}$$

There are models for which (7.26) yields an explicit expression for $\Pi(z)$ up to the unknown vector $\pi(0)$. Then, the entries of $\pi(0)$ are frequently evaluated by using the approach based on the matrix $\mathbf{G}$; see Subsection 7.3.2. In the level dependent case, the matrix $\mathbf{G}$ should be replaced by the sequence $\{\mathbf{G}_j; j \geq 1\}$, where $\mathbf{G}_j$ is recording probabilities, starting from $l(j+1)$, of visiting the states of $l(j)$ in a finite time.

Inspired by the fact that the limit chain is a Markov chain of $M/G/1$-type, Dudin and Klimenok [229] suggest to replace the stationary distribution π of the $AQTMC$ by the stationary probability vector $\overline{\pi}$ of a censored Markov chain with censoring set consisting of all states in and below the level K_f, provided that we take K_f large enough. Therefore, the vector $\overline{\pi}$ is proportional to $(\pi(0), \pi(1), ..., \pi(K_f))$; that is, we have that

$$\overline{\pi}(j) = \frac{\pi(j)}{\sum_{i=0}^{K_f} \pi(i)\mathbf{e}_m}, \quad 0 \leq j \leq K_f. \tag{7.27}$$

Then, it follows that the subvectors $\overline{\pi}(j)$ satisfy (7.24), for $0 \leq j \leq K_f - 1$, and

$$\overline{\pi}(K_f) = \overline{\pi}(0) \left(\mathbf{V}_{K_f} + \sum_{j=K_f+1}^{\infty} \mathbf{V}_j \mathbf{G}_{j-1} \mathbf{G}_{j-2}...\mathbf{G}_{K_f} \right)$$

$$+ \sum_{i=1}^{K_f} \overline{\pi}(i) \sum_{k=1}^{k_0} \mathbf{Q}_k^{(i)} \left(\mathbf{Y}_{K_f-i+1}^{(k)} + \sum_{j=K_f-i+2}^{\infty} \mathbf{Y}_j^{(k)} \mathbf{G}_{j+i-2} \mathbf{G}_{j+i-3}...\mathbf{G}_{K_f} \right).$$

By conditioning on the state visited after the first transition, we find that

$$\mathbf{G}_j = \mathbf{A}_{j+1,j} + \sum_{i=j+1}^{\infty} \mathbf{A}_{j+1,i} \mathbf{G}_{i-1} \mathbf{G}_{i-2}...\mathbf{G}_j, \quad j \geq 0. \tag{7.28}$$

Unless additional assumptions are made about the transition matrix $\mathbf{P}$, it is not evident how to solve this infinite backward recursion in an explicit manner. However, the use of the limit chain allows us to replace the matrices $\mathbf{G}_j$, for $j \geq K$, by the matrix $\mathbf{G}$ related to the limit chain, for a suitably chosen K. To do this, in view of (7.28), we define the value

$$\theta_j = \left\| \mathbf{G} - \mathbf{A}_{j+1,j} - \sum_{i=j+1}^{\infty} \mathbf{A}_{j+1,i} \mathbf{G}^{i-j} \right\|,$$

where $||\mathbf{A}|| = \max_{1 \leq i \leq p} \sum_{j=1}^{q} |a_{ij}|$ for a matrix $\mathbf{A} = (a_{ij})$ of dimension $p \times q$. Then, we notice that θ_j tends to zero as j tends to ∞, since

$$\theta_j \leq C \sum_{k=1}^{k_0} \left\| \mathbf{Q}_k - \mathbf{Q}_k^{(j+1)} \right\|, \qquad (7.29)$$

with the constant $C = \max_{1 \leq k \leq k_0} ||\mathbf{Y}^{(k)}(\mathbf{G})|| < \infty$. Clearly, the right-hand side of (7.29) converges to zero by definition of $AQTMC$. Thus, if $j \geq K$, we may approximate the matrices $\mathbf{G}_j$ by the matrix $\mathbf{G}$, where K is such that $\theta_j < \varepsilon$, for $j \geq K$, with a small predetermined value $\varepsilon > 0$. From Subsection 7.3.2, the matrix $\mathbf{G}$ is here the minimal non-negative solution to

$$\mathbf{G} = \sum_{k=1}^{k_0} \mathbf{Q}_k \mathbf{Y}^{(k)}(\mathbf{G}).$$

If we choose $K_f \geq K$, then the subvector $\overline{\pi}(0)$ is the unique solution to

$$\overline{\pi}(0) \left(\mathbf{V}(\mathbf{G}) - \mathbf{F}_{K_f} \mathbf{G}^{K_f} - \sum_{j=0}^{K_f-1} \mathbf{V}_j \mathbf{G}^j \right.$$

$$\left. + \sum_{i=1}^{K_f} \mathbf{F}_i \sum_{k=1}^{k_0} \mathbf{Q}_k^{(i)} \left(\mathbf{Y}^{(k)}(\mathbf{G}) - \sum_{j=0}^{K_f-i} \mathbf{Y}_j^{(k)} \mathbf{G}^j \right) \mathbf{G}^{i-1} \right) = \mathbf{0}'_m, \quad (7.30)$$

which satisfies $\overline{\pi}(0) \sum_{j=0}^{K_f} \mathbf{F}_j \mathbf{e}_m = 1$. Once $\overline{\pi}(0)$ is computed, we may proceed to evaluate the subvectors $\overline{\pi}(j)$, for $1 \leq j \leq K_f$, from (7.25)

A remaining major question is how to select $K_f \geq K$. Ideally, one would like K_f to be such that the stationary probability of the states in $\cup_{j=K_f+1}^{\infty} l(j)$ is negligible. Then, the mass $\sum_{i=0}^{K_f} \pi(i) \mathbf{e}_m$ in the denominator of the right-hand side of (7.27) should be close to one and, consequently, $\overline{\pi}(j)$ should be close to $\pi(j)$, for $0 \leq j \leq K_f$. Since the tail of the $AQTMC$ is essentially lumped in the subvector $\overline{\pi}(K_f)$, we may progressively increase K until $\overline{\pi}(K_f) \mathbf{e}_m < \varepsilon_f$, for an arbitrary small $\varepsilon_f > 0$.

7.4 Bibliographical Notes

Matrix-analytic methods and, in particular, the notion of a matrix-geometric solution first appeared in [245], where the author considers a QBD structure for which $\mathbf{A}_0$ is a scalar matrix. However, Neuts pioneered their analysis, whose work culminated into two monographs about $GI/M/1$-type processes [544] and $M/G/1$-type processes [546]. These processes can be seen as special cases of a more general structure, which is called of $GI/G/1$-type; see e.g. [329] for a survey on several methodologies to analyze this structure. Markov chains of $GI/G/1$-type are not commonly found in the literature on retrial queues, so that in this book we do not focus on them.

Reviews of the basic properties and modelling applications of Markovian arrivals and PH random variables are found in [173], [464, Chapters 2 and 3], [544, Chapter 2] and [546, Chapter 5]. Aspects of how to fit these models to observed data are analyzed in [116]. The original description of a $BMAP$ [546, Section 5.4] was quite involved because several different types of arrivals were accounted for. In Subsection 7.1.3, we follow a description with simplified notation due to Lucantoni [496]; see also [464, Chapter 3].

In the monographs by Çinlar [197, Chapter 10] and Kulkarni [450, Chapter 9] we find an extensive treatment of the Markov renewal theory with examples applied to queueing theory.

Details about the computational complexity of the algorithm in Theorem 7.1 are given in [307]. The paper [307] also presents numerical methods for evaluating moments of first-passage times in finite QBD processes. In [464, Chapter 10], we may find other approaches that are specifically tailored to the case of a homogeneous finite QBD process, such as the method of folding.

In Section 5.3, we mentioned that there are many direct methods and iterative methods for the solution of a finite system of linear equations. A good summary of both methods can be found in [160, Sections 3.3 and 3.4]. For a major emphasis on procedures oriented to structured Markov chains, see [644].

Two different classes of general-purpose iterative schemes for solving (7.13) might be distinguished. The first class includes the SS method (see [464, Section 8.1] and [544, Section 1.2]) and the LR approach (see [463] and [464, Section 8.4]). This class directly works on the matrices involved in (7.13) and uses probabilistic arguments to derive the matrix-geometric property (7.9). The algorithmic techniques of the second class are not solely based on probabilistic arguments to solve (7.13), but they employ a transformation to some domain to derive a solution of the balanced equations. Among them, the most powerful approaches are the cyclic reduction method [146, Section 7.3], the invariant subspace algorithm [26] and the spectral expansion method [304], which is also termed generalized eigenvalue approach.

A comprehensive and self-contained treatment of algorithms for the QBD and $M/G/1$ structures is the recent book by Bini et al. [146]. We refer the reader to [25] for the use of the invariant subspace approach in Markov chains of $GI/M/1$- and $M/G/1$-types. We also mention the so-called UL-type and LU-type RG-factorizations which provide a unified and algorithmic framework for the study of block structured stochastic models [477, 478].

Contents of Subsection 7.3.3 are mostly based on the paper [229], where the notion of $AQTMC$ first appeared as a tool to solve the $BMAP/SM/1$ retrial queue. In addition to [229], we may cite the recent work by Klimenok and Dudin [411], who introduce a new definition of $AQTMC$, establish stability conditions and propose algorithms for computing the stationary probabilities.

Those unfamiliar with matrix theory should consult a standard textbook; see e.g. [358, Chapter 4]. Properties and applications of Kronecker products and sums can be found in [328].

8
Selected Retrial Queues with QBD Structure

In this chapter, our interest is in a few selected retrial queues with QBD structure which are analyzed through matrix-analytic techniques. In Section 8.1, we analyze the $MAP/PH/1$ retrial queue and discuss the stationary distribution of the system state and the maximal queue length during a busy period. The aim in Section 8.2 is to extend the study to the multiserver case focussing on other performance characteristics. To this end, we study the busy period and the waiting time for the $MAP/M/c$ retrial queue. Finally, in Section 8.3, we focus on a multiserver model with finite population where both the service times and the retrial times follow phase-type distributions.

8.1 The $MAP/PH/1$ Retrial Queue

In this section, we deal with the $MAP/PH/1$ retrial queue. This model provides a natural generalization of the $M/G/1$ retrial queue investigated extensively in previous chapters. We concentrate on the computation of the stationary distribution of the system state and the distribution of the maximum number of customers in orbit during a busy period. The consideration of a generalized truncation of Neuts and Rao type plays again a central role to approximate the distribution of the system state.

8.1.1 Stationary Distribution of the System State

The steady state analysis of the $MAP/PH/1$ model was initially investigated by Diamond and Alfa [210], who propose a numerical treatment based on the generalized truncation models of Falin, and Neuts and Rao type. The analysis here is based on a paper by Artalejo and Chakravarthy [100] where the work [210] is extended and complemented in several directions: refinement of the positive recurrence condition, computation of the rate matrix, determination of the truncation threshold, special cases, etc.

Let $C(t)$, $N(t)$ and $J(t)$ denote, respectively, the phase of the service process, the number of customers in orbit and the phase of the arrival process at time t. Note that $C(t) = 0$ indicates that the server is idle. The process $\mathcal{X} = \{(C(t), N(t), J(t)); t \geq 0\}$ is a continuous-time Markov chain with state space $\mathcal{S} = \{0, ..., r\} \times \mathbb{Z}_+ \times \{1, ..., m\}$. By level $l(j)$, for $j \geq 0$, we denote the set of $(r+1)m$ states given by $l(j) = \{(i, j, k); 0 \leq i \leq r, 1 \leq k \leq m\}$.

Then, the infinitesimal generator $\mathbf{Q}$, in partitioned form, is given by

$$\mathbf{Q} = \begin{pmatrix} \mathbf{A}_{00} & \mathbf{A}_{01} & & & \\ \mathbf{A}_{10} & \mathbf{A}_{11} & \mathbf{A}_{12} & & \\ & \mathbf{A}_{21} & \mathbf{A}_{22} & \mathbf{A}_{23} & \\ & & \ddots & \ddots & \ddots \end{pmatrix}, \tag{8.1}$$

where $\mathbf{A}_{jn}$ are square matrices of order $(r+1)m$. They are given by

$$\mathbf{A}_{j,j-1} = j\mu \begin{pmatrix} \mathbf{0}_{m \times m} & \boldsymbol{\tau} \otimes \mathbf{I}_m \\ \mathbf{0}_{rm \times m} & \mathbf{0}_{rm \times rm} \end{pmatrix}, \quad j \geq 1,$$

$$\mathbf{A}_{jj} = \begin{pmatrix} \mathbf{D}_0 - j\mu \mathbf{I}_m & \boldsymbol{\tau} \otimes \mathbf{D}_1 \\ \mathbf{t} \otimes \mathbf{I}_m & \mathbf{T} \oplus \mathbf{D}_0 \end{pmatrix}, \quad j \geq 0,$$

$$\mathbf{A}_{j,j+1} = \begin{pmatrix} \mathbf{0}_{m \times m} & \mathbf{0}_{m \times rm} \\ \mathbf{0}_{rm \times m} & \mathbf{I}_r \otimes \mathbf{D}_1 \end{pmatrix}, \quad j \geq 0.$$

Let $\mathbf{p}$, partitioned as $\mathbf{p} = (\mathbf{p}(0), \mathbf{p}(1), ...)$, be the stationary probability vector of $\mathbf{Q}$ satisfying that

$$\mathbf{pQ} = \mathbf{0}', \quad \mathbf{pe} = 1. \tag{8.2}$$

In what follows, we assume that $\rho = \lambda \beta_1 < 1$, which is a necessary and sufficient condition for the positive recurrence of the process $\mathcal{X}$; see [100].

The set of equations (8.2) can be solved by employing the truncation methods discussed in Subsections 3.4.1 and 3.4.2 in the context of the $M/M/c$ retrial queue. A comparison of different methods for computing the stationary distribution is not our aim here. Thus, we simply pick the generalized truncation of Neuts and Rao type and discuss its practical implementation. Let K be the threshold associated with the truncated model; that is, if $j \geq K$, then the retrial rate becomes $K\mu$ instead of $j\mu$. In this case, the infinitesimal generator (8.1) is modified as

$$\widetilde{\mathbf{Q}}(K) = \begin{pmatrix} \mathbf{A}_{00} & \mathbf{A}_{01} & & & & & \\ \mathbf{A}_{10} & \mathbf{A}_{11} & \mathbf{A}_{12} & & & & \\ & \ddots & \ddots & \ddots & & & \\ & & \mathbf{A}_{K-1,K-2} & \mathbf{A}_{K-1,K-1} & \mathbf{A}_{K-1,K} & & \\ & & & \mathbf{A}_{K,K-1} & \mathbf{A}_{KK} & \mathbf{A}_{K,K+1} & \\ & & & & \mathbf{A}_{K,K-1} & \mathbf{A}_{KK} & \mathbf{A}_{K,K+1} \\ & & & & & \ddots & \ddots & \ddots \end{pmatrix}.$$

8.1 The $MAP/PH/1$ Retrial Queue

The stationary probability vector $\boldsymbol{\beta}$ of the generator $\mathbf{A}_K = \mathbf{A}_{K,K-1} + \mathbf{A}_{KK} + \mathbf{A}_{K,K+1}$ can be partitioned as $\boldsymbol{\beta} = (\boldsymbol{\beta}_0, \boldsymbol{\beta}_1)$, where $\boldsymbol{\beta}_0$ and $\boldsymbol{\beta}_1$ have m and rm entries, respectively. The following result gives explicit expressions for $\boldsymbol{\beta}_0$ and $\boldsymbol{\beta}_1$.

Theorem 8.1. *The vector $\boldsymbol{\beta}$ is given by*

$$\boldsymbol{\beta}_0 = \boldsymbol{\theta} - \boldsymbol{\beta}_1 (\mathbf{e}_r \otimes \mathbf{I}_m), \tag{8.3}$$

$$\boldsymbol{\beta}_1 = \boldsymbol{\theta}\left(\boldsymbol{\tau} \otimes (\mathbf{D}_1 + K\mu\mathbf{I}_m)\right)\left((\mathbf{e}_r\boldsymbol{\tau}) \otimes (\mathbf{D}_1 + K\mu\mathbf{I}_m) - \mathbf{T} \oplus \mathbf{D}\right)^{-1}. \tag{8.4}$$

Proof. On noting that $\boldsymbol{\beta}_0 + \boldsymbol{\beta}_1(\mathbf{e}_r \otimes \mathbf{I}_m)$ gives the phase of the arrival process in the stationary version, we conclude that $\boldsymbol{\theta} = \boldsymbol{\beta}_0 + \boldsymbol{\beta}_1(\mathbf{e}_r \otimes \mathbf{I}_m)$ as claimed in (8.3).

We also notice that the generator $\mathbf{A}_K$ has the form

$$\mathbf{A}_K = \begin{pmatrix} \mathbf{D}_0 - K\mu\mathbf{I}_m & \boldsymbol{\tau} \otimes (\mathbf{D}_1 + K\mu\mathbf{I}_m) \\ \mathbf{t} \otimes \mathbf{I}_m & \mathbf{T} \oplus \mathbf{D} \end{pmatrix}.$$

Thus, the equations $\boldsymbol{\beta}\mathbf{A}_K = \mathbf{0}'_{(r+1)m}$ can be rewritten as

$$\boldsymbol{\beta}_0(\mathbf{D}_0 - K\mu\mathbf{I}_m) + \boldsymbol{\beta}_1(\mathbf{t} \otimes \mathbf{I}_m) = \mathbf{0}'_m,$$
$$\boldsymbol{\beta}_0(\boldsymbol{\tau} \otimes (\mathbf{D}_1 + K\mu\mathbf{I}_m)) + \boldsymbol{\beta}_1(\mathbf{T} \oplus \mathbf{D}) = \mathbf{0}'_{rm}, \tag{8.5}$$

and the stated equation (8.4) follows immediately by inserting (8.3) into (8.5). □

From formula (7.12), we know that the positive recurrence of the approximate model with generator $\widetilde{\mathbf{Q}}(K)$ is equivalent to the inequality

$$\boldsymbol{\beta}\mathbf{A}_{K,K+1}\mathbf{e}_{(r+1)m} < \boldsymbol{\beta}\mathbf{A}_{K,K-1}\mathbf{e}_{(r+1)m}. \tag{8.6}$$

By noting that

$$\boldsymbol{\beta}\mathbf{A}_{K,K+1}\mathbf{e}_{(r+1)m} = \boldsymbol{\beta}_1(\mathbf{I}_r \otimes \mathbf{D}_1)\mathbf{e}_{rm},$$
$$\boldsymbol{\beta}\mathbf{A}_{K,K-1}\mathbf{e}_{(r+1)m} = K\mu\boldsymbol{\beta}_0\mathbf{e}_m,$$
$$\lambda = -\boldsymbol{\beta}_0\mathbf{D}_0\mathbf{e}_m + \boldsymbol{\beta}_1(\mathbf{I}_r \otimes \mathbf{D}_1)\mathbf{e}_{rm},$$
$$\boldsymbol{\beta}_1(\mathbf{t} \otimes \mathbf{e}_m) = -\boldsymbol{\beta}_0\mathbf{D}_0\mathbf{e}_m + K\mu\boldsymbol{\beta}_0\mathbf{e}_m,$$

we may rewrite (8.6) as $\lambda < \boldsymbol{\beta}_1(\mathbf{t} \otimes \mathbf{e}_m)$. We now post-multiply equation (8.5) by $(-\mathbf{T}^{-1})\mathbf{e}_r \otimes \mathbf{e}_m$ to get that $\boldsymbol{\beta}_1(\mathbf{t} \otimes \mathbf{e}_m) = \beta_1^{-1}\boldsymbol{\beta}_1\mathbf{e}_{rm}$. This finally gives the following expression for the positive recurrence condition of $\widetilde{\mathbf{Q}}(K)$:

$$\rho < \boldsymbol{\beta}_1\mathbf{e}_{rm}. \tag{8.7}$$

Our numerical experience shows that $\boldsymbol{\beta}_1\mathbf{e}_{rm} \to 1$, as $K \to \infty$. Hence, the positive recurrence condition (8.7) of the approximate model approaches the positive recurrence condition for the original process $\mathcal{X}$.

210   8 Selected Retrial Queues with QBD Structure

Under the condition (8.7), the partitioned stationary probability vector $\tilde{\mathbf{p}}$ of $\tilde{\mathbf{Q}}(K)$ is then given by

$$\tilde{\mathbf{p}}(j) = \tilde{\mathbf{p}}(K-1)\mathbf{R}^{j-K+1}, \quad j \geq K-1, \qquad (8.8)$$

where the rate matrix $\mathbf{R}$ satisfies the equation

$$\mathbf{R}^2 \mathbf{A}_{K,K-1} + \mathbf{R}\mathbf{A}_{KK} + \mathbf{A}_{K,K+1} = \mathbf{0}_{(r+1)m \times (r+1)m}, \qquad (8.9)$$

and the vectors $\tilde{\mathbf{p}}(j)$, for $0 \leq j \leq K-1$, are obtained by the following extension of equations (7.10) and (7.11):

$$\tilde{\mathbf{p}}(0)\mathbf{A}_{00} + \tilde{\mathbf{p}}(1)\mathbf{A}_{10} = \mathbf{0}'_{(r+1)m}, \qquad (8.10)$$

$$\tilde{\mathbf{p}}(j-1)\mathbf{A}_{j-1,j} + \tilde{\mathbf{p}}(j)\mathbf{A}_{jj} + \tilde{\mathbf{p}}(j+1)\mathbf{A}_{j+1,j} = \mathbf{0}'_{(r+1)m},$$
$$1 \leq j \leq K-2, \ (8.11)$$

$$\tilde{\mathbf{p}}(K-2)\mathbf{A}_{K-2,K-1} + \tilde{\mathbf{p}}(K-1)(\mathbf{A}_{K-1,K-1} + \mathbf{R}\mathbf{A}_{K,K-1}) = \mathbf{0}'_{(r+1)m}, \ (8.12)$$

subject to the normalizing condition

$$\sum_{j=0}^{K-2} \tilde{\mathbf{p}}(j)\mathbf{e}_{(r+1)m} + \tilde{\mathbf{p}}(K-1)(\mathbf{I}_{(r+1)m} - \mathbf{R})^{-1}\mathbf{e}_{(r+1)m} = 1.$$

Since the matrix $\mathbf{R}$ depends on the value of K, we next make explicit this dependence by writing $\mathbf{R}(K)$ in place of $\mathbf{R}$. On noting that null rows of $\mathbf{A}_{K,K+1}$ are also null rows of $\mathbf{R}(K)$, we obtain

$$\mathbf{R}(K) = \begin{pmatrix} \mathbf{0}_{m \times m} & \mathbf{0}_{m \times rm} \\ \mathbf{V}(K) & \mathbf{W}(K) \end{pmatrix}, \qquad (8.13)$$

where the matrix $\mathbf{V}(K)$ is of dimension $rm \times m$ and $\mathbf{W}(K)$ is a square matrix of order rm.

We may exploit the above form (8.13), and the sparsity of $\mathbf{A}_{K,K-1}$ and $\mathbf{A}_{K,K+1}$ to formulate a variant of the logarithmic reduction algorithm. For full details, we refer the reader to [100].

The following result states the limiting behavior of the rate matrix $\mathbf{R}(K)$, as $K \to \infty$.

Theorem 8.2. *As $K \to \infty$, $\mathbf{V}(K) \to \mathbf{0}_{rm \times m}$ and $\mathbf{W}(K) \to \mathbf{W}$, where $\mathbf{W}$ satisfies*

$$\mathbf{W}^2((\mathbf{t}\boldsymbol{\tau}) \otimes \mathbf{I}_m) + \mathbf{W}(\mathbf{T} \oplus \mathbf{D}_0) + \mathbf{I}_r \otimes \mathbf{D}_1 = \mathbf{0}_{rm \times rm}. \qquad (8.14)$$

Proof. Consider the equation (8.9) in terms of the block matrices $\mathbf{V}(K)$ and $\mathbf{W}(K)$, then

$$\mathbf{V}(K)(\mathbf{D}_0 - K\mu \mathbf{I}_m) + \mathbf{W}(K)(\mathbf{t} \otimes \mathbf{I}_m) = \mathbf{0}_{rm \times m}, \qquad (8.15)$$

$$K\mu \mathbf{W}(K)\mathbf{V}(K)(\boldsymbol{\tau} \otimes \mathbf{I}_m) + \mathbf{V}(K)(\boldsymbol{\tau} \otimes \mathbf{D}_1) + \mathbf{W}(K)(\mathbf{T} \oplus \mathbf{D}_0)$$
$$+ \mathbf{I}_r \otimes \mathbf{D}_1 = \mathbf{0}_{rm \times rm}. \qquad (8.16)$$

8.1 The $MAP/PH/1$ Retrial Queue   211

Pre-multiplying and post-multiplying (8.15) by $\mathbf{W}(K)$ and $\boldsymbol{\tau}\otimes\mathbf{I}_m$, respectively, and using (8.16), we find that

$$\mathbf{W}(K)\mathbf{V}(K)(\boldsymbol{\tau}\otimes\mathbf{D}_0) + \mathbf{W}^2(K)((\mathbf{t}\boldsymbol{\tau})\otimes\mathbf{I}_m) + \mathbf{V}(K)(\boldsymbol{\tau}\otimes\mathbf{D}_1) \\ + \mathbf{W}(K)(\mathbf{T}\oplus\mathbf{D}_0) + \mathbf{I}_r\otimes\mathbf{D}_1 = \mathbf{0}_{rm\times rm}. \quad (8.17)$$

We now observe that $\mathbf{R}(K)\mathbf{A}_{K,K-1}\mathbf{e}_{(r+1)m} = \mathbf{A}_{K,K+1}\mathbf{e}_{(r+1)m} = (\mathbf{e}_{r+1} - \mathbf{e}_{r+1}(1))\otimes(\mathbf{D}_1\mathbf{e}_m)$. This leads to $K\mu\mathbf{V}(K)\mathbf{e}_m = \mathbf{e}_r\otimes(\mathbf{D}_1\mathbf{e}_m)$, which implies that $\mathbf{V}(K)\mathbf{e}_m \to \mathbf{0}_{rm}$, as $K \to \infty$. Since $\mathbf{R}(K)$ is a non-negative matrix, we conclude that $\mathbf{V}(K) \to \mathbf{0}_{rm\times m}$. Using this fact in (8.17), we obtain the desired expression (8.14). □

In the next result we give particular expressions for $\mathbf{V}(K)$ and $\mathbf{W}(K)$ in the models $M/PH/1$ and $PH/M/1$ with retrials. We assume a representation $(\boldsymbol{\gamma},\mathbf{S})$ of order m for the phase-type arrivals. Let $\mathbf{s}$ be such that $\mathbf{Se}_m + \mathbf{s} = \mathbf{0}_m$.

Corollary 8.3. *(i) In the $M/PH/1$ retrial queue, the block matrices $\mathbf{V}(K)$ and $\mathbf{W}(K)$ are explicitly given by*

$$\mathbf{V}(K) = \frac{\lambda}{K\mu}\mathbf{e}_r,$$

$$\mathbf{W}(K) = \lambda\left(\mathbf{I}_r + \frac{\lambda}{K\mu}\mathbf{e}_r\boldsymbol{\tau}\right)(\lambda\mathbf{I}_r - \lambda\mathbf{e}_r\boldsymbol{\tau} - \mathbf{T})^{-1}.$$

(ii) In the $PH/M/1$ retrial queue, we have

$$\mathbf{V}(K) = \beta_1^{-1}\mathbf{sw}\left(K\mu\mathbf{I}_m - \mathbf{S}\right)^{-1},$$
$$\mathbf{W}(K) = \mathbf{sw},$$

where the row vector $\mathbf{w}$ is given by

$$\mathbf{w} = c\boldsymbol{\gamma}\left(\beta_1^{-1}\mathbf{I}_m - \mathbf{S} - K\mu\beta_1^{-1}\eta(K)(K\mu\mathbf{I}_m - \mathbf{S})^{-1}\right)^{-1}$$

and the constant c is

$$c = \left(1 - \beta_1^{-1}\boldsymbol{\gamma}\left(\beta_1^{-1}\mathbf{I}_m - \mathbf{S} - K\mu\beta_1^{-1}\eta(K)(K\mu\mathbf{I}_m - \mathbf{S})^{-1}\right)^{-1}\right. \\ \left. \times (K\mu\mathbf{I}_m - \mathbf{S})^{-1}\mathbf{s}\right)^{-1}.$$

Furthermore, the spectral radius $\eta(K)$ of $\mathbf{R}(K)$ is the unique solution in $(0,1)$ of the equation $\eta(K) = \mathbf{ws}$.

Proof. First, we prove statement (i). The expression for $\mathbf{V}(K)$ is obvious from the identity $K\mu\mathbf{V}(K)\mathbf{e}_m = \mathbf{e}_r\otimes(\mathbf{D}_1\mathbf{e}_m)$ by taking $m = 1$ and $\mathbf{D}_1 = \lambda$. Then, the expression for $\mathbf{W}(K)$ follows easily from (8.16).

The proof of (ii) follows the same lines of the $PH/M/1$ queue without retrials. □

In the matrix-geometric solution (8.8), the stationary probabilities $\tilde{\mathbf{p}}(j)$, with $j \geq K$, depend on the spectral radius $\eta(K)$. To minimize the effect of the approximation caused by the generalized truncation, we would like to choose K such that it satisfies the positive recurrence condition (8.7) and $\eta(K)$ is sufficiently close to the limiting value $\eta = \lim_{K \to \infty} \eta(K)$. From Theorem 8.2, we know that η is the spectral radius of the matrix $\mathbf{W}$ satisfying (8.14). In fact, equation (8.14) corresponds to the matrix-quadratic equation of the standard $MAP/PH/1$ queue. Once $\mathbf{R}(K)$ is evaluated, the spectral radius $\eta(K)$ can be computed using the methods described in Subsection 7.2.2.

To choose an appropriate truncation level, we adopt the following criterion. Starting with an initial minimum threshold, say 100, one can progressively increase the value of K, until at least one of the following conditions is fulfilled:

(i) For a given $\varepsilon_1 > 0$, $|\eta(K) - \eta| < \varepsilon_1$.
(ii) For a given $\varepsilon_2 > 0$, $|\eta(K) - \eta(K-1)| < \varepsilon_2$.

Condition (i) implies condition (ii). However, condition (ii) is considered in order to avoid cases where condition (i) alone results in a very large value of K.

In order to compute the stationary probabilities $\tilde{\mathbf{p}}(j)$, for $0 \leq j \leq K - 1$, we may employ the efficient algorithm given in Theorem 7.1 for systems with a block tridiagonal structure. In what follows, we simply summarize the alternative approach considered in [100].

For the determined threshold K, we can proceed to compute $\tilde{\mathbf{p}}(j)$, for $0 \leq j \leq K - 1$, as follows. First, we partition $\tilde{\mathbf{p}}(j)$ as

$$\tilde{\mathbf{p}}(j) = (\mathbf{q}_0(j), \mathbf{q}(j)) = (\mathbf{q}_0(j), \mathbf{q}_1(j), ..., \mathbf{q}_r(j)),$$

where the row vectors $\mathbf{q}_i(j)$, for $0 \leq i \leq r$, are of dimension m. Then, the equations (8.10)-(8.12) can be rewritten in terms of $\mathbf{q}_i(j)$ as

$$\mathbf{q}_0(j) = \sum_{i=1}^{r} \mathbf{q}_i(j) t_i \left(j \mu \mathbf{I}_m - \mathbf{D}_0\right)^{-1}, \quad 0 \leq j \leq K - 1, \tag{8.18}$$

$$\mathbf{q}(j) = (\mathbf{q}(j-1)(\mathbf{I}_r \otimes \mathbf{D}_1) + \boldsymbol{\tau} \otimes (\mathbf{q}_0(j)\mathbf{D}_1 + (j+1)\mu \mathbf{q}_0(j+1)))$$
$$\times (-\mathbf{T} \oplus \mathbf{D}_0)^{-1}, \quad 0 \leq j \leq K - 2, \tag{8.19}$$

$$\mathbf{q}(K-1) = (\mathbf{q}(K-2)(\mathbf{I}_r \otimes \mathbf{D}_1) + \boldsymbol{\tau} \otimes (\mathbf{q}_0(K-1)\mathbf{D}_1 + K\mu \mathbf{q}(K-1)\mathbf{V}(K)))$$
$$\times (-\mathbf{T} \oplus \mathbf{D}_0)^{-1}, \tag{8.20}$$

where $\mathbf{q}(-1) = \mathbf{0}'_{rm}$ and t_i denotes the ith entry of $\mathbf{t}$.

The set of equations (8.18)-(8.20) is well suited for numerical implementation using block Gauss-Seidel iteration [100]. We notice that expression (8.18) can be modified to minimize the storage requirements of the number of matrices needed. For example, where we need matrices $(j\mu \mathbf{I}_m - \mathbf{D}_0)^{-1}$, for $1 \leq j \leq K - 2$, we simply store $((K-1)\mu \mathbf{I}_m - \mathbf{D}_0)^{-1}$ and modify the corresponding equations as follows:

8.1 The $MAP/PH/1$ Retrial Queue   213

$$\mathbf{q}_0(j) = \left(\sum_{i=1}^{r}\mathbf{q}_i(j)\boldsymbol{t}_i + (K-j-1)\mu\mathbf{q}_0(j)\right)((K-1)\mu\mathbf{I}_m - \mathbf{D}_0)^{-1}.$$

In addition, we may exploit the structure of the Kronecker sum appearing in the inverses. Denote $\mathbf{T} = (t_{ij})$ and $\boldsymbol{\tau} = (\tau_i)$. Let $\theta = \max_{1 \leq i \leq r} |t_{ii}|$, then equation (8.19) in the case $j = 0$ can be expressed in terms of a matrix of order m as follows:

$$\mathbf{q}_i(0) = \left(\sum_{k=1}^{r}\mathbf{q}_k(0)t_{ki} + \theta\mathbf{q}_i(0) + \tau_i(\mathbf{q}_0(0)\mathbf{D}_1 + \mu\mathbf{q}_0(1))\right)(\theta\mathbf{I}_m - \mathbf{D}_0)^{-1},$$

$$1 \leq i \leq r.$$

The other equations in (8.19) and (8.20) can be similarly rewritten. This exploitation of the structure and the storage requirements are especially important if the values of m, r and K are large.

The purpose of the next numerical example is to study the effect of the parameters λ and μ, and the type of arrival process and service time distribution on the truncation level K. For the arrival process, we consider the following five choices of the MAP process:

(i) Erlang (E_5^a)

$$\mathbf{D}_0 = \begin{pmatrix} -5.0 & 5.0 & 0 & 0 & 0 \\ 0 & -5.0 & 5.0 & 0 & 0 \\ 0 & 0 & -5.0 & 5.0 & 0 \\ 0 & 0 & 0 & -5.0 & 5.0 \\ 0 & 0 & 0 & 0 & -5.0 \end{pmatrix}, \quad \mathbf{D}_1 = \begin{pmatrix} 0 & 0 & 0 & 0 & 0 \\ 0 & 0 & 0 & 0 & 0 \\ 0 & 0 & 0 & 0 & 0 \\ 0 & 0 & 0 & 0 & 0 \\ 5.0 & 0 & 0 & 0 & 0 \end{pmatrix}.$$

(ii) Exponential (Exp^a)

$$\mathbf{D}_0 = -1.0, \quad \mathbf{D}_1 = 1.0.$$

(iii) Hyperexponential (H_2^a)

$$\mathbf{D}_0 = \begin{pmatrix} -1.90 & 0 \\ 0 & -0.19 \end{pmatrix}, \quad \mathbf{D}_1 = \begin{pmatrix} 1.710 & 0.190 \\ 0.171 & 0.019 \end{pmatrix}.$$

(iv) MAP with negative correlation (MAP^-)

$$\mathbf{D}_0 = \begin{pmatrix} -1.00221 & 1.00221 & 0 \\ 0 & -1.00221 & 0 \\ 0 & 0 & -225.75 \end{pmatrix}, \quad \mathbf{D}_1 = \begin{pmatrix} 0 & 0 & 0 \\ 0.01002 & 0 & 0.99219 \\ 223.4925 & 0 & 2.2575 \end{pmatrix}.$$

(v) MAP with positive correlation (MAP^+)

$$\mathbf{D}_0 = \begin{pmatrix} -1.00221 & 1.00221 & 0 \\ 0 & -1.00221 & 0 \\ 0 & 0 & -225.75 \end{pmatrix}, \quad \mathbf{D}_1 = \begin{pmatrix} 0 & 0 & 0 \\ 0.99219 & 0 & 0.01002 \\ 2.2575 & 0 & 223.4925 \end{pmatrix}.$$

The five MAP processes have been normalized to fix $\lambda = 1.0$. However, they have different variance and correlation structure. The first three processes, namely E_5^a, Exp^a and H_2^a, correspond to renewal processes so their correlation is 0.0. The processes labelled MAP^- and MAP^+ have correlated arrivals with $c_c = -0.48890$ and 0.48890, respectively. The ratios of the standard deviations of these five processes with respect to E_5^a are, respectively, 1.0, 2.23606, 5.01935, 3.15178 and 3.15178.

On the other hand, for the service time distribution, we consider the following phase-type distributions:

(i) *Erlang* (E_2^s)

$$\boldsymbol{\tau} = (1.0, 0), \quad \mathbf{T} = \begin{pmatrix} -2.0 & 2.0 \\ 0 & -2.0 \end{pmatrix}.$$

(ii) *Exponential* (Exp^s)

$$\boldsymbol{\tau} = 1.0, \quad \mathbf{T} = -1.0.$$

(iii) *Hyperexponential* (H_2^s)

$$\boldsymbol{\tau} = (0.9, 0.1), \quad \mathbf{T} = \begin{pmatrix} -1.90 & 0 \\ 0 & -0.19 \end{pmatrix}.$$

The three phase distributions above have $\beta_1 = 1.0$. We note that the ratios of the standard deviations of these distributions with respect to E_2^s are 1.0, 1.41421 and 3.17451, respectively.

Table 8.1 contains the values of K, which was varied in multiples of 10. An examination of the entries in the table reveals the following observations. As is to be expected, K increases as ρ increases and it appears to decrease as μ increases. K appears to be small for MAP^+ arrivals as compared to the rest of arrival processes. An exception to this observation is the case of Erlang service times with $\rho = 0.25$. It is interesting to note that K is significantly large for MAP^- compared to that of MAP^+, especially when ρ is large. This indicates the important role played by the correlation. In the case of renewal arrivals, there seems to be a pattern depending on the traffic intensity ρ. If $\rho = 0.25$, then we may notice that K increases with increasing variance of the inter-arrival times. However, we observe the opposite behavior for $\rho = 0.95$.

Before concluding this subsection, we point out some results that generalize the corresponding ones in the $M/G/1$ retrial queue. In particular, we have

$$\sum_{j=0}^{\infty} \mathbf{q}(j) \mathbf{e}_{rm} = \rho, \tag{8.21}$$

$$\mu \sum_{j=1}^{\infty} j \mathbf{q}_0(j) \mathbf{e}_m = \sum_{j=0}^{\infty} \mathbf{q}(j) \left(\mathbf{e}_r \otimes (\mathbf{D}_1 \mathbf{e}_m) \right). \tag{8.22}$$

Both expressions are valid for the original $MAP/PH/1$ retrial queue. Note that (8.21) states that the probability that the server is busy equals ρ, whereas formula (8.22) expresses flow conservation across the orbit levels.

8.1 The $MAP/PH/1$ Retrial Queue

Table 8.1. Values of K

	ρ	μ	E_5^a	Exp^a	H_2^a	MAP^-	MAP^+
E_2^s	0.25	0.1	150	950	1460	440	300
		0.5	110	430	660	170	250
		0.75	110	350	540	130	230
		5.0	100	140	220	100	110
		10.0	100	110	160	100	110
		1000.0	100	100	100	100	100
Exp^s	0.25	0.1	320	800	1240	810	140
		0.5	150	360	560	360	110
		0.75	120	300	460	300	110
		5.0	110	120	180	120	100
		10.0	100	110	130	110	100
		1000.0	100	100	100	100	100
H_2^s	0.25	0.1	420	460	580	460	410
		0.5	190	210	270	210	140
		0.75	160	170	220	170	130
		5.0	110	110	110	110	100
		10.0	110	110	110	110	100
		1000.0	100	100	100	100	100
E_2^s	0.95	0.1	3640	3470	1900	2540	330
		0.5	1630	1560	860	1140	160
		0.75	1330	1270	710	930	130
		5.0	520	500	280	370	110
		10.0	370	360	200	260	110
		1000.0	110	110	100	100	100
Exp^s	0.95	0.1	3790	3010	1810	3010	430
		0.5	1700	1350	820	1350	200
		0.75	1390	1110	670	1110	160
		5.0	540	430	270	430	110
		10.0	390	310	190	310	110
		1000.0	110	100	100	100	100
H_2^s	0.95	0.1	1730	1720	1370	1720	470
		0.5	780	770	620	770	440
		0.75	640	630	510	630	350
		5.0	250	250	200	250	120
		10.0	180	180	150	180	110
		1000.0	100	100	100	100	100

8.1.2 Maximal Queue Length in a Busy Period

In this subsection, we concentrate on the computation of the distribution of the maximum number of customers in orbit during a busy period. The arriving customer starting a busy period finds a state of the sub-level $l(0,0) = \{(0,0,k); 1 \leq k \leq m\}$, so we will condition on the distribution of the system state at the beginning of a busy period. This need of considering a stationary version of the maximal queue length is a distinguish feature with respect to the non-stationary analysis performed for other models with Poisson arrivals in Subsections 4.1.3 and 4.2.3.

First, we note that $\{N_{max} < j\}$ corresponds to the event that, starting from level $l(0)$, the process $\mathcal{X}$ hits the sub-level $l(0,0)$ before hitting level $l(j)$, for any $j \geq 1$. Thus, the computation of $P(N_{max} < j)$, for $j \geq 1$, reduces to find an absorbing probability in an auxiliary finite Markov chain with two absorbing states, say 0^* and j^*, whose infinitesimal generator is given by

$$\mathbf{Q}(j) = \begin{pmatrix} 0 & & & & & \\ \mathbf{t} \otimes \mathbf{e}_m & \mathbf{T} \oplus \mathbf{D}_0 & \mathbf{A}_{01}^* & & & \\ & \mathbf{A}_{10}^* & \mathbf{A}_{11} & \mathbf{A}_{12} & & \\ & & \mathbf{A}_{21} & \mathbf{A}_{22} & \mathbf{A}_{23} & \\ & & & \ddots & \ddots & \ddots \\ & & & & \mathbf{A}_{j-1,j-2} & \mathbf{A}_{j-1,j-1} & \mathbf{a}_{j-1,j}^* \\ & & & & & & 0 \end{pmatrix}, \quad (8.23)$$

where the matrices appearing in (8.23) are as seen in (8.1), except $\mathbf{A}_{01}^*$ which is obtained from $\mathbf{A}_{01}$ by deleting the first m rows and $\mathbf{A}_{10}^*$ which is obtained by deleting the first m columns in $\mathbf{A}_{10}$. The column vector $\mathbf{a}_{j-1,j}^*$ is defined as $\mathbf{a}_{j-1,j}^* = (\mathbf{e}_{r+1} - \mathbf{e}_{r+1}(1)) \otimes (\mathbf{D}_1 \mathbf{e}_m)$.

Let $\mathbf{K}(j)$ denote the part of the generator $\mathbf{Q}(j)$ that corresponds to transient states of the auxiliary absorbing Markov chain, and let $\mathbf{k}_0(j)$ be the column vector of dimension $rm + (r+1)(j-1)m$ such that the only non-zero block is the first one, which is given by $\mathbf{t} \otimes \mathbf{e}_m$. Denote also the stationary distribution of the system state, given that a busy period starts, by $\boldsymbol{\delta}(j)$. Then, it is easy to verify that $\boldsymbol{\delta}(j) = d(\boldsymbol{\tau} \otimes (\mathbf{q}_0(0)\mathbf{D}_1), \mathbf{0}'_{(r+1)(j-1)m})$, where d is the normalizing constant such that $\boldsymbol{\delta}(j)\mathbf{e}_{rm+(r+1)(j-1)m} = 1$. The probability $P(N_{max} < j)$ can be expressed as

$$P(N_{max} < j) = -\boldsymbol{\delta}(j)\mathbf{K}^{-1}(j)\mathbf{k}_0(j), \quad j \geq 1. \quad (8.24)$$

It should be pointed out that, in the case of renewal arrivals, there is no need to solve the stationary equations in (8.2). To show this, we consider only PH inter-arrivals. The result can be extended to any arbitrary distribution by a continuity argument, since the class of PH distributions is dense in the class of all non-negative distributions; see also Subsection 7.1.4. In the case of PH inter-arrival times, we notice that $\boldsymbol{\delta}_0(j) = d\boldsymbol{\tau} \otimes (\mathbf{q}_0(0)\mathbf{s}\boldsymbol{\gamma})$, where $\boldsymbol{\delta}_0(j)$ denotes the first non-zero entry of dimension rm of the vector $\boldsymbol{\delta}(j)$. It is easy to note that $d = (\mathbf{q}_0(0)\mathbf{s})^{-1}$ and thus we have $\boldsymbol{\delta}_0(j) = \boldsymbol{\tau} \otimes \boldsymbol{\gamma}$, which is independent of the stationary distribution. This results in a notable simplification when dealing with renewal arrivals.

We next propose a method for computing the desired probability given in (8.24). To this end, we define

$$\mathbf{b}^{(j)} = -\boldsymbol{\delta}(j)\mathbf{K}^{-1}(j), \quad j \geq 1. \quad (8.25)$$

In accordance with (8.23), each entry of $\mathbf{b}^{(j)}$ is equivalent to the amount of time that the auxiliary Markov chain spends in the corresponding transient state before absorption into 0^*, given that the chain starts from one of the transient states. We may partition $\mathbf{b}^{(j)}$ as

$$\mathbf{b}^{(j)} = (\mathbf{b}^{(j)}(0), \mathbf{b}_0^{(j)}(1), \mathbf{b}^{(j)}(1), ..., \mathbf{b}_0^{(j)}(j-1), \mathbf{b}^{(j)}(j-1)),$$

where the vectors $\mathbf{b}_0^{(j)}(n)$, for $1 \leq n \leq j-1$, are of dimension m and the vectors $\mathbf{b}^{(j)}(n)$, for $0 \leq n \leq j-1$, are of dimension rm.

Once the set of equations in (8.25) is solved, we obtain

$$P(N_{max} < j) = \mathbf{b}^{(j)}(0)(\mathbf{t} \otimes \mathbf{e}_m), \quad j \geq 1.$$

One can use a number of numerical methods to solve (8.25) including block Gaussian elimination and a variant of the method described in Subsection 8.1.1 for the stationary probabilities. In what follows, we focus on the latter and store the matrix $((j-1)\mu\mathbf{I}_m - \mathbf{D}_0)^{-1}$ instead of $(n\mu\mathbf{I}_m - \mathbf{D}_0)^{-1}$, for $1 \leq n \leq j-1$. After distinguishing the cases $j = 1$ and $j \geq 2$, the system (8.25) can be rewritten as follows:

Case 1. If $j = 1$, then $P(N_{max} < 1)$ is explicitly given by

$$P(N_{max} < 1) = \delta_0(1)(-\mathbf{T} \oplus \mathbf{D}_0)^{-1}(\mathbf{t} \otimes \mathbf{e}_m).$$

Case 2. If $j \geq 2$, the necessary equations are given by

$$\mathbf{b}^{(j)}(0) = \left(\delta_0(j) + \mu\mathbf{b}_0^{(j)}(1)(\boldsymbol{\tau} \otimes \mathbf{I}_m)\right)(-\mathbf{T} \oplus \mathbf{D}_0)^{-1}$$

and, for $1 \leq n \leq j - 1$,

$$\mathbf{b}_0^{(j)}(n) = \left((j-n-1)\mu\mathbf{b}_0^{(j)}(n) + \mathbf{b}^{(j)}(n)(\mathbf{t} \otimes \mathbf{I}_m)\right)((j-1)\mu\mathbf{I}_m - \mathbf{D}_0)^{-1},$$

$$\mathbf{b}^{(j)}(n) = \Big(\mathbf{b}^{(j)}(n-1)(\mathbf{I}_r \otimes \mathbf{D}_1)$$
$$+ \boldsymbol{\tau} \otimes \left(\mathbf{b}_0^{(j)}(n)\mathbf{D}_1 + (n+1)\mu\mathbf{b}_0^{(j)}(n+1)\right)\Big)(-\mathbf{T} \oplus \mathbf{D}_0)^{-1},$$

where $\mathbf{b}_0^{(j)}(j) = \mathbf{0}'_m$.

We next present some numerical examples showing the performance of N_{max}. In a first example, we fix $\lambda = 0.5$, $\beta_1 = 1.0$ and $\mu = 0.5$. Then, we calculate the probabilities $P(N_{max} = j)$, which are displayed through 99th percentile, for the five arrival processes and the H_2^s distribution given in Subsection 8.1.1. The results for the other two service time distributions exhibit similar characteristics and hence they are omitted. An examination of the probabilities in Table 8.2 reveal that, in the case of renewal arrivals, the probability $P(N_{max} = 0)$ decreases as far as the variability of the arrival process increases. Also the tail of the distribution appears to be longer with increasing arrival variability. In the case of correlated arrivals, we first notice that

Table 8.2. $P(N_{max} = j)$ for H_2^s

j	E_5^a	Exp^a	H_2^a	MAP^-	MAP^+
0	0.87737	0.74003	0.64738	0.03686	0.89916
1	0.05544	0.11416	0.11133	0.51265	0.05565
2	0.01859	0.04814	0.05681	0.16782	0.01552
3	0.01102	0.02564	0.03634	0.08769	0.00766
4	0.00803	0.01596	0.02550	0.04930	0.00427
⋮	⋮	⋮	⋮	⋮	⋮
9	0.00241	0.00440	0.00848	0.01101	0.00034
⋮	—	⋮	⋮	⋮	⋮
13	—	0.00199	0.00483	0.00499	0.00013
⋮	—	—	⋮	⋮	⋮
17	—	—	0.00286	0.00226	0.00010
⋮	—	—	⋮	—	⋮
23	—	—	0.00130	—	0.00009
⋮	—	—	—	—	⋮
27	—	—	—	—	0.00009

Table 8.3. The 99th percentile and the modes of N_{max}

	E_5^a	Exp^a	H_2^a	MAP^-	MAP^+
E_2^s	3,0	7,0	16,0	9,1	21,0
Exp^s	4,0	8,0	17,0	10,1	24,0
H_2^s	9,0	13,0	23,0	17,1	27,0

MAP^+ arrivals have a much larger mass at the origin. In fact, $P(N_{max} = 0)$ for MAP^- is very small even when it is compared to the three renewal arrival processes. Secondly, we observe that MAP^+ presents the heaviest tail among the five arrival processes considered in the example. A plausible explanation might be as follows. The MAP^+ arrival combines large runs of short inter-arrival times and large runs of longer inter-arrival times. The former case results in longer tails for N_{max}, while the latter case explains the large mass at the point $j = 0$.

The next example complements the analysis of N_{max} by illustrating how the variability in the arrival process, as well as the service time distributions, affects the 99th percentile and the mode. We consider $\lambda = 1.0$, $\beta_1 = 0.5$ and $\mu = 1.0$. The first item of each entry in Table 8.3 corresponds to the 99th percentile and the second one refers to the point where the mode occurs. We see that a larger variability, either in the renewal arrival process or in the service time distribution, yields a higher value for the 99th percentile.

The largest value for this percentile always corresponds to positive correlated arrivals. Finally, we observe that the negative correlated arrivals present the mode at $j = 1$, while the rest of arrival processes have the mode at the origin.

8.2 The $MAP/M/c$ Retrial Queue

The purpose of this section is to generalize the model $M/M/c$ with retrials investigated in Part II to Markovian arrivals. We present an exhaustive analysis of the busy period and the waiting time. In general, the study of the queueing descriptors is intractable when dealing with an infinite orbit. Thus, we approximate the model by assuming a direct truncation approach.

8.2.1 Stationary Distribution of the System State

The investigation of the stationary queue length characteristics is the subject matter of most works in the literature [322]. In particular, the computation of the stationary distribution of the system state for the $MAP/M/c$ retrial queue can be obtained as a particular case of the model in [176], where the servers, with a certain probability, search for customers.

Let $C(t)$, $N(t)$ and $J(t)$ be, respectively, the number of busy servers, the number of customers in orbit and the phase of the MAP at time t. The orbit has a finite capacity K. The resulting process $\mathcal{X} = \{(C(t), N(t), J(t)); t \geq 0\}$ is a continuous-time Markov chain with state space $\mathcal{S} = \{0, ..., c\} \times \{0, ..., K\} \times \{1, ..., m\}$. As usual, we partition $\mathcal{S}$ according to the orbit levels $l(j) = \{(i, j, k); 0 \leq i \leq c, 1 \leq k \leq m\}$, for $0 \leq j \leq K$. We also partition $l(0)$ into the sub-levels $l(0,0) = \{(0,0,k); 1 \leq k \leq m\}$ and $l(0) - l(0,0)$.

The infinitesimal generator $\overline{\mathbf{Q}}$ has the form

$$\overline{\mathbf{Q}} = \begin{pmatrix} \mathbf{D}_0 & \mathbf{e}'_c(1) \otimes \mathbf{D}_1 & & & & & \\ \nu \mathbf{e}_c(1) \otimes \mathbf{I}_m & \mathbf{A}^*_{00} & \mathbf{A}^*_{01} & & & & \\ & \mathbf{A}^*_{10} & \mathbf{A}_{11} & \mathbf{A}_{12} & & & \\ & & \mathbf{A}_{21} & \mathbf{A}_{22} & \mathbf{A}_{23} & & \\ & & & \ddots & \ddots & \ddots & \\ & & & & \mathbf{A}_{K-1,K-2} & \mathbf{A}_{K-1,K-1} & \mathbf{A}_{K-1,K} \\ & & & & & \mathbf{A}_{K,K-1} & \overline{\mathbf{A}}_{KK} \end{pmatrix},$$
(8.26)

where the coefficient matrices appearing in (8.26) are given by

$$\mathbf{A}^*_{00} = \begin{pmatrix} \mathbf{D}_0 - \nu \mathbf{I}_m & \mathbf{D}_1 & & & \\ 2\nu \mathbf{I}_m & \mathbf{D}_0 - 2\nu \mathbf{I}_m & \mathbf{D}_1 & & \\ & \ddots & \ddots & \ddots & \\ & & (c-1)\nu \mathbf{I}_m & \mathbf{D}_0 - (c-1)\nu \mathbf{I}_m & \mathbf{D}_1 \\ & & & c\nu \mathbf{I}_m & \mathbf{D}_0 - c\nu \mathbf{I}_m \end{pmatrix},$$

$$\mathbf{A}_{jj} = \begin{pmatrix} \mathbf{D}_0 - j\mu\mathbf{I}_m & \mathbf{D}_1 & & & \\ \nu\mathbf{I}_m & \mathbf{D}_0 - (\nu + j\mu)\mathbf{I}_m & \mathbf{D}_1 & & \\ & \ddots & \ddots & \ddots & \\ & & (c-1)\nu\mathbf{I}_m & \mathbf{D}_0 - ((c-1)\nu + j\mu)\mathbf{I}_m & \mathbf{D}_1 \\ & & & c\nu\mathbf{I}_m & \mathbf{D}_0 - c\nu\mathbf{I}_m \end{pmatrix},$$

$$1 \leq j \leq K,$$

$$\mathbf{A}_{j,j+1} = \begin{pmatrix} \mathbf{0}_{cm \times cm} & \mathbf{0}_{cm \times m} \\ \mathbf{0}_{m \times cm} & \mathbf{D}_1 \end{pmatrix}, \quad 0 \leq j \leq K-1,$$

$$\mathbf{A}_{j,j-1} = j\mu \begin{pmatrix} \mathbf{0}_{m \times m} & \mathbf{I}_m & & & \\ & \mathbf{0}_{m \times m} & \mathbf{I}_m & & \\ & & \ddots & \ddots & \\ & & & \mathbf{0}_{m \times m} & \mathbf{I}_m \\ & & & & \mathbf{0}_{m \times m} \end{pmatrix}, \quad 1 \leq j \leq K,$$

the block matrix $\overline{\mathbf{A}}_{KK}$ is defined as $\mathbf{A}_{KK}$, except for the $(c+1, c+1)$th block entry which is replaced by $\mathbf{D} - c\nu\mathbf{I}_m$, $\mathbf{A}_{01}^*$ is obtained from $\mathbf{A}_{01}$ by removing the first m rows and $\mathbf{A}_{10}^*$ is derived from $\mathbf{A}_{10}$ by deleting the first m columns.

Let $\overline{\mathbf{p}}$, partitioned as $\overline{\mathbf{p}} = (\overline{\mathbf{p}}(0), ..., \overline{\mathbf{p}}(K))$, denote the stationary probability vector of the generator $\overline{\mathbf{Q}}$. That is, $\overline{\mathbf{p}}$ satisfies

$$\overline{\mathbf{p}}\overline{\mathbf{Q}} = \mathbf{0}'_{(c+1)(K+1)m}, \qquad \overline{\mathbf{p}}\mathbf{e}_{(c+1)(K+1)m} = 1.$$

At this point, we notice that the computation of the vector $\overline{\mathbf{p}}$ is reduced to solving a finite block tridiagonal system. This fact will be a common feature along this section when we analyze the busy period and the waiting time characteristics. Once more, we refer the reader to Theorem 7.1 and the techniques mentioned in Section 7.4.

8.2.2 Busy Period

The busy period of the $MAP/M/c$ queue with finite orbit starts when a primary customer sees any state of the sub-level $l(0,0)$ and ends at the first service completion thereafter that the process visits $l(0,0)$ again. Our analysis of the busy period in this subsection includes the calculation of the Laplace-Stieltjes transforms and the generating functions governing the length L of the busy period and the number I of customers served, respectively. We then develop recursive equations for the moments of both characteristics.

Let us denote L_{ijk} as the first-passage time to the sub-level $l(0,0)$, given that the initial state is $(i,j,k) \in \mathcal{S}$. Its Laplace-Stieltjes transform is denoted by $L_{ijk}^*(s)$. The following column vectors comprise the Laplace-Stieltjes transforms in partitioned form:

$$\mathbf{l}_0^*(s) = (L_{101}^*(s), ..., L_{c0m}^*(s))',$$
$$\mathbf{l}_j^*(s) = (L_{0j1}^*(s), ..., L_{cjm}^*(s))', \quad 1 \leq j \leq K,$$
$$\mathbf{l}^*(s) = (\mathbf{l}_0^*(s), ..., \mathbf{l}_K^*(s))'.$$

Moreover, for states in $l(0,0)$, we have $\mathbf{l}_{00}^*(s) = (L_{001}^*(s), ..., L_{00m}^*(s))' = \mathbf{e}_m$.

Theorem 8.4. *The Laplace-Stieltjes transforms $L_{ijk}^*(s)$, for $(i,j,k) \in \mathcal{S} - l(0,0)$, satisfy the following tridiagonal system:*

$$\mathbf{Q}_L(s)\mathbf{l}^*(s) = \mathbf{f}, \tag{8.27}$$

where $\mathbf{Q}_L(s) = \overline{\mathbf{Q}}^* - s\mathbf{I}_{cm+(c+1)Km}$ and

$$\mathbf{f} = -\nu \begin{pmatrix} \mathbf{e}_m \\ \mathbf{0}_{(c-1)m+(c+1)Km} \end{pmatrix}.$$

The matrix $\overline{\mathbf{Q}}^*$ is obtained from $\overline{\mathbf{Q}}$ by deleting the first block column and the first block row.

Proof. We employ first-step analysis to find the following scalar equations:

$$L_{ijk}^*(s) = \sum_{k' \neq k} \frac{d_{kk'}^0}{s + \lambda_k + i\nu + j\mu} L_{ijk'}^*(s) + \sum_{k'=1}^{m} \frac{d_{kk'}^1}{s + \lambda_k + i\nu + j\mu} L_{i+1,j,k'}^*(s)$$

$$+ \frac{i\nu}{s + \lambda_k + i\nu + j\mu} L_{i-1,j,k}^*(s) + \frac{j\mu}{s + \lambda_k + i\nu + j\mu} L_{i+1,j-1,k}^*(s),$$

$$0 \leq i \leq c-1, \ 0 \leq j \leq K, \ 1 \leq k \leq m, \ (i,j) \neq (0,0), \tag{8.28}$$

$$L_{cjk}^*(s) = \sum_{k' \neq k} \frac{d_{kk'}^0}{s + \lambda_k + c\nu} L_{cjk'}^*(s) + \sum_{k'=1}^{m} \frac{d_{kk'}^1}{s + \lambda_k + c\nu} L_{c,j+1,k'}^*(s)$$

$$+ \frac{c\nu}{s + \lambda_k + c\nu} L_{c-1,j,k}^*(s), \quad 0 \leq j \leq K-1, \ 1 \leq k \leq m, \tag{8.29}$$

$$L_{cKk}^*(s) = \sum_{k' \neq k} \frac{d_{kk'}^0 + d_{kk'}^1}{s + \lambda_k' + c\nu} L_{cKk'}^*(s) + \frac{c\nu}{s + \lambda_k' + c\nu} L_{c-1,K,k}^*(s), \quad 1 \leq k \leq m, \tag{8.30}$$

where $\mathbf{D}_0 = (d_{ij}^0)$, $\mathbf{D}_1 = (d_{ij}^1)$, and $\lambda_k = -d_{kk}^0$ and $\lambda_k' = -(d_{kk}^0 + d_{kk}^1)$, for $1 \leq k \leq m$.

By expressing equations (8.28)-(8.30) in matrix form, we automatically obtain the expression (8.27). $\square$

Taking into account that a busy period starts by visiting a state of the sub-level $l(1,0) = \{(1,0,k); 1 \leq k \leq m\}$, we next define the unconditional version with Laplace-Stieltjes transform

$$L^*(s) = \boldsymbol{\pi}_L \mathbf{l}^*(s),$$

where $\boldsymbol{\pi}_L$ is the row vector of dimension $cm + (c+1)Km$ given by

$$\boldsymbol{\pi}_L = \frac{1}{\overline{\mathbf{p}}(0,0)\mathbf{D}_1\mathbf{e}_m} \left(\overline{\mathbf{p}}(0,0)\mathbf{D}_1, \mathbf{0}_{(c-1)m+(c+1)Km}' \right)$$

and $\overline{\mathbf{p}}(0,0)$ is the sub-vector consisting of the m first components of $\overline{\mathbf{p}}(0)$.

We now focus on the moments of L_{ijk}, which are denoted by $L_{ijk}^{(n)}$; that is, we have $L_{ijk}^{(n)} = E[L_{ijk}^n]$, for $n \geq 0$. Let us introduce some notation

$$\mathbf{l}_0^{(n)} = \left(L_{101}^{(n)}, ..., L_{c0m}^{(n)}\right)',$$

$$\mathbf{l}_j^{(n)} = \left(L_{0j1}^{(n)}, ..., L_{cjm}^{(n)}\right)', \quad 1 \leq j \leq K,$$

$$\mathbf{l}^{(n)} = \left(\mathbf{l}_0^{(n)}, ..., \mathbf{l}_K^{(n)}\right)'.$$

Corollary 8.5. *The moments $L_{ijk}^{(n)}$, for $(i,j,k) \in \mathcal{S} - l(0,0)$, verify the following block tridiagonal system:*

$$\mathbf{l}^{(0)} = \mathbf{e}_{cm+(c+1)Km},$$

$$\overline{\mathbf{Q}}^* \mathbf{l}^{(n)} = -n \mathbf{l}^{(n-1)}, \quad n \geq 1.$$

Proof. By differentiating (8.27), we get

$$\mathbf{Q}_L(s) \frac{d^n}{ds^n} \mathbf{l}^*(s) - \frac{d^{n-1}}{ds^{n-1}} \mathbf{l}^*(s) = \mathbf{0}_{cm+(c+1)Km}.$$

Hence, noting that $\mathbf{l}^{(n)} = (-1)^n \frac{d^n}{ds^n} \mathbf{l}^*(s)\big|_{s=0}$ and $\mathbf{Q}_L(0) = \overline{\mathbf{Q}}^*$, we obtain the desired result. □

We also note that the unconditional moments are given by

$$E[L^n] = \pi_L \mathbf{l}^{(n)}, \quad n \geq 0.$$

Now we present numerical results involving the first two moments and the numerical inversion of the unconditional transform $L^*(s)$. First, we fix $c = 5$ and $\lambda = 1.0$, and denote the traffic intensity by $\rho = \lambda/c\nu$. We choose the value of K by increasing it until the first four decimal digits of two successive values of $E[L]$ match. In Table 8.4, we list the resulting values, for several choices of ρ and μ. Each entry in the table gives the truncation levels corresponding to the arrival processes E_5^a, Exp^a and H_2^a defined in Subsection 8.1.1. The last component corresponds to a $MMPP$ with representation

$$\mathbf{D}_0 = \frac{55}{86} \begin{pmatrix} -1.3 & 0.5 & 0.3 \\ 1.0 & -2.5 & 0.5 \\ 2.4 & 0 & -10.4 \end{pmatrix}, \quad \mathbf{D}_1 = \frac{55}{86} \begin{pmatrix} 0.5 & 0 & 0 \\ 0 & 1.0 & 0 \\ 0 & 0 & 8.0 \end{pmatrix}.$$

The $MMPP$ has correlated arrivals and the ratio of its standard deviation with respect to E_5^a is 2.18147.

In Table 8.5, we present the mean value $E[L]$ and the standard deviation $\sigma(L)$ for the three renewal arrival processes. As is to be expected, both measures are increasing functions of ρ and decreasing functions of the retrial rate μ.

8.2 The $MAP/M/c$ Retrial Queue

Table 8.4. Truncation thresholds

ρ		$\mu = 0.05$	$\mu = 0.5$	$\mu = 1.0$	$\mu = 2.5$	$\mu = 5.0$
0.25	E_5^a	3	1	2	2	1
	Exp^a	7	4	5	4	5
	H_2^a	14	9	10	9	9
	$MMPP$	33	18	19	21	19
0.5	E_5^a	11	8	8	9	7
	Exp^a	25	15	15	15	14
	H_2^a	59	40	37	34	32
	$MMPP$	75	49	49	46	48
0.75	E_5^a	54	30	31	28	28
	Exp^a	99	49	50	47	44
	H_2^a	207	138	121	140	125
	$MMPP$	209	145	130	121	129

Table 8.5. Moments of L

ρ			$\mu = 0.05$	$\mu = 0.5$	$\mu = 1.0$	$\mu = 2.5$	$\mu = 5.0$
0.25	E_5^a	$E[L]$	2.28930	2.27990	2.27950	2.27930	2.27924
		$\sigma(L)$	3.15885	3.06292	3.06133	3.06061	3.06040
	Exp^a	$E[L]$	2.94830	2.52350	2.50500	2.46622	2.49410
		$\sigma(L)$	6.24091	3.15530	3.09491	3.07062	3.06522
	H_2^a	$E[L]$	7.33213	3.50993	3.37880	3.31692	3.30102
		$\sigma(L)$	20.21505	4.43020	4.08798	3.94726	3.91341
0.5	E_5^a	$E[L]$	20.84490	14.02606	13.76183	13.62571	13.58560
		$\sigma(L)$	40.33858	20.31492	19.75304	19.47443	19.39336
	Exp^a	$E[L]$	41.75191	12.71971	12.01540	11.67232	11.57512
		$\sigma(L)$	95.75959	17.77823	16.31635	15.63480	15.44442
	H_2^a	$E[L]$	155.75531	15.69120	14.07320	13.32370	13.10752
		$\sigma(L)$	313.02446	20.54201	17.60480	16.27244	15.88795
0.75	E_5^a	$E[L]$	2334.59355	115.37530	98.44500	89.81731	87.15331
		$\sigma(L)$	3817.40231	165.90911	138.96143	125.35888	121.17625
	Exp^a	$E[L]$	3794.35110	76.19950	62.66100	56.24280	54.35280
		$\sigma(L)$	6417.54554	113.16275	92.27115	79.48115	76.30562
	H_2^a	$E[L]$	2588.76030	57.98240	49.04640	44.83870	43.56590
		$\sigma(L)$	4195.64390	89.40936	74.05324	66.67907	64.41793

Using numerical inversion algorithms, we obtain the unconditional density $f_L(x)$ associated with $L^*(s)$. In Figures 8.1 and 8.2, we illustrate the influence of ρ and μ for the case of a $MMPP$. Firstly, in Figure 8.1, we fix $\mu = 1.0$ and display three curves corresponding to $\rho = 0.25$, 0.5 and 0.75. We notice that the value at the origin equals $\nu = (5\rho)^{-1}$, in agreement with the Tauberian result $f_L(0) = \lim_{s \to \infty} sL^*(s)$. Since

$$\lim_{s\to\infty} sL^*_{ijk}(s) = \begin{cases} \nu, & \text{if } (i,j,k) = (1,0,k), \\ 0, & \text{otherwise,} \end{cases}$$

it easily follows that $f_L(0) = \nu$. The three curves exhibit decreasing shapes with heavier tails for higher values of ρ.

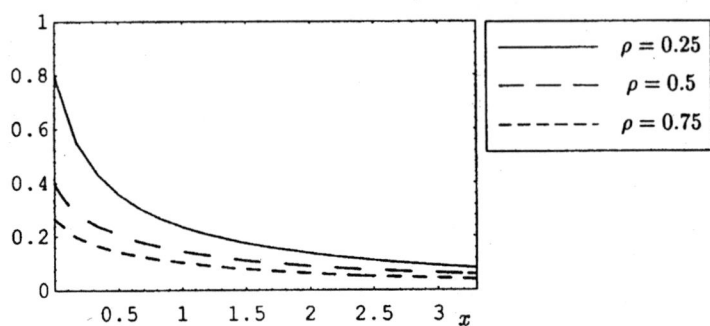

Fig. 8.1. Density functions of L as ρ varies

In Figure 8.2, we plot the density $f_L(x)$ for $\rho = 0.5$ and $\mu = 0.05$, 1.0 and 5.0. The three curves in the figure seem to be undistinguished. However, the tail of the distribution becomes heavier as far as μ decreases. The expectations $E[L] = 124.42693$, for the case $\mu = 0.05$, and $E[L] = 9.95433$, when $\mu = 5.0$, corroborate this fact.

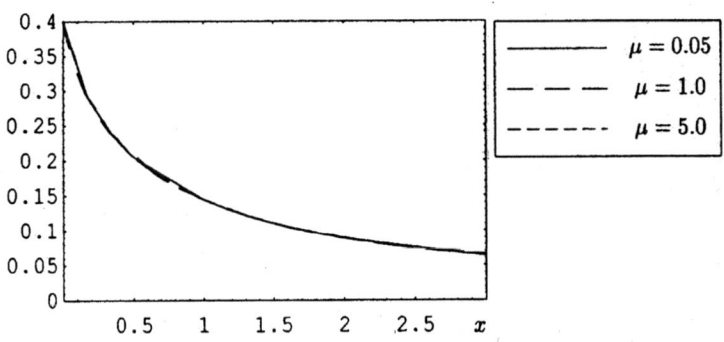

Fig. 8.2. Density functions of L as μ varies

To finish this subsection, we consider the number of customers served during a busy period. The methodology is similar to that employed for the length of the busy period, hence in the sequel we omit repetitive details.

We next summarize the necessary definitions and notations. For the number of customers served during L_{ijk}, denoted by I_{ijk}, we have

$$I_{ijk}(z) = E\left[z^{I_{ijk}}\right], \quad |z| \le 1, \ (i,j,k) \in \mathcal{S},$$
$$\mathbf{i}_0(z) = (I_{101}(z), ..., I_{c0m}(z))',$$
$$\mathbf{i}_j(z) = (I_{0j1}(z), ..., I_{cjm}(z))', \quad 1 \le j \le K,$$
$$\mathbf{i}(z) = (\mathbf{i}_0(z), ..., \mathbf{i}_K(z))',$$
$$I_{ijk}^{(n)} = E\left[I_{ijk}(I_{ijk}-1)...(I_{ijk}-n+1)\right], \quad n \ge 1,$$
$$\mathbf{i}_0^{(n)} = \left(I_{101}^{(n)}, ..., I_{c0m}^{(n)}\right)', \quad n \ge 1,$$
$$\mathbf{i}_j^{(n)} = \left(I_{0j1}^{(n)}, ..., I_{cjm}^{(n)}\right)', \quad 1 \le j \le K, \ n \ge 1,$$
$$\mathbf{i}^{(n)} = \left(\mathbf{i}_0^{(n)}, ..., \mathbf{i}_K^{(n)}\right)', \quad n \ge 1.$$

An appropriate first-step analysis leads to the following results for the generating functions and the factorial moments.

Theorem 8.6. *The generating functions $I_{ijk}(z)$, for $(i,j,k) \in \mathcal{S} - l(0,0)$, verify the block tridiagonal system*

$$\mathbf{Q}_I(z)\mathbf{i}(z) = \mathbf{f}(z), \tag{8.31}$$

where $\mathbf{Q}_I(z) = \overline{\mathbf{Q}}^* - \mathbf{A}(z)$, $\mathbf{f}(z) = z\mathbf{f}$ and

$$\mathbf{A}(z) = (1-z)\nu \begin{pmatrix} \mathbf{W}_0 & \mathbf{0}_{c \times (c+1)K} \\ \mathbf{0}_{(c+1)K \times c} & \mathbf{I}_K \otimes \mathbf{W} \end{pmatrix} \otimes \mathbf{I}_m.$$

Here, the matrix $\mathbf{W}_0 = (w_{ij}^0)$ is square of order c with entries

$$w_{ij}^0 = \begin{cases} i, & \text{if } 2 \le i \le c, \ j = i-1, \\ 0, & \text{otherwise}, \end{cases}$$

and the matrix $\mathbf{W}$ was defined in Theorem 6.13.

Corollary 8.7. *The factorial moments $I_{ijk}^{(n)}$, for $(i,j,k) \in \mathcal{S} - l(0,0)$, can be obtained as the solution of the block tridiagonal system*

$$\mathbf{i}^{(0)} = \mathbf{e}_{cm+(c+1)Km},$$
$$\overline{\mathbf{Q}}^*\mathbf{i}^{(n)} = \delta_{1n}\mathbf{f} - n\mathbf{A}(0)\mathbf{i}^{(n-1)}, \quad n \ge 1. \tag{8.32}$$

Needless to say, the unconditional version leads to the factorial moments

$$E[I(I-1)...(I-n+1)] = \pi_L \mathbf{i}^{(n)}, \quad n \ge 1.$$

Because of the spatial heterogeneity caused by the retrials, it seems impossible to solve the system (8.31) for the generating functions $I_{ijk}(z)$, or the system (8.32) for the moments $I_{ijk}^{(n)}$, in the case $K = \infty$. In contrast, we next present a recursive scheme for computing the probability mass function of the number of customers served, which is valid for the model with infinite orbit capacity.

For $(i, j, k) \in S - l(0, 0)$ and $n \geq 1$, we define $x_{ijk}^n = P(I_{ijk} = n)$. The partition accordingly the orbit levels yields

$$\mathbf{x}_0^n = (x_{101}^n, ..., x_{c0m}^n)',$$
$$\mathbf{x}_j^n = (x_{0j1}^n, ..., x_{cjm}^n)',$$
$$\mathbf{x}^n = (\mathbf{x}_0^n, \mathbf{x}_1^n, ...)'.$$

We may extend the definition to the boundary cases $n = 0$ and $(i, j) = (0, 0)$ as follows:

$$x_{ijk}^0 = \begin{cases} 1, & \text{if } (i, j, k) \in l(0, 0), \\ 0, & \text{otherwise,} \end{cases}$$
$$x_{00k}^n = 0, \quad 1 \leq k \leq m, \ n \geq 1.$$

Theorem 8.8. *For every fixed $n \geq 1$, the probability vector $\mathbf{x}^n$ satisfies the following block tridiagonal system:*

$$(\mathbf{Q}_I^n - \mathbf{A}^n)\mathbf{x}^n(n) = \delta_{1n}\mathbf{f}^n - \mathbf{A}^n\mathbf{x}^{n-1}(n), \tag{8.33}$$

where $\mathbf{x}^n(n) = (\mathbf{x}_0^n, ..., \mathbf{x}_n^n)'$, $\mathbf{x}^{n-1}(n) = (\mathbf{x}_0^{n-1}, ..., \mathbf{x}_{n-1}^{n-1}, \mathbf{0}_{(c+1)m})'$, $\mathbf{Q}_I^n$ is the square matrix of order $cm + (c+1)nm$ obtained from $\overline{\mathbf{Q}}^*$ by replacing the $(n+1, n+1)$th block entry by $\mathbf{A}_{nn}$ and setting $K = n$, $\mathbf{A}^n$ is obtained from $\mathbf{A}(z)$ by setting $K = n$ and $z = 0$, and $\mathbf{f}^n$ is obtained from $\mathbf{f}$ by taking $K = n$.

Proof. Once more, we employ first-step analysis to find that

$$x_{ijk}^n = \sum_{k' \neq k} \frac{d_{kk'}^0}{\lambda_k + i\nu + j\mu} x_{ijk'}^n + \sum_{k'=1}^m \frac{d_{kk'}^1}{\lambda_k + i\nu + j\mu} x_{i+1,j,k'}^n$$
$$+ \frac{i\nu}{\lambda_k + i\nu + j\mu} x_{i-1,j,k}^{n-1} + \frac{j\mu}{\lambda_k + i\nu + j\mu} x_{i+1,j-1,k}^n,$$
$$0 \leq i \leq c-1, \ j \geq 0, \ 1 \leq k \leq m, \ n \geq 1, \ (i, j) \neq (0, 0), \tag{8.34}$$

$$x_{cjk}^n = \sum_{k' \neq k} \frac{d_{kk'}^0}{\lambda_k + c\nu} x_{cjk'}^n + \sum_{k'=1}^m \frac{d_{kk'}^1}{\lambda_k + c\nu} x_{c,j+1,k'}^n + \frac{c\nu}{\lambda_k + c\nu} x_{c-1,j,k}^{n-1},$$
$$j \geq 0, \ 1 \leq k \leq m, \ n \geq 1. \tag{8.35}$$

For every $n \geq 1$, we notice that $x_{ijk}^n = 0$, for $i + j > n$. As a result, the system in (8.34) and (8.35) involves only a finite number of unknowns. The equation (8.33) follows after appropriate matrix formulation. □

Table 8.6. $P(I=n)$ for queues with Exp^a and $MMPP$ arrivals

n		$\mu = 0.05$	$\mu = 0.5$	$\mu = 1.0$	$\mu = 5.0$
1	Exp^a	0.28571	0.28571	0.28571	0.28571
	$MMPP$	0.32720	0.32720	0.32720	0.32720
2	Exp^a	0.09070	0.09070	0.09070	0.09070
	$MMPP$	0.09908	0.09908	0.09908	0.09908
3	Exp^a	0.05628	0.05628	0.05628	0.05628
	$MMPP$	0.05958	0.05958	0.05958	0.05958
4	Exp^a	0.04260	0.04260	0.04260	0.04260
	$MMPP$	0.04401	0.04401	0.04401	0.04401
5	Exp^a	0.03533	0.03533	0.03533	0.03533
	$MMPP$	0.03575	0.03575	0.03575	0.03575
6	Exp^a	0.03034	0.03055	0.03063	0.03072
	$MMPP$	0.02874	0.02989	0.03020	0.03046
7	Exp^a	0.02642	0.02703	0.02724	0.02745
	$MMPP$	0.02277	0.02547	0.02615	0.02668
8	Exp^a	0.02315	0.02427	0.02463	0.02495
	$MMPP$	0.01801	0.02212	0.02306	0.02377
9	Exp^a	0.02036	0.02204	0.02251	0.02292
	$MMPP$	0.01432	0.01953	0.02064	0.02143
10	Exp^a	0.01797	0.02018	0.02076	0.02123
	$MMPP$	0.01149	0.01749	0.01867	0.01949

Conditioning on the distribution of the first state visited when the busy period starts, we derive the unconditional distribution

$$P(I = n) = \frac{\overline{\mathbf{p}}(0,0)\mathbf{D}_1 \mathbf{x}_{(1,0)}^n}{\overline{\mathbf{p}}(0,0)\mathbf{D}_1 \mathbf{e}_m}, \quad n \geq 1, \tag{8.36}$$

where $\mathbf{x}_{(1,0)}^n = (x_{101}^n, ..., x_{10m}^n)'$.

By the right-hand expression in (8.36), we notice that the computation of the unconditional probability $P(I = n)$ implies to know $\overline{\mathbf{p}}(0,0)$. Thus, we need to truncate the orbit capacity.

Numerical experiments not included here indicate that the expected value $E[I]$ and the standard deviation $\sigma(I)$ decrease with increasing values of μ and increase as function of ρ. The same behavior was observed in Table 8.5 for L.

In Table 8.6, we compare the probability mass function of I for two models with Poisson and $MMPP$ arrivals. We fix $\rho = 0.5$ and employ the arrival representation described earlier when dealing with L. Since repeated attempts occur only when the c servers are busy, the probabilities $P(I = n)$, for $1 \leq n \leq 5$, do not depend on the retrial rate. For the four choices of μ, we observe that the queue with $MMPP$ arrivals has a larger mass at the origin (entries for $1 \leq n \leq 5$), whereas the model with Poisson arrivals presents a heavier tail.

8.2.3 Waiting Time

In this subsection, we study the waiting time which is defined as the sojourn time of a tagged customer in orbit.

Let W_{ijk} be the conditional waiting time of the tagged customer, given that the system state is $(i,j,k) \in \mathcal{S}-l(0)$. We also denote the Laplace-Stieltjes transform of W_{ijk} by $W_{ijk}^*(s)$. This yields

$$\mathbf{w}_j^*(s) = \left(W_{0j1}^*(s), ..., W_{cjm}^*(s)\right)', \quad 1 \le j \le K,$$
$$\mathbf{w}^*(s) = (\mathbf{w}_1^*(s), ..., \mathbf{w}_K^*(s))'.$$

The following result provides the block tridiagonal nature of the system governing the Laplace-Stieltjes transforms.

Theorem 8.9. *The Laplace-Stieltjes transforms $W_{ijk}^*(s)$, for $(i,j,k) \in \mathcal{S} - l(0)$, verify the following system:*

$$\mathbf{Q}_W(s)\mathbf{w}^*(s) = \mathbf{g}, \tag{8.37}$$

where $\mathbf{Q}_W(s) = \hat{\mathbf{Q}} - s\mathbf{I}_{(c+1)Km}$, $\mathbf{g} = -\mu \mathbf{e}_K \otimes (\mathbf{e}_{c+1} - \mathbf{e}_{c+1}(c+1)) \otimes \mathbf{e}_m$,

$$\hat{\mathbf{Q}} = \begin{pmatrix} \mathbf{A}_{11} & \mathbf{A}_{12} & & & & \\ \hat{\mathbf{A}}_{21} & \mathbf{A}_{22} & \mathbf{A}_{23} & & & \\ & \ddots & \ddots & \ddots & & \\ & & \hat{\mathbf{A}}_{K-1,K-2} & \mathbf{A}_{K-1,K-1} & \mathbf{A}_{K-1,K} \\ & & & \hat{\mathbf{A}}_{K,K-1} & \mathbf{A}_{KK} \end{pmatrix},$$

and $\hat{\mathbf{A}}_{j,j-1} = \frac{j-1}{j}\mathbf{A}_{j,j-1}$, *for* $2 \le j \le K$.

Proof. Using again a first-step analysis, we obtain

$$W_{ijk}^*(s) = \sum_{k' \ne k} \frac{d_{kk'}^0}{s + \lambda_k + i\nu + j\mu} W_{ijk'}^*(s) + \sum_{k'=1}^{m} \frac{d_{kk'}^1}{s + \lambda_k + i\nu + j\mu} W_{i+1,j,k'}^*(s)$$
$$+ \frac{i\nu}{s + \lambda_k + i\nu + j\mu} W_{i-1,j,k}^*(s) + \frac{(j-1)\mu}{s + \lambda_k + i\nu + j\mu} W_{i+1,j-1,k}^*(s)$$
$$+ \frac{\mu}{s + \lambda_k + i\nu + j\mu}, \quad 0 \le i \le c-1, \ 1 \le j \le K, \ 1 \le k \le m, \tag{8.38}$$

$$W_{cjk}^*(s) = \sum_{k' \ne k} \frac{d_{kk'}^0}{s + \lambda_k + c\nu} W_{cjk'}^*(s) + \sum_{k'=1}^{m} \frac{d_{kk'}^1}{s + \lambda_k + c\nu} W_{c,j+1,k'}^*(s)$$
$$+ \frac{c\nu}{s + \lambda_k + c\nu} W_{c-1,j,k}^*(s), \quad 1 \le j \le K-1, \ 1 \le k \le m, \tag{8.39}$$

$$W_{cKk}^*(s) = \sum_{k' \ne k} \frac{d_{kk'}^0 + d_{kk'}^1}{s + \lambda_k' + c\nu} W_{cKk'}^*(s) + \frac{c\nu}{s + \lambda_k' + c\nu} W_{c-1,K,k}^*(s), \quad 1 \le k \le m. \tag{8.40}$$

The last term on the right-hand side of formula (8.38) corresponds to the case where the first event is a successful reattempt for service made by the tagged customer. In contrast, if another retrial customer applies for service, we get the previous term in the formula. Similar arguments explain the terms due to primary arrivals and service completions. Finally, after routine block identification, we may rewrite (8.38)-(8.40) as given in (8.37). □

The tagged customer joins the orbit if he finds the system at any state belonging to the subset $\mathcal{S}_W = \{(c, j, k); 0 \leq j \leq K-1, 1 \leq k \leq m\}$. Thus, the unconditional version of the waiting time has the Laplace-Stieltjes transform

$$W^*(s) = 1 - \pi_W \mathbf{e}_{(c+1)Km} + \pi_W \mathbf{w}^*(s),$$

where π_W denotes the row vector of dimension $(c+1)Km$ defined by

$$\pi_W = \frac{1}{\omega} \left(\mathbf{0}'_{cm}, \overline{\mathbf{p}}(c,0)\mathbf{D}_1, ..., \mathbf{0}'_{cm}, \overline{\mathbf{p}}(c, K-1)\mathbf{D}_1 \right),$$

and $\overline{\mathbf{p}}(c, j)$ is the sub-vector consisting of the m last components of $\overline{\mathbf{p}}(j)$, for $0 \leq j \leq K-1$. The normalizing constant ω is given by

$$\omega = \sum_{j=0}^{K} \overline{\mathbf{p}}(j) \left(\mathbf{e}_{c+1} \otimes (\mathbf{D}_1 \mathbf{e}_m) \right).$$

The probability of non-waiting $P(W = 0) = 1 - \pi_W \mathbf{e}_{(c+1)Km}$ is the discrete contribution to the unconditional waiting time. It occurs either when the tagged customer finds a free server or when he sees states of the sub-level $l(c, K)$ and becomes a lost customer. On the other hand, $\pi_W \mathbf{w}^*(s)$ is the transform of the continuous contribution with density $f_{W^c}(x)$ on $(0, \infty)$.

Let $W_{ijk}^{(n)}$ be the nth moment of W_{ijk}. Then, in partitioned form, we have

$$\mathbf{w}_j^{(n)} = \left(W_{0j1}^{(n)}, ..., W_{cjm}^{(n)} \right)', \quad 1 \leq j \leq K, \; n \geq 0,$$
$$\mathbf{w}^{(n)} = \left(\mathbf{w}_1^{(n)}, ..., \mathbf{w}_K^{(n)} \right)', \quad n \geq 0.$$

The following result follows straightforward from Theorem 8.9.

Corollary 8.10. *The moments $W_{ijk}^{(n)}$, for $(i, j, k) \in \mathcal{S} - l(0)$, satisfy the following block tridiagonal system:*

$$\mathbf{w}^{(0)} = \mathbf{e}_{(c+1)Km},$$
$$\hat{\mathbf{Q}} \mathbf{w}^{(n)} = -n \mathbf{w}^{(n-1)}, \quad n \geq 1.$$

The nth moment of the unconditional waiting time is given by

$$E[W^n] = \pi_W \mathbf{w}^{(n)}, \quad n \geq 0.$$

In agreement with the intuitive expectations, the numerical experience shows that $E[W]$ and $\sigma(W)$ increase with increasing values of ρ and, in contrast, they decrease as far as μ increases [103].

We next present numerical results regarding the inversion of the distribution function $F_W(x)$. The curves displayed in Figures 8.3 and 8.4 correspond to those orbit levels that guarantee at least four decimal places of $E[W]$ are matched.

In Figure 8.3 we deal with a queue with a $MMPP$ with $\lambda = 1.0$, $c = 5$ and $\mu = 1.0$. The traffic intensity takes the values $\rho = 0.25$, 0.5 and 0.75. The jump at the point $x = 0.0$ amounts the probability $P(W = 0)$, and it becomes higher with smaller values of ρ. Obviously, for $\rho = 0.75$, the distribution function exhibits the heaviest tail.

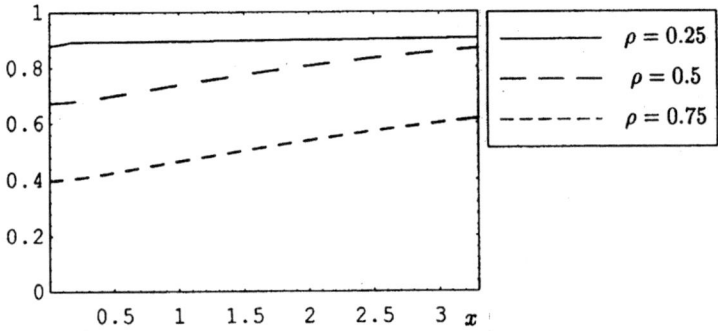

Fig. 8.3. Distribution functions of W as ρ varies

The influence of μ is illustrated in Figure 8.4. We keep the same arrival process and $c = 5$. Then, we fix $\rho = 0.5$ and plot the distribution function $F_W(x)$, for $\mu = 0.05$, 1.0 and 5.0. We observe that the distribution becomes sparser when μ decreases. Moreover, the increasing competition for occupying free servers explains the decrease of $P(W = 0)$ as μ increases.

Finally, we concentrate on the distribution of the number of repeated attempts made by a tagged customer until he reaches a free server.

Let R_{ijk} be the number of retrials that a tagged customer will make, given that the system state is (i, j, k). The corresponding generating functions are denoted by $R_{ijk}(z)$. Then, partitioning according to orbit levels yields

$$\mathbf{r}_j(z) = (R_{0j1}(z), ..., R_{cjm}(z))', \quad 1 \leq j \leq K,$$
$$\mathbf{r}(z) = (\mathbf{r}_1(z), ..., \mathbf{r}_K(z))'.$$

Theorem 8.11. *The generating functions $R_{ijk}(z)$, for $(i, j, k) \in S - l(0)$, satisfy the following block tridiagonal system:*

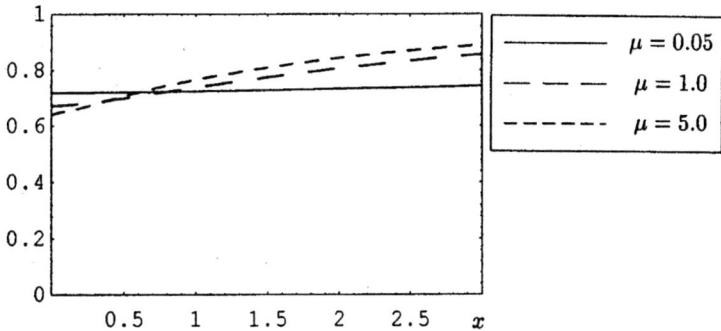

Fig. 8.4. Distribution functions of W as μ varies

$$\mathbf{Q}_R(z)\mathbf{r}(z) = \mathbf{g}(z), \tag{8.41}$$

where $\mathbf{Q}_R(z) = \hat{\mathbf{Q}} + (1-z)\mathbf{h}$, $\mathbf{h} = -\mu \mathbf{I}_K \otimes (\mathbf{e}_{c+1}(c+1)\mathbf{e}'_{c+1}(c+1)) \otimes \mathbf{I}_m$ and $\mathbf{g}(z) = z\mathbf{g}$.

Proof. The proof is similar to that given for the waiting time and thus it is omitted. We just point out that it is convenient to avoid the consideration of vain retrials made by non-tagged customers which neither affect the event under study nor modify the current system state. □

If we denote the nth factorial moment of R_{ijk} by $R_{ijk}^{(n)}$, then we have in partitioned form the following

$$\mathbf{r}_j^{(n)} = \left(R_{0j1}^{(n)}, ..., R_{cjm}^{(n)}\right)', \quad 1 \leq j \leq K, \ n \geq 0,$$

$$\mathbf{r}^{(n)} = \left(\mathbf{r}_1^{(n)}, ..., \mathbf{r}_K^{(n)}\right)', \quad n \geq 0.$$

By differentiating (8.41) in Theorem 8.11, we readily find the following result for the factorial moments.

Corollary 8.12. *The factorial moments $R_{ijk}^{(n)}$, for $(i,j,k) \in \mathcal{S} - l(0)$, verify the block tridiagonal system*

$$\mathbf{r}^{(0)} = \mathbf{e}_{(c+1)Km},$$

$$\hat{\mathbf{Q}}\mathbf{r}^{(n)} = \delta_{1n}\mathbf{g} + n\mathbf{h}\mathbf{r}^{(n-1)}, \quad n \geq 1.$$

The unconditional version of the number of retrials made by a customer leads to the factorial moments

$$E[R(R-1)...(R-n+1)] = \boldsymbol{\pi}_W \mathbf{r}^{(n)}, \quad n \geq 1.$$

Table 8.7. Moments of R

ρ			$\mu = 0.05$	$\mu = 0.5$	$\mu = 1.0$	$\mu = 2.5$	$\mu = 5.0$
0.25	E_5^a	$E[L]$	0.00019	0.00022	0.00026	0.00035	0.00050
		$\sigma(L)$	0.01417	0.01720	0.02041	0.02957	0.04425
	Exp^a	$E[L]$	0.00776	0.00942	0.01130	0.01651	0.02490
		$\sigma(L)$	0.09004	0.11398	0.14095	0.21829	0.34446
	H_2^a	$E[L]$	0.04080	0.05433	0.06820	0.10720	0.16951
		$\sigma(L)$	0.20782	0.28657	0.37078	0.61507	1.01375
0.5	E_5^a	$E[L]$	0.02860	0.04435	0.05991	0.10263	0.17032
		$\sigma(L)$	0.18411	0.29991	0.41842	0.75490	1.30115
	Exp^a	$E[L]$	0.10522	0.17301	0.24391	0.44780	0.77930
		$\sigma(L)$	0.35730	0.63143	0.92948	1.80771	3.25581
	H_2^a	$E[L]$	0.30191	0.63720	0.99150	2.01590	3.68880
		$\sigma(L)$	0.61823	1.43764	2.33362	4.98086	9.35609
0.75	E_5^a	$E[L]$	0.34060	0.66210	0.98970	1.92423	3.44221
		$\sigma(L)$	0.79291	1.67478	2.60948	5.35854	9.90257
	Exp^a	$E[L]$	0.57930	1.25420	1.97510	4.09041	7.57670
		$\sigma(L)$	1.03684	2.58730	4.29800	9.38993	17.84378
	H_2^a	$E[L]$	1.06930	3.56840	6.28670	14.33750	27.67170
		$\sigma(L)$	1.54204	6.17578	11.27336	26.47920	51.75798

Table 8.7 lists the values of $E[R]$ and $\sigma(R)$, for the model with $c = 5$ and the three renewal arrival processes with $\lambda = 1.0$. The numerical results show that both descriptors are increasing functions of ρ and μ. We notice that a rapid reattempt for service has a significant chance of being blocked. This explains that $E[R]$ and $\sigma(R)$ increase for increasing values of μ.

The numerical inversion of the system in (8.41) gives a first possibility for getting the probability mass function of R. An alternative approach may be attained introducing the probabilities $z_{l,(i,j,k)}^r$ defined as the probabilities that the tagged customer makes r retrials before entering service, given that he has accumulated l retrials and the system state is (i, j, k), for $(i, j, k) \in S - l(0)$.

The probability mass function $P(R = r)$, for $r \geq 0$, satisfies the equations

$$P(R = 0) = 1 - \pi_W \mathbf{e}_{(c+1)Km},$$
$$P(R = r) = \pi_W \mathbf{z}_0^r, \quad r \geq 1,$$

where $\mathbf{z}_l^r$ is the column vector of dimension $(c+1)Km$ containing the unknowns $z_{l,(i,j,k)}^r$.

A generalization of the arguments given in Subsection 5.2.2 leads to

$$\mathbf{Q}_R \mathbf{z}_{r-1}^r = \mathbf{g}, \quad r \geq 1, \tag{8.42}$$
$$\mathbf{Q}_R \mathbf{z}_l^r = \mathbf{h} \mathbf{z}_{l+1}^r, \quad r \geq 1, \ 0 \leq l \leq r - 2, \tag{8.43}$$

where $\mathbf{Q}_R = \hat{\mathbf{Q}} + \mathbf{h}$. For any fixed value of r, equations (8.42) and (8.43) can be solved backward, from $l = r - 1$ to 0, to get $\mathbf{z}_l^r$ and, consequently, the probabilities $P(R = r)$, for $r \geq 0$.

Table 8.8. $P(R = r)$ for queues with Exp^a and $MMPP$ arrivals

r		$\mu = 0.05$	$\mu = 0.5$	$\mu = 1.0$	$\mu = 5.0$
0	Exp^a	0.90912	0.89748	0.89124	0.87803
	$MMPP$	0.72031	0.68845	0.67337	0.64173
1	Exp^a	0.07856	0.06247	0.05151	0.02300
	$MMPP$	0.02240	0.13638	0.09757	0.03547
2	Exp^a	0.01056	0.02339	0.02573	0.01786
	$MMPP$	0.04406	0.07364	0.06574	0.03014
3	Exp^a	0.00148	0.00935	0.01349	0.01405
	$MMPP$	0.00904	0.04041	0.04457	0.02605
4	Exp^a	0.00021	0.00395	0.00738	0.01117
	$MMPP$	0.00196	0.02308	0.03084	0.02279
5	Exp^a	0.00003	0.00175	0.00418	0.00897
	$MMPP$	0.00044	0.01370	0.02181	0.02014
6	Exp^a	5.0×10^{-6}	0.00080	0.00244	0.00727
	$MMPP$	0.00010	0.00840	0.01573	0.01792
7	Exp^a	7.7×10^{-7}	0.00038	0.00146	0.00595
	$MMPP$	0.00002	0.00530	0.01154	0.01604
8	Exp^a	1.2×10^{-7}	0.00018	0.00090	0.00490
	$MMPP$	6.4×10^{-6}	0.00341	0.00860	0.01443
9	Exp^a	1.9×10^{-8}	0.00009	0.00056	0.00407
	$MMPP$	1.6×10^{-6}	0.00224	0.00648	0.01303
10	Exp^a	3.1×10^{-9}	0.00004	0.00035	0.00340
	$MMPP$	4.3×10^{-7}	0.00150	0.00495	0.01180

Finally, in Table 8.8, we display the probabilities $P(R = r)$, for $0 \leq r \leq 10$, for models with Poisson and $MMPP$ arrivals, $c = 5$ and $\rho = 0.5$. We observe that, for several choices of μ, the mass function always decreases. The queue with Poisson arrivals has a larger mass at $r = 0$. However, $MMPP$ arrivals have heavier tails when compared with the case of Poisson arrivals.

8.3 A Queue with Finite Population and PH Service and Retrial Times

In this section, we generalize the multiserver model with finite population studied in Section 3.5 by allowing the service times and the retrial times of each customer to follow phase-type distributions. Since phase-type distributions can be used to approximate general distributions and fit observed data, the resulting model is more appropriate for practical purposes.

Let $C(t)$ and $N(t)$ be, respectively, the number of busy servers and the number of customers in orbit at time t. We also define the random vectors $\mathbf{i}(t) = (I_1, ..., I_{C(t)})(t)$ and $\mathbf{j}(t) = (J_1, ..., J_{N(t)})(t)$ containing the phases of the customers in service and the customers in orbit, if any. We assume that the inter-retrial times of any arbitrary customer in orbit have a phase-type distribution of order s with representation $(\boldsymbol{\delta}, \mathbf{U})$ and $\mathbf{u} = -\mathbf{U}\mathbf{e}_s$. For the service times, we hold the representation $(\boldsymbol{\tau}, \mathbf{T})$ of order r.

The state space $\mathcal{S}$ can be partitioned according to the orbit levels $l(j)$ and the sub-levels $l(j, i)$ defined by

$$\mathcal{S} = \bigcup_{j=0}^{M-c} l(j) = \bigcup_{j=0}^{M-c} \bigcup_{i=0}^{c} l(j, i),$$

where, for $1 \leq j \leq M - c$ and $1 \leq i \leq c$, we have

$l(0,0) = \{(0,0)\}$,
$l(0,i) = \{(0,i,I_1,...,I_i); \ 1 \leq I_1,...,I_i \leq r\}$,
$l(j,0) = \{(j,0,J_1,...,J_j); \ 1 \leq J_1,...,J_j \leq s\}$,
$l(j,i) = \{(j,i,J_1,...,J_j,I_1,...,I_i); \ 1 \leq J_1,...,J_j \leq s, \ 1 \leq I_1,...,I_i \leq r\}$.

Therefore, we have that the cardinality of the sub-level $l(j, i)$ is $|l(j, i)| = s^j r^i$ and, consequently, the cardinality of $l(j)$ is $|l(j)| = s^j(r^{c+1} - 1)(r - 1)^{-1}$.

Then, the infinitesimal generator of $\mathcal{X} = \{(N(t), C(t), \mathbf{j}(t), \mathbf{i}(t)); \ t \geq 0\}$ has the following finite QBD structure:

$$\mathbf{Q} = \begin{pmatrix} \mathbf{A}_{00} & \mathbf{A}_{01} & & & & \\ \mathbf{A}_{10} & \mathbf{A}_{11} & \mathbf{A}_{12} & & & \\ & \ddots & \ddots & \ddots & & \\ & & \mathbf{A}_{M-c-1,M-c-2} & \mathbf{A}_{M-c-1,M-c-1} & \mathbf{A}_{M-c-1,M-c} \\ & & & \mathbf{A}_{M-c,M-c-1} & \mathbf{A}_{M-c,M-c} \end{pmatrix},$$

where the matrices $\mathbf{A}_{jj'}$ are of dimension $|l(j)| \times |l(j')|$.

The matrices $\mathbf{A}_{j,j-1}$, for $1 \leq j \leq M - c$, represent the case in which the orbit size decreases one unit due to a successful repeated attempt. They are given by

$$\mathbf{A}_{j,j-1} = \begin{pmatrix} 0 & \mathbf{A}_{j,j-1}(0,1) & & & \\ & 0 & \mathbf{A}_{j,j-1}(1,2) & & \\ & & \ddots & \ddots & \\ & & & 0 & \mathbf{A}_{j,j-1}(c-1,c) \\ & & & & 0 \end{pmatrix},$$

where $\mathbf{A}_{j,j-1}(i, i+1)$, for $0 \leq i \leq c-1$, gives the infinitesimal motion between the sub-levels $l(j, i)$ and $l(j-1, i+1)$. Thus, we have

$$\mathbf{A}_{j,j-1}(i, i+1) = \left(\oplus_{l=1}^{j} \mathbf{u}\right) \otimes \mathbf{I}_{r^i} \otimes \boldsymbol{\tau}, \quad 0 \leq i \leq c-1.$$

We notice that $\mathbf{A}_{j,j-1}(i, i+1)$ consists of the Kronecker product of three matrices, where the Kronecker sum of order j of the vector $\mathbf{u}$ is defined by

8.3 A Queue with Finite Population and PH Service and Retrial Times 235

$\oplus_{l=1}^{j}\mathbf{u} = (\mathbf{u}\otimes\mathbf{I}_s\otimes...\otimes\mathbf{I}_s)+...+(\mathbf{I}_s\otimes...\otimes\mathbf{I}_s\otimes\mathbf{u})$. The first matrix allows us to identify the retrial unit performing the successful attempt. The second factor $\mathbf{I}_{r^i}$ reflects that there is no change in the service phases currently in progress. The last factor $\boldsymbol{\tau}$ assigns a service phase to the retrial unit just starting its service time. Other matrices $\mathbf{A}_{jj'}$ in what follows can be interpreted in a similar manner.

The matrices $\mathbf{A}_{j,j+1}$, for $0 \le j \le M-c-1$, correspond to the case where an arriving customer finds the c servers busy, so the number of customers in orbit increases by one. Hence, we have

$$\mathbf{A}_{j,j+1} = \begin{pmatrix} 0 & 0 \\ 0 & (M-c-j)\alpha\mathbf{I}_{s^j}\otimes\boldsymbol{\delta}\otimes\mathbf{I}_{r^c} \end{pmatrix}.$$

Finally, the matrices $\mathbf{A}_{jj}$, for $0 \le j \le M-c$, describe the motion associated with those transitions where there is no change in the orbit size. This may occur due to a variety of possibilities: service completions, arrivals finding empty servers, blocked retrials followed by a change of phase, and service and retrial phase changes. Thus, the matrices $\mathbf{A}_{jj}$ are also structured in block tridiagonal form:

$$\mathbf{A}_{jj} = \begin{pmatrix} \mathbf{A}_{jj}(0,0) & \mathbf{A}_{jj}(0,1) & & & \\ \mathbf{A}_{jj}(1,0) & \mathbf{A}_{jj}(1,1) & \mathbf{A}_{jj}(1,2) & & \\ \ddots & \ddots & \ddots & & \\ & & \mathbf{A}_{jj}(c-1,c-2) & \mathbf{A}_{jj}(c-1,c-1) & \mathbf{A}_{jj}(c-1,c) \\ & & & \mathbf{A}_{jj}(c,c-1) & \mathbf{A}_{jj}(c,c) \end{pmatrix},$$

where

$$\mathbf{A}_{jj}(i,i-1) = \mathbf{I}_{s^j}\otimes\left(\oplus_{l=1}^{i}\mathbf{t}\right),\quad 1 \le i \le c,$$
$$\mathbf{A}_{jj}(i,i+1) = (M-i-j)\alpha\mathbf{I}_{s^j}\otimes\mathbf{I}_{r^i}\otimes\boldsymbol{\tau},\quad 0 \le i \le c-1,$$
$$\mathbf{A}_{jj}(i,i) = -(M-i-j)\alpha\mathbf{I}_{s^j}\otimes\mathbf{I}_{r^i}+\left(\oplus_{l=1}^{j}\mathbf{U}+\delta_{ic}\left(\oplus_{l=1}^{j}\mathbf{u}\right)\otimes\boldsymbol{\delta}\right)\otimes\mathbf{I}_{r^i}$$
$$+\mathbf{I}_{s^j}\otimes\left(\oplus_{l=1}^{i}\mathbf{T}\right),\quad 0 \le i \le c.$$

Since the state space is finite and $\mathbf{Q}$ is irreducible, the stationary probability vector of the generator $\mathbf{Q}$ always exists and satisfies

$$\mathbf{p}\mathbf{Q} = \mathbf{0}'_d,\quad \mathbf{p}\mathbf{e}_d = 1,$$

where $d = (s^{M-c+1}-1)(s-1)^{-1}(r^{c+1}-1)(r-1)^{-1}$ is the cardinality of the state space $\mathcal{S}$. For the computation of the vector $\mathbf{p}$, we refer the reader to Theorem 7.1 and information given in Sections 5.3 and 7.4.

Finally, we numerically illustrate the main performance measures of the model. The following figures represent the mean number of customers in orbit, the mean waiting time, and the blocking probability, for the case $M = 6$ and $c = 3$. We assume hyperexponential service times and Erlang retrial times with representations

$$\tau = (0.75, 0.25), \quad \mathbf{T} = \begin{pmatrix} -1.5 & 0 \\ 0 & -0.5 \end{pmatrix},$$

$$\delta = (1.0, 0), \quad \mathbf{U} = \begin{pmatrix} -\mu & \mu \\ 0 & -\mu \end{pmatrix},$$

that is, we fix the hyperexponential parameters to fit $\beta_1 = 1.0$ and $c_v = 1.29099$.

Taking into account that the population is finite, we have

$$E[N] = \overline{\lambda} E[W],$$

where $\overline{\lambda}$ is the mean arrival rate defined in Section 3.5.

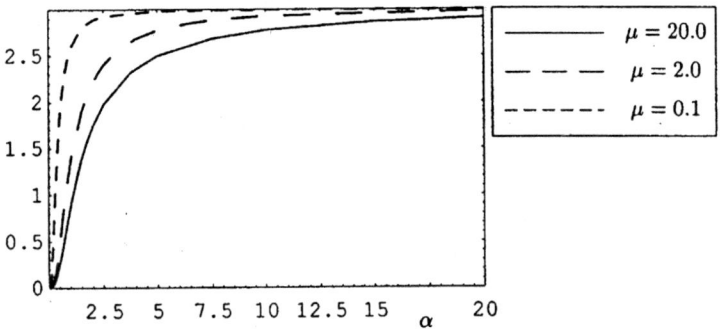

Fig. 8.5. The effect of α on $E[N]$

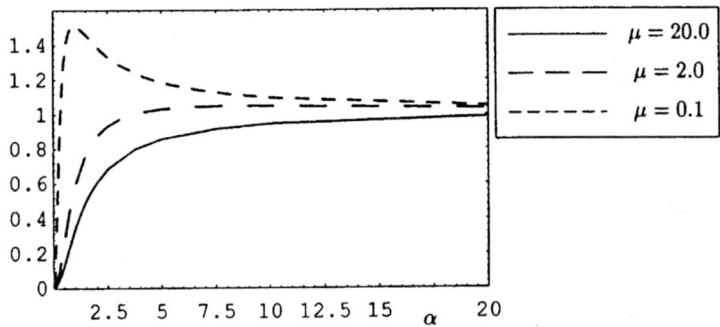

Fig. 8.6. The effect of α on $E[W]$

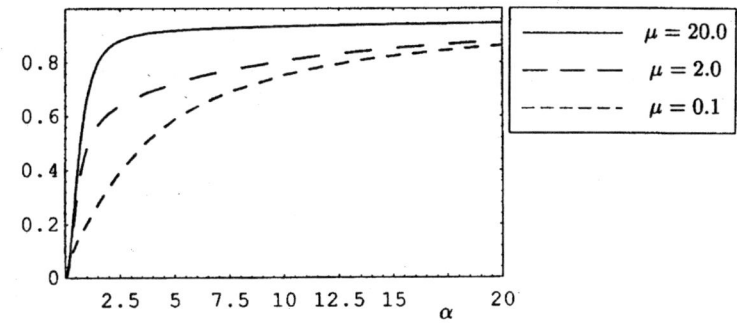

Fig. 8.7. The effect of α on B

The effect of the arrival rate is shown in Figures 8.5-8.7, where we have plotted the three performance measures versus α, for $\mu = 0.1$, 2.0 and 20.0. The mean number of customers in orbit and the blocking probability are strictly increasing.

An interesting feature of the quasi-random input is that the arrival rates $\lambda_{ij} = (M - i - j)\alpha$ are non-homogeneous. As a result, sometimes the interpretation of the effect of the system parameters on performance measures becomes more complicated. This occurs in Figure 8.6 for $E[W]$, which shows a maximum in the figure for $\mu = 0.1$. For ease of visualization, we have plotted the curves in the domain $(0, 20]$. By expanding this domain, we could visualize the maximum corresponding to the cases $\mu = 2.0$ and 20.0. The numerical results show that $E[W] = \overline{\lambda}^{-1} E[N]$ is obtained as the product of $\overline{\lambda}^{-1}$, which decreases with increasing values of α, and $E[N]$, which is increasing with α. The equilibrium between both opposite monotonic behaviors yields the observed maximum.

In Figures 8.8-8.10, we choose the values $\alpha = 0.1$, 2.0 and 20.0 to represent the three descriptors as functions of the retrial rate μ. We observe that, as is to be expected, $E[N]$ and $E[W]$ decrease with increasing values of μ, but B is an increasing function of μ with α fixed.

In the case $\alpha = 0.1$, the descriptors take values in a narrow domain. Indeed, the curve associated with $\alpha = 0.1$ and the horizontal axis are almost undistinguished.

8.4 Bibliographical Notes

The paper [322] provides a bibliographical guide to the use of matrix-analytic methods in retrial queueing systems. This guide consists of an author index and a subject index of papers written in English and published in journals or

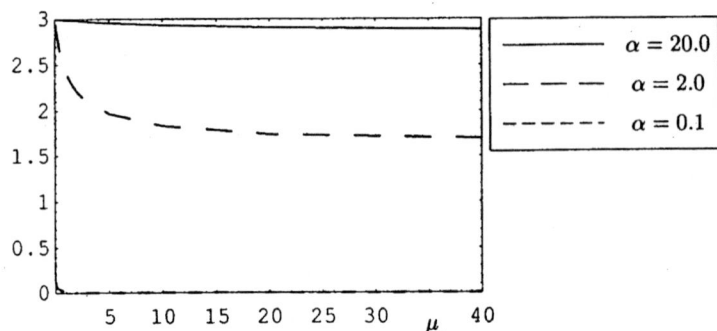

Fig. 8.8. The effect of μ on $E[N]$

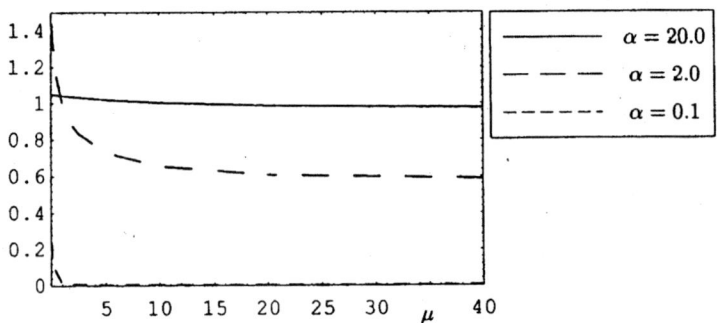

Fig. 8.9. The effect of μ on $E[W]$

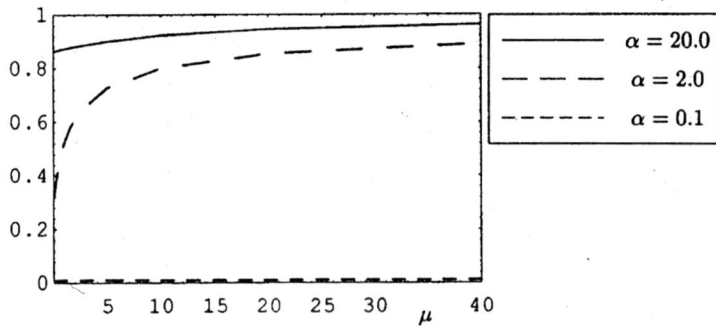

Fig. 8.10. The effect of μ on B

collective publications. In particular, the subject list includes several keywords closely related to the QBD structure under study in the present chapter.

The MAP has been widely used to model the arrival input of retrial queues with QBD structure. Alfa and Li [28] consider a personal communication services network where the arrival of customers is modelled by MAP flows. The rest of random variables (service times, retrial times, etc.) involved have PH distributions. Chakravarthy and Dudin [175] study a model with full access rule in which two types of customers arrive according to a MAP. Avram and Gómez-Corral [140] investigate a finite buffer retrial queue with MAP processes of positive and negative arrivals. Customers are served in groups following requirements of phase-type. These models and many others [322] provide examples of the use of the MAP preserving the QBD matrix structure.

In Section 8.1, we presented the $MAP/PH/1$ retrial queue as a natural generalization of the $M/G/1$ retrial queue. The $MAP/G/1$ retrial queue investigated by Shin and Pearce [602] provides an alternative; see also the orbit truncated version in [158, Section 9.4]. At this point, we remark that the consideration of the $MAP/PH/1$ case preserves the analysis at Markovian level. In contrast, the study of the $MAP/G/1$ retrial queue leads to an embedded Markov chain of $M/G/1$-type.

In Corollary 8.3, we showed how the $PH/M/1$ retrial queue can be obtained as a particular case of the results for the $MAP/PH/1$ case. We notice that the proof follows the same lines of the standard $PH/M/1$ queue without retrials. Interested readers are referred to the paper [100] and the book [548]. The analysis of the maximal queue length in the $MAP/PH/1$ retrial queue is based on the paper [100]. For the study of this descriptor for the $MAP/M/c$ retrial queue, see the paper by Artalejo and Chakravarthy [98]. Unlike the study of the stationary system state, the analysis of extreme values can be done for general block structures. In [99], the authors develop an algorithmic analysis of the maximum level visited by a two-dimensional continuous-time Markov chain with a general block level dependent structure.

Among the literature on multiserver retrial queues with QBD structure, we next comment on a few selected papers. Choi and Chang [191] consider a priority queue with finite orbit, where both priority and non-priority customers arrive according to MAP flows. Artalejo et al. [96] study a multiserver retrial model in which the servers are switched on and off according to a multi-threshold strategy that depends on the number of customers in the system. Efrosinin and Breuer [239] investigate another controlled retrial queue with Markovian arrivals and heterogeneous servers providing service of PH type. They prove that a threshold policy provides an optimal allocation of customers among servers, when the objective is to minimize the mean number of customers in the system.

The assumption of PH service times is widely extended in the retrial literature [322]. However, the number of papers considering PH retrial times is very limited [28, 107, 211, 352]. The problem of using a PH random variable to describe the retrial time of each individual customer in orbit is to find appropriate representations of the system state to work in practice. Section 8.3 follows the paper [107] which generalizes the study by Alfa and Sapna Isotupa [29] who investigated the model with exponential retrial times. A number of misprints in [29] are now corrected.

The three retrial queues presented in this chapter are described by means of continuous-time stochastic processes. As examples of discrete-time retrial queues, with QBD structure and studied through the matrix-analytic formalism, we mention the $PH/Geo/1$ queue [474] and the approximating approach of the $GI/G/1$ queue [30]. In both papers, the retrial times are assumed to follow a geometric law.

9

Selected Retrial Queues with $GI/M/1$ and $M/G/1$ Structures

In this last chapter, we turn the attention to the $GI/M/1$ and $M/G/1$ structures. In order to illustrate the former, we present an exhaustive analysis of the $Geo/Geo/c$ retrial queue, which includes the study of the stationary distribution of the system state, the busy period and the waiting time analysis. In Section 9.2, we discuss the $BMAP/SM/1$ retrial queue. We first analyze the system state at the departure epochs. Then, we extend the study at any arbitrary time.

9.1 The $Geo/Geo/c$ Retrial Queue

In Section 3.3, we studied the limiting probabilities of the $Geo/G/1$ retrial queue. The aim of the present section is to investigate the $Geo/Geo/c$ queue with geometric retrials; that is, the discrete-time analogue of the $M/M/c$ retrial queue. We propose several algorithmic procedures for the efficient computation of the main performance measures. More specifically, we focus on the stationary distribution of the system state, the busy period and the waiting time.

Although the efficient computational analysis of non-homogeneous $GI/M/1$-type processes remains still an open problem, we shall show that it is possible to exploit the matrix structure and sparsity of the underlying blocks to proceed algorithmically. There are several methods for an efficient solution of the stationary probabilities and the rate matrix $\mathbf{R}$. For the sake of brevity, we present in detail only those methods that we use for obtaining the numerical results.

9.1.1 Stationary Distribution of the System State

We start with the description of the mathematical model. We assume that the time axis has been divided into equal intervals of unit length called slots. In the $Geo/Geo/c$ retrial queue, primary customers arrive to the system according to a geometric arrival process with probability p. We denote the complementary probability $1 - p$ by $\bar{p}$. Service is rendered by c identical servers with service times geometrically distributed with probability mass function $q\bar{q}^{k-1}$, for $k \geq 1$, where $\bar{q} = 1 - q$. This means that any busy server finishes the undergoing service in the next slot with probability q. Customers in orbit behave independently of each other and retry with probability s at every time slot; that is, their retrial times are independent geometric random variables with probability mass function $s\bar{s}^{k-1}$, for $k \geq 1$, where $\bar{s} = 1 - s$. Any arriving customer who finds a free server begins immediately to be served. Otherwise, he joins the orbit and reapplies for service later.

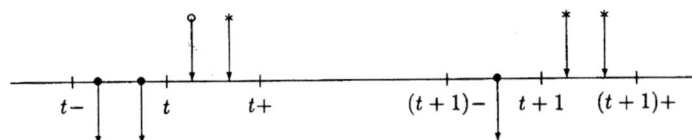

Fig. 9.1. Various epochs in *G-EAS*

We consider a *G-EAS* policy in which all events occur around the slot boundaries. We recall that, at a given slot boundary t, departures occur in $(t-, t)$, while primary arrivals and retrials occur in $(t, t+)$, see Figure 9.1. For analyzing the waiting time distribution, we must be even more specific about which customers are selected for service at a given slot. If we have a primary arrival and/or many retrials that exceed the number of free servers, then we assume that primary arrivals have priority over retrials. We also assume the random order discipline among the customers in orbit; that is, they are selected randomly to occupy the free servers.

The state of the system at time $t+$ can be described by the process $X_t = (C_t, N_t)$, where C_t denotes the number of busy servers and N_t represents the number of customers in orbit. The process $\{X_t; t \geq 0\}$ is an irreducible Markov chain with state space $S = \{0, ..., c\} \times \mathbb{Z}_+$ and one-step transition probabilities $P_{(i,j)(m,n)}$ given as in the following theorem.

Theorem 9.1. *The non-zero transition probabilities $P_{(i,j)(m,n)}$ of the process $\{X_t; t \geq 0\}$ are as follows:*
  (i) If $n = j + 1$ and $i = m = c$, then

$$P_{(c,j)(c,j+1)} = p\bar{q}^c. \tag{9.1}$$

9.1 The *Geo/Geo/c* Retrial Queue

(ii) If $\max(0, j-c) \leq n \leq j$, $j-n \leq m \leq c$ and $\max(0, m+n-j-1) \leq i \leq c$, then

$$P_{(i,j)(m,n)} = \left((1 - \delta_{m,i+j-n+1})\bar{p}\binom{i}{i+j-m-n}q^{i+j-m-n}\bar{q}^{m+n-j}\right.$$
$$+ (1 - \delta_{j-n,m})p\binom{i}{i+j+1-m-n}q^{i+j+1-m-n}\bar{q}^{m+n-j-1}\right)$$
$$\times \left(\binom{j}{j-n}s^{j-n}\bar{s}^n + \delta_{mc}\sum_{r=j-n+1}^{j}\binom{j}{r}s^r\bar{s}^{j-r}\right). \quad (9.2)$$

Proof. Suppose that we denote by (i,j) and (m,n) the current state of the process and the next state to be visited, respectively. For any fixed j, the range for n is clearly from $\max(0, j-c)$ to $j+1$. Indeed, the extreme case $n = j+1$ occurs only when all servers are busy, there are no departures and one primary arrival occurs and joins the orbit. So we have $n = j+1$ only in the case $i = m = c$, with the probability given in formula (9.1).

The other extreme case $n = j - c$ occurs if all the servers complete their service times, no primary arrivals occur and at least c retrial customers reattempt for service. This case implies that $m = c$ and accounts for the last term of (9.2). In general, in order to have $m < c$ and $n \leq j$ in the next slot, we must have $r = j - n$ retrials which correspond to obtain $j - n$ successful trials in a binomial distribution with parameters j (number of trials) and s (probability of having a successful trial).

Let d be the number of departures and a the number of primary arrivals at the slot under consideration. Given the states (i, j) and (m, n), we have necessarily that $d - a = i + j - m - n$. This happens either if $(d, a) = (i + j - m - n, 0)$ or if $(d, a) = (i + j + 1 - m - n, 1)$, which lead respectively to the two first terms in (9.2). Note that, for given values of j and n with $\max(0, j - c) \leq n \leq j$, the range for m varies from $j - n$ to c. Indeed, we have at least $j - n$ retrial customers who will occupy servers at the next slot. For given j, n and m, the range for i is from $\max(0, m + n - j - 1)$ to c since $0 \leq d \leq i$. Finally, we notice that $d = -1$ has no sense so transitions with $(d, a) = (i+j-m-n, 0)$ happen only for $i \neq m+n-j-1$. On the other hand, $d = i + 1$ is also impossible so transitions with $(d, a) = (i + j + 1 - m - n, 1)$ occur only for $m \neq j - n$. This explains the use of Kronecker's function in (9.2). □

Let $S_{(i,j)}$ be the set of one-step accessible states from the initial state (i, j). This set can be decomposed as follows:

$$S_{(i,j)} = \begin{cases} \cup_{k=0}^{i+1} D_j^k, & \text{if } 0 \leq i \leq c-1, \\ \cup_{k=0}^{c} D_j^k \cup \{(c, j+1)\}, & \text{if } i = c, \end{cases}$$

where $D_j^k = \{(k, j), (k+1, j-1), ..., (k + \min(c-k, j), j - \min(c-k, j))\}$. The superscript k in D_j^k represents the number of servers that remain busy just after the departures and the primary arrival occurring in the next slot. The sets D_j^k are disjoint with cardinality $|D_j^k| = \min(c-k, j) + 1$. Thus, it follows that

$$|S_{(i,j)}| = \begin{cases} (i+2)(j+1), & \text{if } 0 \leq i \leq c-1,\ 0 \leq j \leq c-i-1, \\ (c-j+1)(j+1) + \delta_{ic} \\ \quad + \frac{(j+c-i)(i+j+1-c)}{2}, & \text{if } 1 \leq i \leq c,\ c-i \leq j \leq c-1, \\ \delta_{ic} + \frac{(i+2)(2c+1-i)}{2}, & \text{if } 0 \leq i \leq c,\ j \geq c. \end{cases} \quad (9.3)$$

Equation (9.3) shows that the number of immediately accessible states can be large. Of course, this number increases as c grows. This is a crucial difference between the $Geo/Geo/c$ retrial queue and the $M/M/c$ retrial queue.

Because the number j of customers in orbit may increase at most by 1 or decrease at most by $\min(c, j)$ at each slot, the one-step transition probability matrix $\mathbf{P} = \left(P_{(i,j)(m,n)}\right)$ has a particular level dependent structure of $GI/M/1$-type. Let $l(j) = \{(0, j), ..., (c, j)\}$ denote the jth level of the process. Then, as a slight variant of (7.3), the matrix $\mathbf{P}$ can be partitioned in square blocks $\mathbf{A}_{jn}$ of dimension $c+1$ containing the transition probabilities from $l(j)$ to $l(n)$. We notice that $\mathbf{A}_{jn} = \mathbf{0}_{(c+1) \times (c+1)}$, for $n < j - c$ and $n > j + 1$. The blocks $\mathbf{A}_{j,j+1}$ have only one non-zero element, $P_{(c,j)(c,j+1)}$, while $\mathbf{A}_{jn}$, for $j - c \leq n \leq j$, have $c + 1 + n - j$ non-zero columns. These features of the matrix structure are the basis for the algorithmic analysis in the sequel.

We expect to have the same positive recurrence condition for the standard and retrial $Geo/Geo/c$ queues because, as the number of customers in orbit grows, the model with retrials approaches the standard $Geo/Geo/c$ queue. A formal proof uses Foster's mean drift criterion giving that the discrete-time Markov chain $\{X_t; t \geq 0\}$ is positive recurrent if and only if $p < cq$ (see [112, Theorem 2]). From now on, we suppose that this condition holds.

We next concentrate on the computation of the stationary probabilities

$$\pi_{ij} = \lim_{t \to \infty} P(C_t = i, N_t = j), \quad (i, j) \in \mathcal{S}.$$

Because the transition matrix $\mathbf{P}$ is level dependent, it seems impossible to compute exactly the probabilities π_{ij}. Thus, in what follows, we consider two approximate methods: direct truncation and generalized truncation of Neuts and Rao type. In Section 3.4, both methods have been proved simple and effective for dealing with the $M/M/c$ retrial queue.

In the method of direct truncation, we put a fictitious limit K on the orbit capacity. For convenience, we assume $K \geq c + 1$. The resulting transition matrix $\overline{\mathbf{P}}(K)$ is identical to the submatrix of $\mathbf{P}$ describing the transitions among levels $l(j)$, for $0 \leq j \leq K$, except the block $\mathbf{A}_{KK}$ which is replaced by $\mathbf{A}_{KK} + \mathbf{A}_{K,K+1}$. This ensures that $\overline{\mathbf{P}}(K)$ is stochastic. Thus, the matrix $\overline{\mathbf{P}}(K)$ is given by

$$\begin{pmatrix} \mathbf{A}_{00} & \mathbf{A}_{01} & & & & & \\ \mathbf{A}_{10} & \mathbf{A}_{11} & \mathbf{A}_{12} & & & & \\ \vdots & \vdots & \vdots & \ddots & & & \\ \mathbf{A}_{c0} & \mathbf{A}_{c1} & \mathbf{A}_{c2} & \cdots & \mathbf{A}_{c,c+1} & & \\ & \mathbf{A}_{c+1,1} & \mathbf{A}_{c+1,2} & \cdots & \mathbf{A}_{c+1,c+1} & \mathbf{A}_{c+1,c+2} & \\ & \ddots & \ddots & & & \ddots & \\ & & \mathbf{A}_{K-1,K-1-c} & \mathbf{A}_{K-1,K-c} & \cdots & \mathbf{A}_{K-1,K-1} & \mathbf{A}_{K-1,K} \\ & & & \mathbf{\Phi}_c(K) & \cdots & \mathbf{\Phi}_1(K) & \mathbf{\Phi}_0(K) \end{pmatrix},$$
(9.4)

where

$$\mathbf{\Phi}_0(K) = \mathbf{A}_{KK} + \mathbf{A}_{K,K+1}, \tag{9.5}$$
$$\mathbf{\Phi}_i(K) = \mathbf{A}_{K,K-i}, \quad 1 \leq i \leq c. \tag{9.6}$$

Let $\{\bar{\pi}_{ij}; 0 \leq i \leq c, 0 \leq j \leq K\}$ be the stationary distribution of $\overline{\mathbf{P}}(K)$ and $\bar{\pi}(j) = (\bar{\pi}_{0j}, ..., \bar{\pi}_{cj})$ be the row vector associated with the level $l(j)$, for $0 \leq j \leq K$. The vectors $\bar{\pi}(j)$ can be determined by solving the system of balance equations

$$(\bar{\pi}(0), \ldots, \bar{\pi}(K)) \left(\mathbf{I}_{(c+1)(K+1)} - \overline{\mathbf{P}}(K) \right) = \mathbf{0}'_{(c+1)(K+1)} \tag{9.7}$$

and the normalization equation

$$\sum_{j=0}^{K} \bar{\pi}(j) \mathbf{e}_{c+1} = 1.$$

If we introduce the vectors

$$\mathbf{z}_0(K) = (\bar{\pi}(0), ..., \bar{\pi}(K-1)),$$
$$\mathbf{z}_1(K) = \bar{\pi}(K),$$

and partition $\mathbf{I}_{(c+1)(K+1)} - \overline{\mathbf{P}}(K)$ according to $\mathbf{z}_0(K)$ and $\mathbf{z}_1(K)$, then we may express (9.7) in the form

$$(\mathbf{z}_0(K), \mathbf{z}_1(K)) \begin{pmatrix} \mathbf{B}_{00}(K) & \mathbf{B}_{01}(K) \\ \mathbf{B}_{10}(K) & \mathbf{B}_{11}(K) \end{pmatrix} = \left(\mathbf{0}'_{(c+1)K}, \mathbf{0}'_{c+1} \right), \tag{9.8}$$

where $\mathbf{B}_{00}(K)$, $\mathbf{B}_{01}(K)$, $\mathbf{B}_{10}(K)$ and $\mathbf{B}_{11}(K)$ are block structured matrices with $K \times K$, $K \times 1$, $1 \times K$ and 1×1 square blocks of dimension $c+1$, respectively. Furthermore, we note that the block matrix $\mathbf{B}_{00}(K)$ can be indeed written by using blocks associated with the previous level $K-1$ as follows:

$$\mathbf{B}_{00}(K) = \begin{pmatrix} \mathbf{B}_{00}(K-1) & \mathbf{B}_{01}(K-1) \\ \mathbf{C}_0(K-1) & \mathbf{C}_1(K-1) \end{pmatrix}, \tag{9.9}$$

where

$$C_0(K-1) = \left(\mathbf{0}_{(c+1)\times(c+1)(K-1-c)}, -\mathbf{A}_{K-1,K-1-c}, ..., -\mathbf{A}_{K-1,K-2}\right),$$
$$C_1(K-1) = \mathbf{I}_{c+1} - \mathbf{A}_{K-1,K-1}.$$

The next result provides an algorithm for computing the stationary distribution of $\overline{\mathbf{P}}(K_f)$, where K_f is a suitably chosen threshold. Further details about how to choose K_f will be given in the numerical work.

Theorem 9.2. *The stationary vectors $\mathbf{z}_0(K_f)$ and $\mathbf{z}_1(K_f)$ can be computed from the following steps:*

Step 1. Put $K = c + 1$.
Step 2. Compute $\mathbf{B}_{00}^{-1}(K)$ as

$$\mathbf{B}_{00}^{-1}(K) = \begin{pmatrix} \mathbf{D}_{00} & \mathbf{D}_{01} \\ \mathbf{D}_{10} & \mathbf{D}_{11} \end{pmatrix}, \quad (9.10)$$

where

$$\mathbf{D}_{00} = \left(\mathbf{B}_{00}(K-1) - \mathbf{B}_{01}(K-1)\mathbf{C}_1^{-1}(K-1)\mathbf{C}_0(K-1)\right)^{-1},$$
$$\mathbf{D}_{10} = -\mathbf{C}_1^{-1}(K-1)\mathbf{C}_0(K-1)\mathbf{D}_{00},$$
$$\mathbf{D}_{11} = \left(\mathbf{C}_1(K-1) - \mathbf{C}_0(K-1)\mathbf{B}_{00}^{-1}(K-1)\mathbf{B}_{01}(K-1)\right)^{-1},$$
$$\mathbf{D}_{01} = -\mathbf{B}_{00}^{-1}(K-1)\mathbf{B}_{01}(K-1)\mathbf{D}_{11}.$$

Compute $\mathbf{z}_1(K)$ as the solution of

$$\mathbf{z}_1(K)\left(\mathbf{B}_{11}(K) - \mathbf{B}_{10}(K)\mathbf{B}_{00}^{-1}(K)\mathbf{B}_{01}(K)\right) = \mathbf{0}'_{c+1}, \quad (9.11)$$
$$\mathbf{z}_1(K)\left(\mathbf{e}_{c+1} - \mathbf{B}_{10}(K)\mathbf{B}_{00}^{-1}(K)\mathbf{e}_{(c+1)K}\right) = 1. \quad (9.12)$$

Compute $\mathbf{z}_0(K)$ by

$$\mathbf{z}_0(K) = -\mathbf{z}_1(K)\mathbf{B}_{10}(K)\mathbf{B}_{00}^{-1}(K). \quad (9.13)$$

Store $\mathbf{B}_{00}(K)$, $\mathbf{B}_{00}^{-1}(K)$, $\mathbf{B}_{01}(K)$ and $\mathbf{B}_{10}(K)$.
Put $K = K + 1$.
Step 3. While $K \leq K_f$,
  Compute $\mathbf{B}_{00}^{-1}(K)$ by (9.10).
  Compute $\mathbf{z}_1(K)$ by (9.11) and (9.12), and $\mathbf{z}_0(K)$ by (9.13).
  Update $\mathbf{B}_{00}(K)$, $\mathbf{B}_{00}^{-1}(K)$, $\mathbf{B}_{01}(K)$ and $\mathbf{B}_{10}(K)$.
  Put $K = K + 1$.

Proof. The proof is based on the partition of $\mathbf{B}_{00}(K)$ given in (9.9). Specifically, since $\mathbf{C}_0(K-1) = \mathbf{B}_{10}(K-1)$, we can store up the matrices $\mathbf{B}_{00}(K-1)$, $\mathbf{B}_{01}(K-1)$ and $\mathbf{B}_{10}(K-1)$ and notice that an appeal to [358, Theorem 4.2.4] yields the equation (9.10). Equations (9.11)-(9.13) readily follow from (9.8). □

We may exploit the matrix structure and notice that the sparse block forms of $\mathbf{B}_{01}(K-1)$ and $\mathbf{C}_0(K-1)$ simplify the calculations. In particular, note that the matrix $\mathbf{B}_{01}(K-1)$ has only one non-zero element $-p\bar{q}^c$ in its bottom-right entry. This implies that only the last column of $\mathbf{B}_{00}^{-1}(K-1)\mathbf{B}_{01}(K-1)$ is non-zero and it is obtained by multiplying the last column of $\mathbf{B}_{00}^{-1}(K-1)$ by $-p\bar{q}^c$. Therefore, the key step is the computation of $\mathbf{D}_{00}$. The rest of operations only involve multiplications of known simple matrices and the inversion of matrices of order $c+1$.

The inverse in the definition of $\mathbf{D}_{00}$ can be computed by using small-rank adjustment [358]. Let us assume that we have computed the inverse of a matrix $\mathbf{A}$ and we want the inverse of its adjustment $\mathbf{B} = \mathbf{A} + \mathbf{XWY}$, where $\mathbf{W}$ is a matrix of smaller order than $\mathbf{A}$. Then, we have

$$\mathbf{B}^{-1} = \left(\mathbf{I} - \mathbf{A}^{-1}\mathbf{X}\left(\mathbf{W}^{-1} + \mathbf{Y}\mathbf{A}^{-1}\mathbf{X}\right)^{-1}\mathbf{Y}\right)\mathbf{A}^{-1}.$$

The above formula shows that only the computation of inverses of smaller matrices $\mathbf{W}$ and $\mathbf{W}^{-1} + \mathbf{Y}\mathbf{A}^{-1}\mathbf{X}$ is needed. In our case, we have $\mathbf{A} = \mathbf{B}_{00}(K-1)$, $\mathbf{X} = -\mathbf{B}_{01}(K-1)$, $\mathbf{W} = \mathbf{C}_1^{-1}(K-1)$ and $\mathbf{Y} = \mathbf{C}_0(K-1)$. Thus, we obtain that $\mathbf{D}_{00} = \mathbf{B}^{-1} = \left(\mathbf{I}_{(c+1)(K-1)} - \mathbf{D}_{01}\mathbf{C}_0(K-1)\right)\mathbf{B}_{00}^{-1}(K-1)$, so $\mathbf{D}_{00}$ is obtained by multiplications and additions of already computed matrices.

In what follows, we focus on a second approximation for the stationary distribution of the system state, which is based on the use of generalized truncation of Neuts and Rao type. According to our comments in Subsection 3.4.2, from a level $l(K)$ and up, only K customers in orbit are permitted to conduct retrials. This implies that the new transition matrix $\widetilde{\mathbf{P}}(K)$ can be thought of as a level independent Markov chain of $GI/M/1$-type, with a large number of boundary states corresponding to levels $l(j)$, for $0 \leq j \leq K-1$.

Let $\{\tilde{\pi}_{ij}; 0 \leq i \leq c, j \geq 0\}$ be the stationary distribution of $\widetilde{\mathbf{P}}(K)$ and $\tilde{\boldsymbol{\pi}}(j) = (\tilde{\pi}_{0j}, ..., \tilde{\pi}_{cj})$ be the row vector with the stationary probabilities of the jth level. It is well-known that the stationary distribution of $\widetilde{\mathbf{P}}(K)$ has the matrix-geometric form

$$\tilde{\boldsymbol{\pi}}(j) = \tilde{\boldsymbol{\pi}}(K)\mathbf{R}^{j-K}, \quad j \geq K+1, \tag{9.14}$$

where $\mathbf{R}$ is the minimal solution to the matrix equation

$$\mathbf{R} = \sum_{j=0}^{c+1} \mathbf{R}^j \mathbf{A}_{K,K+1-j},$$

in the set of non-negative matrices.

Following Subsection 7.3.1, by starting somewhat arbitrarily with $\mathbf{R}_0 = \mathbf{0}_{(c+1)\times(c+1)}$, we may iteratively evaluate the sequence defined by (7.19), which becomes

$$\mathbf{R}_{k+1} = \left(\mathbf{A}_{K,K+1} + \sum_{j=2}^{c+1} \mathbf{R}_k^j \mathbf{A}_{K,K+1-j}\right)(\mathbf{I}_{c+1} - \mathbf{A}_{KK})^{-1}, \quad k \geq 0.$$

To define a stopping criterion, the standard practice is to keep track of two successive matrices in each iteration and stop iterations when $||\mathbf{R}_k - \mathbf{R}_{k-1}||_\infty < \varepsilon$.

Nevertheless, we can take advantage of the special matrix structure for $\mathbf{A}_{K,K+1}$, which has the form

$$\mathbf{A}_{K,K+1} = p\bar{q}^c \mathbf{e}_{c+1}(c+1)\mathbf{e}'_{c+1}(c+1).$$

Straightforward algebra, which mostly repeats arguments of [464, Theorem 8.5.2], allows us to prove that $\mathbf{R}$ can be explicitly determined, once its spectral radius $\eta(K)$ is known, as

$$\mathbf{R} = p\bar{q}^c \begin{pmatrix} \mathbf{0}_{c\times(c+1)} \\ \mathbf{u} \end{pmatrix},$$

where $\mathbf{u}$ is the left eigenvector of $\mathbf{R}$ corresponding to $\eta(K)$. Indeed, we notice that $\eta(K) = p\bar{q}^c u_{c+1}$, where u_{c+1} is the $(c+1)$th entry of $\mathbf{u}$. The matrix $\mathbf{R}$ can be also expressed as

$$\mathbf{R} = \mathbf{A}_{K,K+1}\left(\mathbf{I}_{c+1} - \sum_{j=1}^{c+1}\eta^{j-1}(K)\mathbf{A}_{K,K+1-j}\right)^{-1}.$$

Thus, the problem reduces to determine the spectral radius $\eta(K)$ and its eigenvector $\mathbf{u}$. One procedure for computing $\eta(K)$ uses Elsner's algorithm. An alternative scheme, appropriate to the present case, is to solve the equation

$$\det\left(\eta(K)\mathbf{I}_{c+1} - \sum_{j=0}^{c+1}\eta^j(K)\mathbf{A}_{K,K+1-j}\right) = 0,$$

which leads to the computation of $\eta(K)$ as a root in $(0,1)$ of a polynomial equation by applying an elementary procedure such as the bisection or the secant method. Once $\eta(K)$ is numerically computed, $\mathbf{u}$ is determined as the solution to

$$\mathbf{u}\left(\eta(K)\mathbf{I}_{c+1} - \sum_{j=0}^{c+1}\eta^j(K)\mathbf{A}_{K,K+1-j}\right) = \mathbf{0}'_{c+1},$$

which satisfies $\eta(K) = p\bar{q}^c u_{c+1}$.

It remains to compute the vectors $\tilde{\pi}(j)$, for $0 \leq j \leq K$. To this end, we remark that the censored Markov chain in the set $\cup_{j=0}^{K}l(j)$ has a transition matrix of the form (9.4) with $\mathbf{\Phi}_i(K)$ given by

$$\mathbf{\Phi}_i(K) = \sum_{j=0}^{c-i}\mathbf{R}^j\mathbf{A}_{K,K-i-j}, \quad 0 \leq i \leq c. \tag{9.15}$$

Therefore, we can follow the arguments used to derive Theorem 9.2. Then, the algorithm is still valid with four modifications. First, we have to compute $\Phi_i(K)$, for $0 \leq i \leq c$, by (9.15) instead of (9.5) and (9.6). Second, we notice that $\mathbf{C}_0(K-1) \neq \mathbf{B}_{10}(K-1)$. As a result, we need to compute $\mathbf{C}_0(K-1)$ instead of keeping $\mathbf{B}_{10}(K-1)$. Third, we must consider the normalization condition given by

$$\tilde{\pi}(K)\left((\mathbf{I}_{c+1} - \mathbf{R})^{-1}\mathbf{e}_{c+1} - \mathbf{B}_{10}(K)\mathbf{B}_{00}^{-1}(K)\mathbf{e}_{(c+1)K}\right) = 1,$$

instead of (9.12). Finally, we have to compute $\tilde{\pi}(j)$, for $j \geq K+1$, by (9.14), up to any desired level of accuracy.

It should be noted that the positive recurrence of $\widetilde{\mathbf{P}}(K)$ implies that $\eta(K) < 1$. Thus, in order to select K_f, we should start with an initial value $K \geq c+1$ such that $\eta(K) < 1$.

Next we discuss the criterion for the selection of K_f. It is clear that a suitably chosen level K_f for a particular performance measure could be inappropriate for another measure. Thus, we suggest here to focus on a unified criterion based on the spectral radius $\eta(K)$. We propose to start with an initial value $K \geq c+1$ and progressively increase the value of K until the change in $\eta(K)$ is sufficiently small. To be more precise, for an arbitrary small $\varepsilon > 0$, we choose the smallest value K_f with relative error

$$\left|1 - \frac{\eta(K_f - 1)}{\eta(K_f)}\right| < \varepsilon. \tag{9.16}$$

With this selection criterion, Tables 9.1 and 9.2 list the values of K_f, the probability B that the server is busy and the mean number of customers in orbit $E[N]$, for queues with $c = 5$, $q = 0.2$, different values of p and s, and $\varepsilon = 10^{-3}$. In both tables, $\overline{R}(K_f)$ and $\widetilde{R}(K_f)$ denote respectively the relative errors associated with the direct truncation and the generalized truncation. Thus, similarly to (9.16), $\overline{R}(K_f)$ and $\widetilde{R}(K_f)$ are evaluated for those approximate values of B and $E[N]$ obtained from $\overline{\mathbf{P}}(K)$ and $\widetilde{\mathbf{P}}(K)$ by setting $K = K_f - 1$ and K_f. In all cases, we observe that $\widetilde{R}(K_f) < \overline{R}(K_f)$, so that we conclude that the generalized truncation seems to provide a better approximation for the stationary characteristics of the $Geo/Geo/c$ retrial queue. Hence, the values for B and $E[N]$ listed in the tables are those obtained by applying a generalized truncation at the level K_f. Note that, for fixed values of s, B and $E[N]$ are increasing functions of p. For fixed values of p, B increases with increasing values of s and, in contrast, $E[N]$ is a decreasing function of s.

9.1.2 Busy Period

Let L_{ij} be a random variable representing the first-passage time to $(0,0)$, starting from a state (i,j). Then, the busy period is defined as $L = L_{10}$. We

Table 9.1. The probability that the server is busy

s		$p=0.2$	$p=0.4$	$p=0.6$	$p=0.8$
0.1	K_f	11	18	23	29
	$\overline{R}(K_f)$	1.112×10^{-11}	5.652×10^{-12}	1.348×10^{-9}	2.062×10^{-6}
	$\widetilde{R}(K_f)$	7.563×10^{-13}	1.071×10^{-13}	1.269×10^{-11}	1.171×10^{-8}
	B	0.00151	0.03010	0.14869	0.43449
0.3	K_f	7	9	11	13
	$\overline{R}(K_f)$	9.847×10^{-8}	1.555×10^{-6}	0.00002	0.00069
	$\widetilde{R}(K_f)$	5.830×10^{-9}	3.670×10^{-8}	2.831×10^{-7}	4.862×10^{-6}
	B	0.00154	0.03134	0.15562	0.44921
0.5	K_f	7	7	7	8
	$\overline{R}(K_f)$	1.025×10^{-7}	0.00002	0.00081	0.00545
	$\widetilde{R}(K_f)$	2.133×10^{-9}	4.186×10^{-7}	0.00001	0.00005
	B	0.00156	0.03212	0.15986	0.45804
0.7	K_f	7	7	7	7
	$\overline{R}(K_f)$	1.016×10^{-7}	0.00002	0.00075	0.00794
	$\widetilde{R}(K_f)$	4.540×10^{-10}	8.641×10^{-8}	2.232×10^{-6}	0.00002
	B	0.00158	0.03266	0.16278	0.46391
0.9	K_f	7	7	7	7
	$\overline{R}(K_f)$	1.003×10^{-7}	0.00002	0.00072	0.00755
	$\widetilde{R}(K_f)$	1.669×10^{-11}	3.111×10^{-9}	7.719×10^{-8}	8.240×10^{-7}
	B	0.00159	0.03307	0.16492	0.46804

are interested in obtaining the probabilities $P(L_{ij} = k)$, for $k \geq 1$, and the moments of L. Conditioning on the first transition of the process yields

$$P(L_{ij} = 1) = P_{(i,j)(0,0)}, \quad 0 \leq i \leq c,\ j \geq 0,\ (i,j) \neq (0,0), \quad (9.17)$$

$$P(L_{ij} = k) = \sum_{(m,n)\in S_{(i,j)}-\{(0,0)\}} P_{(i,j)(m,n)} P(L_{mn} = k-1),$$

$$0 \leq i \leq c,\ j \geq 0,\ (i,j) \neq (0,0),\ k \geq 2. \quad (9.18)$$

For every $k \geq 1$, we define the infinite column vector $\mathbf{l}(k) = (l_0(k), l_1(k), ...)'$ where

$$l_0(k) = (P(L_{10} = k), ..., P(L_{c0} = k))',$$
$$l_j(k) = (P(L_{0j} = k), ..., P(L_{cj} = k))', \quad j \geq 1.$$

Further, we denote by $\mathbf{P}_L$ the matrix that results by removing the first row and the first column of the transition matrix $\mathbf{P}$. Let $\mathbf{p}_0$ be the column vector created by removing the first element of the first column of the matrix $\mathbf{P}$ and $\mathbf{p}_0(0)$ be the column vector containing the c first entries of $\mathbf{p}_0$. The equations (9.17) and (9.18) may be rewritten as

$$\mathbf{l}(1) = \mathbf{p}_0,$$
$$\mathbf{l}(k) = \mathbf{P}_L \mathbf{l}(k-1), \quad k \geq 2. \quad (9.19)$$

9.1 The *Geo/Geo/c* Retrial Queue

Table 9.2. The mean number of customers in orbit

s		$p=0.2$	$p=0.4$	$p=0.6$	$p=0.8$
0.1	K_f	11	18	23	29
	$\overline{R}(K_f)$	3.235×10^{-10}	8.510×10^{-11}	1.466×10^{-8}	0.00001
	$\widetilde{R}(K_f)$	1.907×10^{-11}	1.656×10^{-12}	1.944×10^{-10}	3.055×10^{-7}
	$E[N]$	0.00108	0.04660	0.41150	2.41272
0.3	K_f	7	9	11	13
	$\overline{R}(K_f)$	3.093×10^{-6}	0.00002	0.00028	0.00594
	$\widetilde{R}(K_f)$	1.130×10^{-7}	4.928×10^{-7}	4.040×10^{-6}	0.00010
	$E[N]$	0.00042	0.01997	0.19030	1.19400
0.5	K_f	7	7	7	8
	$\overline{R}(K_f)$	4.163×10^{-6}	0.00043	0.00782	0.03843
	$\widetilde{R}(K_f)$	4.674×10^{-8}	5.500×10^{-6}	0.00013	0.00081
	$E[N]$	0.00029	0.01448	0.14506	0.95982
0.7	K_f	7	7	7	7
	$\overline{R}(K_f)$	5.083×10^{-6}	0.00050	0.00847	0.05742
	$\widetilde{R}(K_f)$	1.048×10^{-8}	1.141×10^{-6}	0.00002	0.00034
	$E[N]$	0.00023	0.01209	0.12574	0.86713
0.9	K_f	7	7	7	7
	$\overline{R}(K_f)$	5.850×10^{-6}	0.00055	0.00898	0.05909
	$\widetilde{R}(K_f)$	3.665×10^{-10}	3.614×10^{-8}	7.229×10^{-7}	8.598×10^{-6}
	$E[N]$	0.00020	0.01077	0.11543	0.82221

Then, the probability $P(L=k)$ equals the first element of $\mathbf{l}_0(k)$; that is, $P(L=k) = \mathbf{e}'_c(1)\mathbf{l}_0(k)$.

For any fixed k, the vector $\mathbf{l}_0(k)$ can be computed recursively by exploiting (9.19). Indeed, because of the $GI/M/1$ structure of $\mathbf{P}_L$, we see that (9.19) leads to

$$\mathbf{l}_0(k) = \mathbf{A}_{00}^*\mathbf{l}_0(k-1) + \mathbf{A}_{01}^*\mathbf{l}_1(k-1), \tag{9.20}$$

$$\mathbf{l}_j(k) = \mathbf{A}_{j0}^*\mathbf{l}_0(k-1) + \sum_{i=1}^{j+1}\mathbf{A}_{ji}\mathbf{l}_i(k-1), \quad 1\leq j\leq c, \tag{9.21}$$

$$\mathbf{l}_j(k) = \sum_{i=j-c}^{j+1}\mathbf{A}_{ji}\mathbf{l}_i(k-1), \quad j\geq c+1, \tag{9.22}$$

where $\mathbf{A}_{00}^*$ is obtained from $\mathbf{A}_{00}$ by deleting the first row and column. Analogously, $\mathbf{A}_{01}^*$ follows from $\mathbf{A}_{01}$ by deleting the first row, while the matrices $\mathbf{A}_{j0}^*$, for $1\leq j\leq c$, are obtained from $\mathbf{A}_{j0}$ by removing the first column.

We next organize the above facts to formulate an algorithm for the computation of $P(L=k)$.

Theorem 9.3. *The probability $P(L=k)$, for any given $k\geq 1$, can be computed as follows:*

Step 1. Set $l_0(1) = p_0(0)$ and $l_j(1) = 0_{c+1}$, for $j = 1, ..., k-1$.
Step 2. For $i = 2, ..., k$, calculate $l_j(i)$ using (9.20)-(9.22), for $j = 0, ..., \min(k - i, (i-1)c)$. For $j = \min(k-i, (i-1)c) + 1, ..., k-i$, set $l_j(i) = 0_{c+1}$.
Step 3. Calculate $P(L = k)$ as

$$P(L = k) = e'_c(1)l_0(k).$$

Proof. From (9.20) and (9.21), we observe that $l_0(k)$ depends on $l_0(k-1)$ and $l_1(k-1)$, but $l_1(k-1)$ depends on $l_0(k-2)$, $l_1(k-2)$ and $l_2(k-2)$, and so on. It readily follows that we need to start by computing the vectors $l_0(1), ..., l_{k-1}(1)$. If the orbit contains at least one customer, then it is impossible to move in one transition to the state $(0, 0)$. Hence, $l_j(1) = 0_{c+1}$, for $j \geq 1$, and using (9.22) we conclude that $l_j(i) = 0_{c+1}$, for $i \geq 1$ and $j \geq (i-1)c + 1$.

After these specifications, the algorithm reduces to a straightforward iteration of formulas (9.20)-(9.22). □

For a given level of accuracy $\varepsilon > 0$, we may compute $P(L = k)$ until the $(1 - \varepsilon)$th percentile of L is achieved, thus stopping computations at a level K such that $P(L \leq K - 1) \leq 1 - \varepsilon < P(L \leq K)$. An alternative option is to choose K as K_f. Then, the idea is to approximate the moment $E[L^n]$ by the truncated moment $\sum_{k=0}^{K} k^n P(L = k)$. However, this method neglects small probabilities $P(L = k)$ which are multiplied by large numbers k^n. An alternative approach may be attained by considering the direct truncation model with transition matrix $\overline{\mathbf{P}}(K)$. If we denote by L^K the length of a busy period in the truncated model, then we have

$$P(L = k) = P(L^K = k), \quad 0 \leq k \leq K.$$

This happens because the event $\{L = k\}$ with $k \leq K$ implies that the process $\{X_t; t \geq 0\}$ does not visit $l(K)$ during the busy period. Then, the transition probabilities of the paths that satisfy the event $\{L = k\}$ are equal either if we use the matrix $\mathbf{P}$ or its truncation $\overline{\mathbf{P}}(K)$. It seems numerically advantageous to approximate the moments of L by the moments of L^K. The reason is that we can think of L^K as a discrete PH distribution with representation $(e'_{c+(c+1)K}(1), \overline{\mathbf{P}}_L(K))$, where $\overline{\mathbf{P}}_L(K)$ is obtained by neglecting the first row and column of $\overline{\mathbf{P}}(K)$. Using standard results for the moments of a discrete PH distribution [464], we have that the factorial moments of L^K are given by

$$E[L^K(L^K - 1)...(L^K - n + 1)]$$
$$= n! e'_{c+(c+1)K}(1)(\mathbf{I}_{c+(c+1)K} - \overline{\mathbf{P}}_L(K))^{-n} \overline{\mathbf{P}}_L^{n-1}(K) e_{c+(c+1)K}, \quad (9.23)$$

for $n \geq 1$. In order to use (9.23), we need to compute the column vector $(\mathbf{I}_{c+(c+1)K} - \overline{\mathbf{P}}_L(K))^{-n} \overline{\mathbf{P}}_L^{n-1}(K) e_{c+(c+1)K}$. It amounts to successively solve linear systems of the form $(\mathbf{I}_{c+(c+1)K} - \overline{\mathbf{P}}_L(K))\mathbf{x} = \mathbf{a}$. This can be done at low

9.1 The $Geo/Geo/c$ Retrial Queue   253

Table 9.3. $P(L = k)$ for several choices of (p, s)

k	$(p,s) = (0.2, 0.1)$	$(p,s) = (0.2, 0.9)$	$(p,s) = (0.8, 0.1)$	$(p,s) = (0.8, 0.9)$
1	0.16000	0.16000	0.03999	0.03999
2	0.11392	0.11392	0.01792	0.01792
3	0.08730	0.08730	0.01068	0.01068
4	0.07038	0.07038	0.00741	0.00741
5	0.05877	0.05877	0.00566	0.00566
6	0.05029	0.05029	0.00462	0.00462
7	0.04380	0.04380	0.00394	0.00395
8	0.03863	0.03863	0.00348	0.00349
9	0.03439	0.03439	0.00315	0.00317
10	0.03081	0.03081	0.00290	0.00293
⋮	⋮	⋮	⋮	⋮
146	1.602×10^{-7}	1.370×10^{-7}	6.115×10^{-4}	1.353×10^{-3}
147	1.467×10^{-7}	1.252×10^{-7}	6.106×10^{-4}	1.351×10^{-3}
148	1.343×10^{-7}	1.144×10^{-7}	6.096×10^{-4}	1.348×10^{-3}
149	1.230×10^{-7}	1.045×10^{-7}	6.087×10^{-4}	1.345×10^{-3}
150	1.126×10^{-7}	9.552×10^{-8}	6.079×10^{-4}	1.342×10^{-3}
$P(L \leq 150)$	0.99999	0.99999	0.22767	0.32783

Table 9.4. Moments of L

s		$p = 0.2$	$p = 0.4$	$p = 0.6$	$p = 0.8$
0.1	$E[L]$	9.46730	22.64749	86.88340	1135.99651
	$Var(L)$	109.54890	754.33596	11530.15448	1682179.59557
0.3	$E[L]$	9.46014	22.15289	74.22936	560.08671
	$Var(L)$	108.99381	692.91946	7659.65627	383643.55288
0.5	$E[L]$	9.45913	22.07519	72.19682	487.87305
	$Var(L)$	108.93164	684.78211	7148.27291	286688.58022
0.7	$E[L]$	9.45878	22.04737	71.44458	461.90924
	$Var(L)$	108.91167	681.97266	6967.03397	255490.27774
0.9	$E[L]$	9.45862	22.03391	71.07639	450.61853
	$Var(L)$	108.90239	680.63252	6879.93031	242618.81403

computational cost by exploiting the $GI/M/1$ structure of the coefficient matrix $\mathbf{I}_{c+(c+1)K} - \overline{\mathbf{P}}_L(K)$. In particular, we can use a block Gaussian elimination procedure.

In Table 9.3, we present the exact mass function of L for various choices of the pair (p, s). In the light of our numerical experience including examples that are not reported here, we point out that $P(L = k)$ seems to be a decreasing function of k, irrespective of the magnitudes of p and s. Nevertheless, Table 9.3 shows that the rate of convergence of $P(L = k)$ towards zero does notably depend on the value of p. In particular, for fixed values of s, the slower convergence is associated with the choice of higher values of p. In Table 9.4, we approximate the moments $E[L]$ and $Var(L)$ by $E[L^{K_f}]$ and $Var(L^{K_f})$, with those levels K_f listed in Table 9.1. As was expected, both descriptors increase with increasing values of p, when we fix s. In contrast, they decrease with increasing values of s, with differences of magnitude more apparent for higher values of p.

9.1.3 Waiting Time

Let W be the waiting time that a tagged customer spends in orbit. We remark that a primary arrival has priority over customers coming from the orbit and that the random order policy among retrying customers is assumed. In what follows, we approximate the analysis of W in the $Geo/Geo/c$ retrial queue by the parallel study for a truncated model with sufficiently large K.

Since we deal with the G-EAS policy, the probability that a primary arriving customer finds the state (c,j) is $\bar{q}^c \bar{\pi}_{cj}$. Let W_{ij} be the conditional waiting time for a customer in orbit, given that the current system state is (i,j). We now may condition on the state viewed by the tagged customer. This yields the following.

Theorem 9.4. *The approximate probability mass function of W is given by*

$$P(W=0) = 1 - \bar{q}^c \sum_{j=0}^{K-1} \bar{\pi}_{cj},$$

$$P(W=k) = \bar{q}^c \sum_{j=0}^{K-1} \bar{\pi}_{cj} P(W_{c,j+1} = k), \quad k \geq 1,$$

where

$$P(W_{ij} = 1) = \sum_{(m,n) \in S^*_{(i,j)}} P_{(i,j)(m,n)} \frac{j-n}{j}, \quad 0 \leq i \leq c, \ 1 \leq j \leq K, \quad (9.24)$$

$$P(W_{ij} = k) = \sum_{(m,n) \in S^*_{(i,j)}} P_{(i,j)(m,n)} \frac{n}{j} P(W_{mn} = k-1)$$
$$+ \delta_{ic}(1 - \delta_{jK}) P_{(c,j)(c,j+1)} P(W_{c,j+1} = k-1)$$
$$+ \delta_{ic}\delta_{jK} P_{(c,K)(c,K+1)} P(W_{cK} = k-1),$$
$$0 \leq i \leq c, \ 1 \leq j \leq K, \ k \geq 2, \quad (9.25)$$

with $S^*_{(i,j)} = S_{(i,j)}$, for $0 \leq i \leq c-1$, and $S^*_{(c,j)} = S_{(c,j)} - \{(c,j+1)\}$.

Equations (9.24) and (9.25) provide a simple recursive scheme to obtain $P(W=k)$ up to any given level of accuracy. In particular, they can be expressed in matrix form as in the case of (9.17) and (9.18) for $P(L_{ij} = k)$. Thus, a similar algorithm could be developed.

We next give recursive relations for the factorial conditional waiting time moments $W_{ij}^{(k)} = E[W_{ij}(W_{ij} - 1)...(W_{ij} - k + 1)]$, for $k \geq 1$.

Corollary 9.5. *The factorial moments* $W_{ij}^{(k)}$, *for* $k \geq 1$, *satisfy that*

$$W_{ij}^{(k)} = \sum_{(m,n) \in S^*_{(i,j)}} P_{(i,j)(m,n)} \left(\delta_{1k} \frac{j-n}{j} + \frac{n}{j} \left(W_{mn}^{(k)} + kW_{mn}^{(k-1)} \right) \right)$$
$$+ \delta_{ic}(1 - \delta_{jK}) P_{(c,j)(c,j+1)} \left(W_{c,j+1}^{(k)} + kW_{c,j+1}^{(k-1)} \right)$$
$$+ \delta_{ic} \delta_{jK} P_{(c,K)(c,K+1)} \left(W_{cK}^{(k)} + kW_{cK}^{(k-1)} \right),$$
$$0 \leq i \leq c, \ 1 \leq j \leq K, \ k \geq 1.$$

Proof. By multiplying (9.24) by z and (9.25) by z^k, and adding for all k, we obtain recursive relations for the corresponding probability generating functions $W_{ij}(z) = \sum_{k=1}^{\infty} z^k P(W_{ij} = k)$:

$$W_{ij}(z) = \sum_{(m,n) \in S^*_{(i,j)}} P_{(i,j)(m,n)} \left(\frac{j-n}{j} z + \frac{n}{j} z W_{mn}(z) \right)$$
$$+ \delta_{ic}(1 - \delta_{jK}) P_{(c,j)(c,j+1)} z W_{c,j+1}(z)$$
$$+ \delta_{ic} \delta_{jK} P_{(c,K)(c,K+1)} z W_{cK}(z), \quad 0 \leq i \leq c, \ 1 \leq j \leq K.$$

Then, by differentiating at $z = 1$, we obtain the desired result. □

Consequently, we may compute $E[W]$ as

$$E[W] = \bar{q}^c \sum_{j=0}^{K-1} \bar{\pi}_{cj} W_{c,j+1}^{(1)}.$$

Tables 9.5 and 9.6 show the influence of p and s on the mass function, the mean and the variance of W. To this end, we approximate these measures by the corresponding characteristics derived from the truncated model with levels K_f given in Table 9.1. Table 9.5 lists values of $P(W = k)$ for four choices of the pair (p, s). The mass function is a decreasing function of k in the cases $(p, s) \in \{(0.2, 0.9), (0.8, 0.9)\}$ and exhibits a two-modal behavior when dealing with $(p, s) \in \{(0.2, 0.1), (0.8, 0.1)\}$. The first mode is always located at $k = 0$. We observe from Table 9.6 that, for fixed values of s, the mean and the variance of W increase with increasing values of p. Both descriptors are decreasing functions of s, for each fixed value of p. Heavier tails correspond to increase p and decrease s, thus implying that the case $(p, s) = (0.8, 0.1)$ with a higher congestion has heavier tails.

9.2 The $BMAP/SM/1$ Retrial Queue

In this section, the model under consideration is an extension of the $M/G/1$ retrial queue to include batch Markovian arrivals and semi-Markovian service

Table 9.5. $P(W = k)$ for several choices of (p, s)

k	$(p,s) = (0.2, 0.1)$	$(p,s) = (0.2, 0.9)$	$(p,s) = (0.8, 0.1)$	$(p,s) = (0.8, 0.9)$
0	0.99950	0.99947	0.85762	0.85134
1	2.920×10^{-5}	2.687×10^{-4}	0.00436	0.03169
2	3.682×10^{-5}	1.353×10^{-4}	0.00559	0.02347
3	3.770×10^{-5}	6.168×10^{-5}	0.00595	0.01723
4	3.617×10^{-5}	2.869×10^{-5}	0.00597	0.01308
5	3.375×10^{-5}	1.382×10^{-5}	0.00584	0.01020
6	3.107×10^{-5}	6.880×10^{-6}	0.00564	0.00813
7	2.837×10^{-5}	3.515×10^{-6}	0.00540	0.00659
8	2.579×10^{-5}	1.836×10^{-6}	0.00515	0.00541
9	2.338×10^{-5}	9.777×10^{-7}	0.00491	0.00450
10	2.115×10^{-5}	5.288×10^{-7}	0.00466	0.00377
11	1.911×10^{-5}	2.899×10^{-7}	0.00443	0.00319
12	1.725×10^{-5}	1.609×10^{-7}	0.00420	0.00271
13	1.556×10^{-5}	9.025×10^{-8}	0.00399	0.00232
14	1.403×10^{-5}	5.109×10^{-8}	0.00379	0.00200
15	1.264×10^{-5}	2.917×10^{-8}	0.00359	0.00173
$P(W \leq 15)$	0.99988	0.99999	0.93117	0.98743

Table 9.6. Moments of W

s		$p = 0.2$	$p = 0.4$	$p = 0.6$	$p = 0.8$
0.1	$E[W]$	0.00541	0.11650	0.68584	3.01579
	$Var(W)$	0.10517	2.39483	16.72482	115.10389
0.3	$E[W]$	0.00212	0.04994	0.31710	1.47391
	$Var(W)$	0.01400	0.38694	3.33642	28.33867
0.5	$E[W]$	0.00146	0.03621	0.24019	1.10087
	$Var(W)$	0.00619	0.19862	1.94009	16.13409
0.7	$E[W]$	0.00117	0.03023	0.20810	0.95268
	$Var(W)$	0.00392	0.14044	1.51183	12.47816
0.9	$E[W]$	0.00101	0.02692	0.19095	0.90038
	$Var(W)$	0.00294	0.11469	1.32479	11.55164

times. Our interest is in the process that describes the system state at arbitrary times. The main feature of the underlying Markov renewal process is the block structure of its embedded Markov chain. Specifically, the embedded Markov chain at departure epochs can be thought of as an asymptotically quasi-Toeplitz Markov chain.

Subsection 9.2.1 focusses on the system state at departure epochs. Here, we will use notation and results in Subsection 7.3.3 as part of our analysis. Moreover, we also denote the kth derivative of a matrix function, say $\mathbf{A}(z)$, at the point $z = 1$ by $\mathbf{A}^{(k)}$. In Subsection 9.2.2, we use the preceding analysis to obtain the limiting distribution of the system state at arbitrary times. To illustrate our analysis, we conclude with some numerical examples.

9.2.1 Embedded Markov Chain at Departure Epochs

The steady state analysis of the $BMAP/SM/1$ retrial queue has been investigated by Dudin and Klimenok [229], who introduce $AQTMCs$ to derive an algorithmic solution for the embedded Markov chain at departure epochs. Their approach is mainly based on the replacement of a non-homogeneous Markov chain by a suitably defined censored Markov chain, whose behavior is similar when the number of customers in orbit is sufficiently large.

Let N_n, J_n and K_n denote the number of customers in orbit, the phase of the arrival process and the phase of the service process at the epoch immediately after the nth service completion. The sequence $\{(N_n, J_n, K_n); n \geq 1\}$ constitutes an irreducible Markov chain with state space

$$\mathcal{S} = \bigcup_{j=0}^{\infty} l(j),$$

where the jth level $l(j)$ is defined as $\{(j, k, l); 1 \leq k \leq m, 1 \leq l \leq r\}$, for $j \geq 0$.

We note that the one-step transition probability matrix $\mathbf{P}$ of $\{(N_n, J_n, K_n); n \geq 1\}$ is given by

$$\mathbf{P} = \begin{pmatrix} \mathbf{A}_{00} & \mathbf{A}_{01} & \mathbf{A}_{02} & \mathbf{A}_{03} & \cdots \\ \mathbf{A}_{10} & \mathbf{A}_{11} & \mathbf{A}_{12} & \mathbf{A}_{13} & \cdots \\ & \mathbf{A}_{21} & \mathbf{A}_{22} & \mathbf{A}_{23} & \cdots \\ & & \ddots & \ddots & \ddots \end{pmatrix}, \quad (9.26)$$

where the square blocks $\mathbf{A}_{jn}$ of dimension mr consist of the transition probabilities from states of $l(j)$ to states in $l(n)$, for $n \geq \max(0, j-1)$.

The following result gives expressions for the above blocks $\mathbf{A}_{jn}$.

Theorem 9.6. *For $n \geq \max(0, j-1)$, the matrices $\mathbf{A}_{jn}$ are given by*

$$\mathbf{A}_{jn} = j\mu \left(j\mu \mathbf{I}_{mr} - \widetilde{\mathbf{D}}_0\right)^{-1} \mathbf{\Omega}_{n-j+1} + \sum_{k=1}^{n-j+1} \left(j\mu \mathbf{I}_{mr} - \widetilde{\mathbf{D}}_0\right)^{-1} \widetilde{\mathbf{D}}_k \mathbf{\Omega}_{n-j-k+1}, \quad (9.27)$$

where $\widetilde{\mathbf{D}}_k = \mathbf{D}_k \otimes \mathbf{I}_r$, for $k \geq 0$, and the matrices $\mathbf{\Omega}_l$, for $l \geq 0$, are defined as the coefficients in the series expansion of the matrix

$$\hat{\beta}(z) = \int_0^\infty \exp\{\mathbf{D}(z)x\} \otimes d\mathbf{B}(x),$$

with $\mathbf{D}(z) = \sum_{k=0}^{\infty} \mathbf{D}_k z^k$.

Proof. From the matrix expansion $\hat{\beta}(z) = \sum_{l=0}^{\infty} \Omega_l z^l$, we may remark that the matrix Ω_l records transition probabilities of the joint distribution of the arrival and service phases corresponding to the arrival of l primary customers during the length of a service time. Thus, in the right-hand side of (9.27), the first term corresponds to the case where a successful repeated attempt occurs. In contrast, the second term represents the cases in which a primary customer belonging to an arriving group of size k occupies the server. □

The matrices $\mathbf{A}_{jn}$ in (9.27) depend not only on $n-j$, but they also depend on j. This implies that the one-step transition probability matrix $\mathbf{P}$ in (9.26) has a particular level dependent structure of $M/G/1$-type.

It can be shown that the Markov chain $\{(N_n, J_n, K_n); n \geq 1\}$ is indeed an AQTMC with $k_0 = 2$ and

$$\mathbf{Q}_1^{(j)} = j\mu \left(j\mu \mathbf{I}_{mr} - \widetilde{\mathbf{D}}_0\right)^{-1}, \quad j \geq 1, \tag{9.28}$$

$$\mathbf{Q}_2^{(j)} = -\widetilde{\mathbf{D}}_0 \left(j\mu \mathbf{I}_{mr} - \widetilde{\mathbf{D}}_0\right)^{-1}, \quad j \geq 1, \tag{9.29}$$

$$\mathbf{V}_j = -\widetilde{\mathbf{D}}_0^{-1} \sum_{k=1}^{j+1} \widetilde{\mathbf{D}}_k \Omega_{j-k+1}, \quad j \geq 0, \tag{9.30}$$

$$\mathbf{Y}_j^{(1)} = \Omega_j, \quad j \geq 0, \tag{9.31}$$

$$\mathbf{Y}_0^{(2)} = \mathbf{0}_{mr \times mr}, \tag{9.32}$$

$$\mathbf{Y}_j^{(2)} = -\widetilde{\mathbf{D}}_0^{-1} \sum_{k=1}^{j} \widetilde{\mathbf{D}}_k \Omega_{j-k}, \quad j \geq 1, \tag{9.33}$$

$$\mathbf{Q}_1 = \mathbf{I}_{mr}, \tag{9.34}$$

$$\mathbf{Q}_2 = \mathbf{0}_{mr \times mr}. \tag{9.35}$$

Expressions (9.28)-(9.33) readily follow from (7.21)-(7.23) by direct verification. The proof of (9.34) and (9.35) is an algebraic proof based on [358, Theorem 4.5.6].

Following [229, Theorem 3] we notice that a sufficient condition for the Markov chain $\{(N_n, J_n, K_n); n \geq 1\}$ to be positive recurrent is the inequality $\rho < 1$, where the traffic load of the system is defined by $\rho = \lambda \beta_1$. In the case of SM service times, the mean value β_1 is given by $\beta_1 = \mathbf{b}\boldsymbol{\delta}$, where $\mathbf{b}$ is the stationary probability vector of $\mathbf{B}(\infty)$ and the column vector $\boldsymbol{\delta} = (\delta_i)$ has entries $\delta_i = \int_0^{\infty}(1 - \sum_{j=1}^{r} b_{ij}(x))dx$, for $1 \leq i \leq r$. In the rest of the chapter, we assume that $\rho < 1$.

Our objective is to investigate the stationary probabilities

$$\pi_{jkl} = \lim_{n \to \infty} P(N_n = j, J_n = k, K_n = l), \quad (j, k, l) \in \mathcal{S}.$$

To this end, we introduce the vectors $\boldsymbol{\pi}(j,k) = (\pi_{jk1}, ..., \pi_{jkr})$, for $j \geq 0$ and $1 \leq k \leq m$, and define the vector generating function

$$\Pi(z) = \sum_{j=0}^{\infty} \pi(j) z^j, \quad |z| \leq 1,$$

where $\pi(j) = (\pi(j,1), ..., \pi(j,m))$ assembles the stationary probabilities of states in the level $l(j)$, for $j \geq 0$.

From Subsection 7.3.3, we establish the following result.

Theorem 9.7. *If $\rho < 1$, then the vector generating function $\Pi(z)$ satisfies the integral equation*

$$\Pi(z) = -\frac{1}{\mu} \int_0^z \Pi(u) e^{-(\mathbf{I}_{mr} + \mu^{-1}\widetilde{\mathbf{D}}_0)\ln u} du \, e^{\mu^{-1}\widetilde{\mathbf{D}}_0 \ln z} \widetilde{\mathbf{D}}_0 \Phi(z), \quad (9.36)$$

where $\Phi(z) = -\widetilde{\mathbf{D}}_0^{-1} \widetilde{\mathbf{D}}(z) \hat{\beta}(z)(z\mathbf{I}_{mr} - \hat{\beta}(z))^{-1}$ and $\widetilde{\mathbf{D}}(z) = \mathbf{D}(z) \otimes \mathbf{I}_r$.

Proof. First, we write (7.26) in the particular case of (9.28)-(9.33) as follows:

$$\Pi(z) \left(z\mathbf{I}_{mr} - \hat{\beta}(z)\right) = \sum_{j=0}^{\infty} \pi(j) z^j \left(j\mu\mathbf{I}_{mr} - \widetilde{\mathbf{D}}_0\right)^{-1} \widetilde{\mathbf{D}}(z) \hat{\beta}(z).$$

The equation (9.36) is then obtained from the equality

$$\sum_{j=0}^{\infty} \pi(j) z^j \left(j\mu\mathbf{I}_{mr} - \widetilde{\mathbf{D}}_0\right)^{-1} = \int_0^{\infty} \Pi(ze^{-\mu v}) e^{\widetilde{\mathbf{D}}_0 v} dv$$

and the change of variable $u = ze^{-\mu v}$. □

Differentiating (9.36) with respect to z gives

$$\frac{d\Pi(z)}{dz} = \Pi(z) \widetilde{\mathbf{S}}(z), \quad (9.37)$$

where the matrix function $\widetilde{\mathbf{S}}(z)$ is defined by

$$\widetilde{\mathbf{S}}(z) = \Phi^{-1}(z) \frac{d\Phi(z)}{dz} - \frac{1}{\mu z} \left(\mathbf{I}_{mr} - \Phi^{-1}(z)\right) \widetilde{\mathbf{D}}_0 \Phi(z).$$

Dudin and Klimenok [229] point out that the problem of solving the linear differential equation (9.37) seems to be notably difficult since $\widetilde{\mathbf{S}}(z)$ is a singular matrix in some points of the unit disk. Thus, they suggest to proceed as in Subsection 7.3.3 and adapt the recursion given by the equation (7.25) to formulas (9.28)-(9.33).

At this stage, we recall that the distribution $\{\pi_{jkl}; (j,k,l) \in \mathcal{S}\}$ is then replaced by the stationary distribution $\{\bar{\pi}_{jkl}; 0 \leq j \leq K_f, 1 \leq k \leq m, 1 \leq l \leq r\}$ of the censored Markov chain with censoring set consisting of all states in and below the level $l(K_f)$, provided we take K_f large enough. To evaluate the probabilities of the censored Markov chain, we note that the vectors $\bar{\pi}(j)$

satisfy (7.25), for $1 \leq j \leq K_f$, where $\bar{\pi}(0)$ is computed as the solution to the set of linear equations (7.30).

From (9.28)-(9.35), it becomes easy to observe that the limit chain is indeed a Markov chain of $M/G/1$-type with one-step transition probability matrix

$$\widetilde{\mathbf{P}} = \begin{pmatrix} \mathbf{V}_0 & \mathbf{V}_1 & \mathbf{V}_2 & \mathbf{V}_3 & \cdots \\ \mathbf{\Omega}_0 & \mathbf{\Omega}_1 & \mathbf{\Omega}_2 & \mathbf{\Omega}_3 & \cdots \\ & \mathbf{\Omega}_0 & \mathbf{\Omega}_1 & \mathbf{\Omega}_2 & \cdots \\ & & \ddots & \ddots & \ddots \end{pmatrix}.$$

It can be verified that $\widetilde{\mathbf{P}}$ coincides with the one-step transition probability matrix of the embedded Markov chain at departure epochs in the standard $BMAP/SM/1$ queue. The underlying matrix $\mathbf{G}$ is thus the minimal solution to

$$\mathbf{G} = \int_0^\infty \exp\{\mathbf{D}(\mathbf{G})x\} \otimes d\mathbf{B}(x), \qquad (9.38)$$

in the set of non-negative matrices.

Following Subsection 7.3.3, we remember that, to choose a suitable value of K_f, we may take for any $\varepsilon > 0$ an initial value K to obtain a global error $\theta_j < \varepsilon$, for all $j \geq K$, where θ_j was defined to approximate the level dependent matrices $\mathbf{G}_j$, for $j \geq K$, as accurately as possible by means of the solution $\mathbf{G}$ to (9.38). Then, we start with this value of K and progressively increase K until $\bar{\pi}(K_f)\mathbf{e}_{mr} < \varepsilon_f$, for an arbitrary small $\varepsilon_f > 0$.

As an alternative proposal, we may use residuals. More specifically, straightforward algebra based on (9.37) leads to

$$\Pi^{(1)}\mathbf{e}_{mr} = \Pi(1)\mathbf{\Phi}_0^{-1}\left(\mathbf{\Phi}_1 + \frac{1}{\mu}(\mathbf{I}_{mr} - \mathbf{\Phi}_0)\widetilde{\mathbf{D}}_0\mathbf{\Phi}_0\right)\mathbf{e}_{mr}, \qquad (9.39)$$

where the matrices $\mathbf{\Phi}_0$ and $\mathbf{\Phi}_1$ are defined as the coefficients in the matrix expansion

$$\mathbf{\Phi}(z) = \mathbf{\Phi}_0 + \mathbf{\Phi}_1(z-1) + o(z-1).$$

Thus, by substituting the probabilities $\{\bar{\pi}_{jkl}; 0 \leq j \leq K, 1 \leq k \leq m, 1 \leq l \leq r\}$ instead of the exact values $\{\pi_{jkl}; (j,k,l) \in \mathcal{S}\}$ into (9.39), we may define the residual

$$g(K) = \bar{\Pi}^{(1)}\mathbf{e}_{mr} - \bar{\Pi}(1)\mathbf{\Phi}_0^{-1}\left(\mathbf{\Phi}_1 + \frac{1}{\mu}(\mathbf{I}_{mr} - \mathbf{\Phi}_0)\widetilde{\mathbf{D}}_0\mathbf{\Phi}_0\right)\mathbf{e}_{mr},$$

where $\bar{\Pi}(z) = \sum_{j=0}^K \bar{\pi}(j)z^j$. Hence, we may successively increase the level K until finding a value K_f such that $|g(K_f)| < \varepsilon_f$.

To finish this subsection we show how to derive expressions for $\mathbf{\Phi}_0$ and $\mathbf{\Phi}_1$. We know that $\mathbf{\Phi}(z)$ satisfies the matrix equation

$$\Phi(z)\left(z\mathbf{I}_{mr} - \hat{\beta}(z)\right) = -\tilde{\mathbf{D}}_0^{-1}\tilde{\mathbf{D}}(z)\hat{\beta}(z), \tag{9.40}$$

from which it follows that

$$\Phi_0\left(\hat{\beta}(1) - \mathbf{I}_{mr}\right) = \tilde{\mathbf{D}}_0^{-1}\tilde{\mathbf{D}}(1)\hat{\beta}(1), \tag{9.41}$$

$$\Phi_1\left(\hat{\beta}(1) - \mathbf{I}_{mr}\right) = \tilde{\mathbf{D}}_0^{-1}\left(\tilde{\mathbf{D}}^{(1)}\hat{\beta}(1) + \tilde{\mathbf{D}}(1)\hat{\beta}^{(1)}\right) - \Phi_0\left(\hat{\beta}^{(1)} - \mathbf{I}_{mr}\right). \tag{9.42}$$

Since the matrix $\hat{\beta}(1) - \mathbf{I}_{mr}$ is singular, Φ_0 and Φ_1 shall be evaluated as the solution of (9.41) and (9.42) subject to

$$\Phi_0\left(\hat{\beta}^{(1)} - \mathbf{I}_{mr}\right)\mathbf{e}_{mr} = \tilde{\mathbf{D}}_0^{-1}\left(\tilde{\mathbf{D}}^{(1)} + \tilde{\mathbf{D}}(1)\hat{\beta}^{(1)}\right)\mathbf{e}_{mr},$$

$$2\Phi_1\left(\hat{\beta}^{(1)} - \mathbf{I}_{mr}\right)\mathbf{e}_{mr} = \tilde{\mathbf{D}}_0^{-1}\left(\tilde{\mathbf{D}}^{(2)} + 2\tilde{\mathbf{D}}^{(1)}\hat{\beta}^{(1)} + \tilde{\mathbf{D}}(1)\hat{\beta}^{(2)}\right)\mathbf{e}_{mr}$$
$$- \Phi_0\hat{\beta}^{(2)}\mathbf{e}_{mr}.$$

The proof of these equations is by differentiation of both sides in (9.40).

In matrix form, Φ_0 and Φ_1 are given by

$$\Phi_l = \mathbf{H}_l\mathbf{C}^{-1}, \quad l \in \{0,1\},$$

where $\mathbf{C}$ is obtained from the matrix $\hat{\beta}(1) - \mathbf{I}_{mr}$ by replacing its first column by the vector $(\hat{\beta}^{(1)} - \mathbf{I}_{mr})\mathbf{e}_{mr}$. To define $\mathbf{H}_0$, we replace the first column of the matrix $\tilde{\mathbf{D}}_0^{-1}\tilde{\mathbf{D}}(1)\hat{\beta}(1)$ by the vector

$$\tilde{\mathbf{D}}_0^{-1}\left(\tilde{\mathbf{D}}^{(1)} + \tilde{\mathbf{D}}(1)\hat{\beta}^{(1)}\right)\mathbf{e}_{mr}.$$

In a similar manner, $\mathbf{H}_1$ is obtained from

$$\tilde{\mathbf{D}}_0^{-1}\left(\tilde{\mathbf{D}}^{(1)}\hat{\beta}(1) + \tilde{\mathbf{D}}(1)\hat{\beta}^{(1)}\right) - \Phi_0\left(\hat{\beta}^{(1)} - \mathbf{I}_{mr}\right),$$

by replacing its first column by

$$\frac{1}{2}\left(\tilde{\mathbf{D}}_0^{-1}\left(\tilde{\mathbf{D}}^{(2)} + 2\tilde{\mathbf{D}}^{(1)}\hat{\beta}^{(1)} + \tilde{\mathbf{D}}(1)\hat{\beta}^{(2)}\right) - \Phi_0\hat{\beta}^{(2)}\right)\mathbf{e}_{mr}.$$

9.2.2 Limiting Distribution of the System State

We have seen in the previous subsection how to derive the stationary distribution of the embedded Markov chain at departure epochs. Now we want to use this in order to obtain an expression for the limiting probabilities of the process $\mathcal{X} = \{(C(t), N(t), J(t), K(t)); t \geq 0\}$ defined by

$$P_{ijkl} = \lim_{t\to\infty} P\left(C(t) = i, N(t) = j, J(t) = k, K(t) = l\right),$$

for $i \in \{0, 1\}$, $j \geq 0$, $1 \leq k \leq m$ and $1 \leq l \leq r$.

Let $\eta_0 = 0 < \eta_1 < \eta_2 < ...$ be the instants of successive service completions. Obtaining the limiting distribution of the process $\mathcal{X}$ is a simple matter if we conceive this process as a semi-regenerative process with embedded Markov renewal process $\{(N_n, J_n, K_n, \eta_n); n \geq 1\}$. The process is regular, since $\rho < 1$.

We begin with some notation. Let $\mathbf{p}_0(j)$ (respectively, $\mathbf{p}_1(j)$) be the row vector of dimension mr with $((k-1)r + l)$th entry recording the limiting probability that the server is idle (respectively, busy), the orbit size equals j, and the arrival and the service phases are k and l, respectively, for $j \geq 0$, $1 \leq k \leq m$ and $1 \leq l \leq r$. Let $\mathbf{p}_i(z)$ be the vector generating function defined by $\sum_{j=0}^{\infty} \mathbf{p}_i(j) z^j$, for $i \in \{0, 1\}$.

The following theorem is the main result of this subsection.

Theorem 9.8. *The vector generating function of the orbit size at an arbitrary time is given by*

$$(\mathbf{p}_0(z) + \mathbf{p}_1(z))\mathbf{e}_{mr} = -\frac{1}{\lambda}(1-z)\boldsymbol{\Pi}(z)\hat{\boldsymbol{\beta}}^{-1}(z)\widetilde{\mathbf{D}}^{-1}(z)\mathbf{e}_{mr}. \quad (9.43)$$

Moreover, the generating function of the number of customers in the system at an arbitrary time satisfies

$$(\mathbf{p}_0(z) + z\mathbf{p}_1(z))\mathbf{e}_{mr} = -\frac{1}{\lambda}(1-z)\boldsymbol{\Pi}(z)\widetilde{\mathbf{D}}^{-1}(z)\mathbf{e}_{mr}. \quad (9.44)$$

Proof. Since we deal with an irreducible aperiodic recurrent Markov renewal process, an appeal to [197, Theorem 10.6.12] yields

$$\mathbf{p}_0(j) = \frac{1}{\pi \mathbf{m}} \boldsymbol{\pi}(j) \left(j\mu \mathbf{I}_{mr} - \widetilde{\mathbf{D}}_0\right)^{-1}, \quad j \geq 0,$$

where $\boldsymbol{\pi} = (\boldsymbol{\pi}(0), \boldsymbol{\pi}(1), ...)$ and $\mathbf{m}$ is a column vector consisting of the expected sojourn times in each state $(j, k, l) \in \mathcal{S}$.

Post-multiplying both sides in this equality by $z^j \mathbf{e}_{mr}$ and adding, we get

$$\mathbf{p}_0(z)\mathbf{e}_{mr} = -\frac{1}{\pi \mathbf{m}} \boldsymbol{\Pi}(z) \boldsymbol{\Phi}^{-1}(z) \widetilde{\mathbf{D}}_0^{-1} \mathbf{e}_{mr}. \quad (9.45)$$

In a similar manner, we find that

$$\mathbf{p}_1(z)\mathbf{e}_{mr} = -\frac{1}{\pi \mathbf{m}} \boldsymbol{\Pi}(z) \hat{\boldsymbol{\beta}}^{-1}(z) \widetilde{\mathbf{D}}^{-1}(z)(\mathbf{I}_{mr} - \hat{\boldsymbol{\beta}}(z))\mathbf{e}_{mr}. \quad (9.46)$$

Then, in view of (9.45) and (9.46), expressions (9.43) and (9.44) are readily derived, since $\lambda = \pi \mathbf{m}$ [229]. □

Formulas (9.43) and (9.44) are appropriate to obtain the moments of the orbit size and the number of customers in the system in terms of those moments corresponding to the embedded Markov chain at departures.

It is important to observe that due to the nature of the input process, which is not Poisson, the limiting probabilities of the process $\mathcal{X}$ will not be necessarily the steady state probabilities at arrivals. We give an illustration of this fact in the corollary below. Other probabilistic descriptors at arrival epochs can be found in [229, Section 6].

Corollary 9.9. *(i) The vector* $\mathbf{p}_0 = \mathbf{p}_0(1)$ *describing, at an arbitrary time, the system state when the server is idle is given by*

$$\mathbf{p}_0 = -\frac{1}{\lambda}\boldsymbol{\Pi}(1)\boldsymbol{\Phi}_0^{-1}\widetilde{\mathbf{D}}_0^{-1}.$$

(ii) The probability that a batch of customers finds the server idle upon arrival is given by

$$p_0^{(a)} = -\frac{\mathbf{p}_0 \widetilde{\mathbf{D}}_0 \mathbf{e}_{mr}}{\lambda}.$$

(iii) The conditional probability that, at an arbitrary time, the server is idle, given that the orbit size is j, has the form

$$q_0(j) = \frac{\mathbf{p}_0(j)\mathbf{e}_{mr}}{(\mathbf{p}_0(j) + \mathbf{p}_1(j))\mathbf{e}_{mr}}, \quad j \geq 0.$$

To finish, we present some numerical results for $BMAP/SM/1$ retrial queues where the maximum batch size equals three. In a first set of experiments, we assume that the semi-Markov kernel describing the service process is characterized by

$$\mathbf{B}(x) = \mathrm{diag}(B_1(x), ..., B_r(x))\mathbf{B}(\infty),$$

where $B_i(x)$ is the probability distribution function of the Erlang law with k_i phases and rate ν_i, for $1 \leq i \leq r$, and the matrix $\mathbf{B}(\infty)$ gives the transition probabilities among phases of the SM service process. Such an assumption facilitates the representation of $\boldsymbol{\Omega}_l$ as

$$\boldsymbol{\Omega}_l = \mathbf{U}\mathrm{diag}(\boldsymbol{\Omega}_l^{(1)}, ..., \boldsymbol{\Omega}_l^{(r)})\mathbf{U}'\left(\mathbf{I}_m \otimes \mathbf{B}(\infty)\right), \quad l \geq 0,$$

where the matrices $\boldsymbol{\Omega}_l^{(i)}$, for $l \geq 0$, are defined as the coefficients in the series expansion

$$\sum_{l=0}^{\infty} \boldsymbol{\Omega}_l^{(i)} z^l = \int_0^{\infty} \exp\{\mathbf{D}(z)x\}\, dB_i(x), \quad 1 \leq i \leq r,$$

and $\mathbf{U}$ denotes the square permutation matrix of order mr defined by

$$\mathbf{U} = \sum_{k=1}^{m}\sum_{n=1}^{r} \mathbf{E}_{mr}(k,n) \otimes \mathbf{E}_{rm}(n,k),$$

with $\mathbf{E}_{mr}(k,n) = \mathbf{e}_m(k)\mathbf{e}'_r(n)$.

In Figure 9.2, we illustrate the influence of ρ and μ on $E[N]$ in the case $m = 2$ and $r = 3$. In particular, the above semi-Markov kernel has been normalized to fix $\beta_1 = 1.0$ by choosing the pairs $(k_1, \nu_1) = (10, 100.0)$, $(k_2, \nu_2) = (3, 3.0)$ and $(k_3, \nu_3) = (2, 1.05263)$, and the matrix

$$\mathbf{B}(\infty) = \frac{1}{6}\begin{pmatrix} 3.0 & 3.0 & 0 \\ 2.0 & 2.0 & 2.0 \\ 0 & 3.0 & 3.0 \end{pmatrix}.$$

We plot three curves that correspond to the values of the traffic load $\rho = 0.25$, 0.5 and 0.75. To get these values, we consider the following three choices of the $BMAP$ process, with a coefficient of correlation $c_c = 0.1$:

(i) $\rho = 0.25$

$$\mathbf{D}_0 = \begin{pmatrix} -0.14686 & 4.353 \times 10^{-8} \\ 4.353 \times 10^{-8} & -0.00321 \end{pmatrix}, \quad \mathbf{D}_1 = \begin{pmatrix} 0.04392 & 0.00013 \\ 0.00075 & 0.00021 \end{pmatrix},$$

$$\mathbf{D}_2 = \begin{pmatrix} 0.05856 & 0.00018 \\ 0.00100 & 0.00028 \end{pmatrix}, \quad \mathbf{D}_3 = \mathbf{D}_1.$$

(ii) $\rho = 0.5$

$$\mathbf{D}_0 = \begin{pmatrix} -0.29373 & 8.707 \times 10^{-8} \\ 8.707 \times 10^{-8} & -0.00643 \end{pmatrix}, \quad \mathbf{D}_1 = \begin{pmatrix} 0.08785 & 0.00027 \\ 0.00150 & 0.00042 \end{pmatrix},$$

$$\mathbf{D}_2 = \begin{pmatrix} 0.11713 & 0.00036 \\ 0.00200 & 0.00056 \end{pmatrix}, \quad \mathbf{D}_3 = \mathbf{D}_1.$$

(iii) $\rho = 0.75$

$$\mathbf{D}_0 = \begin{pmatrix} -0.44060 & 1.306 \times 10^{-7} \\ 1.306 \times 10^{-7} & -0.00965 \end{pmatrix}, \quad \mathbf{D}_1 = \begin{pmatrix} 0.13177 & 0.00040 \\ 0.00225 & 0.00063 \end{pmatrix},$$

$$\mathbf{D}_2 = \begin{pmatrix} 0.17570 & 0.00054 \\ 0.00300 & 0.00085 \end{pmatrix}, \quad \mathbf{D}_3 = \mathbf{D}_1.$$

From Figure 9.2, it is observed that the mean orbit size decreases as μ increases, for ρ fixed, and increases as ρ increases, for μ fixed. Furthermore, it is worth noting that the approach yields values for the probability B that the server is busy which fit the five first decimal digits of the corresponding value of ρ, hence we may surmise that B equals ρ in the $BMAP/SM/1$ retrial queue.

In Table 9.7, we show how the variability in the arrival and service processes affects the mean orbit size in retrial queues with $\mu = 2.0$. For the service process, we consider simple distributions, such as deterministic, E_3 and H_2 laws with $\beta_1 = 1.0$. In the case H_2, the value $\beta_1 = 1.0$ is obtained from the choice $p_1 = 0.1$, $\nu_1 = 18.1$ and $\nu_2 = 0.905$. For the arrival stream, we define

9.2 The $BMAP/SM/1$ Retrial Queue

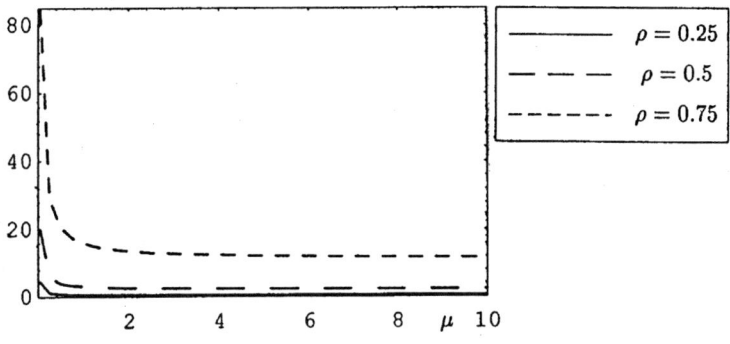

Fig. 9.2. $E[N]$ versus μ

$BMAP$ processes to fit the values $c_c = 0.0, 0.1$ and 0.2. Specifically, the arrival processes with null correlation are Poisson processes with batch arrivals and the following representations:

(i) $\rho = 0.25$

$$\mathbf{D}_0 = -0.13888, \quad \mathbf{D}_1 = 0.05555, \quad \mathbf{D}_2 = \mathbf{D}_1, \quad \mathbf{D}_3 = 0.02777.$$

(ii) $\rho = 0.5$

$$\mathbf{D}_0 = -0.27777, \quad \mathbf{D}_1 = 0.11111, \quad \mathbf{D}_2 = \mathbf{D}_1, \quad \mathbf{D}_3 = 0.05555.$$

(iii) $\rho = 0.75$

$$\mathbf{D}_0 = -0.41666, \quad \mathbf{D}_1 = 0.16666, \quad \mathbf{D}_2 = \mathbf{D}_1, \quad \mathbf{D}_3 = 0.08333.$$

The choices of the $BMAP$ with $c_c = 0.1$ are defined as those used in our preceding numerical example. To define the cases of the $BMAP$ with $c_c = 0.2$, we choose the following representations:

(i) $\rho = 0.25$

$$\mathbf{D}_0 = \begin{pmatrix} -0.16863 & 1.363 \times 10^{-7} \\ 1.363 \times 10^{-7} & -0.00548 \end{pmatrix}, \quad \mathbf{D}_1 = \begin{pmatrix} 0.05025 & 0.00033 \\ 0.00091 & 0.00072 \end{pmatrix},$$

$$\mathbf{D}_2 = \begin{pmatrix} 0.06700 & 0.00044 \\ 0.00122 & 0.00097 \end{pmatrix}, \quad \mathbf{D}_3 = \mathbf{D}_1.$$

(ii) $\rho = 0.5$

$$\mathbf{D}_0 = \begin{pmatrix} -0.33726 & 2.727 \times 10^{-7} \\ 2.727 \times 10^{-7} & -0.01097 \end{pmatrix}, \quad \mathbf{D}_1 = \begin{pmatrix} 0.10051 & 0.00067 \\ 0.00183 & 0.00145 \end{pmatrix},$$

$$\mathbf{D}_2 = \begin{pmatrix} 0.13401 & 0.00089 \\ 0.00244 & 0.00194 \end{pmatrix}, \quad \mathbf{D}_3 = \mathbf{D}_1.$$

Table 9.7. Values of $E[N]$ for several arrival and service processes

ρ		Poisson	$c_c = 0.1$	$c_c = 0.2$
0.25	D	0.52277	0.59281	0.62922
	E_3	0.57340	0.65538	0.70467
	H_2	0.53686	0.61024	0.65024
0.5	D	1.57193	2.02069	2.49877
	E_3	1.87591	2.44081	3.07044
	H_2	1.65661	2.13795	2.65915
0.75	D	4.73364	9.46050	21.61340
	E_3	6.09063	12.04450	24.82200
	H_2	5.11148	10.19380	22.54070

(iii) $\rho = 0.75$

$$\mathbf{D}_0 = \begin{pmatrix} -0.50590 & 4.090 \times 10^{-7} \\ 4.090 \times 10^{-7} & -0.01645 \end{pmatrix}, \quad \mathbf{D}_1 = \begin{pmatrix} 0.15076 & 0.00100 \\ 0.00275 & 0.00218 \end{pmatrix},$$

$$\mathbf{D}_2 = \begin{pmatrix} 0.20102 & 0.00134 \\ 0.00367 & 0.00291 \end{pmatrix}, \quad \mathbf{D}_3 = \mathbf{D}_1.$$

An examination of the entries in Table 9.7 reveals that $E[N]$ appears to be an increasing function of the correlation value, irrespective of the service time distribution. If we fix the arrival pattern, then we see that higher and smaller values are always related to E_3 and deterministic service times, respectively, thus showing that the mean orbit size seems to have a non-monotone behavior as a function of the service time coefficient of variation. Similarly to Figure 9.2, we observe that $E[N]$ increases with increasing values of ρ.

9.3 Bibliographical Notes

The materials in Section 9.1 are drawn mostly from the paper [112]. In such a paper, the proof of the positive recurrence of $\{X_t; t \geq 0\}$ is based on mean drifts and the limit behavior of the one-step transition probabilities $P_{(i,j)(m,n)}$ as $n \to \infty$.

In the literature on retrial queues, the $GI/M/1$ structure commonly arises from the presence of negative arrivals [42, 610] and the use of a full access rule [73]. In [42], the averaging principle and the matrix-analitic formalism are combined in order to analyze a Markovian multiserver queue with negative arrivals, which have the effect of removing a random batch of customers from the retrial group. Shin [610] considers exponential service and retrial patterns, and several types of arrivals governed by a single MAP with marked transitions. The distinguishing feature in [73] is that a successful retrial transfers a batch of customers from the orbit to the idle servers.

9.3 Bibliographical Notes

The presentation of Section 9.2 is largely based on [229], where the notion of $AQTMC$ first appeared. $AQTMC$s are treated in their full generality in [411]. We refer to equations (29)-(31) in [411] for an alternative recursion to evaluate the stationary vector at departure epochs. As noted in [229], the most laborious part in implementation of the solution given in Subsection 9.2.1 is the calculation of the matrices $\boldsymbol{\Omega}_l$, for $l \geq 0$, which should be derived by exploiting the special nature of the $BMAP$ and the SM service process under study. As the reader may verify, the matrix function $\boldsymbol{\Phi}(z)$ defined in Theorem 9.7 yields the matrix functional equation $\boldsymbol{\Pi}(z) = \boldsymbol{\pi}(0)\boldsymbol{\Phi}(z)$ for the corresponding vector generating function of the standard $BMAP/SM/1$ queue; see e.g. [497, 498].

Without any claim to an exhaustive enumeration, we remark that $AQTMC$s are applied to a variety of queues of type $BMAP/PH/c$ [167, 401, 412], the $BMAP/G/1$ queue with search of customers from the orbit [237] and the $BMAP/SM/1$ retrial queue with multi-threshold control [393].

As an alternative to the use of $AQTMC$s, Li et al. [479] apply RG-factorizations of a level dependent $M/G/1$ structure to the $BMAP/G/1$ retrial queue with unreliable server. To that end, they use the supplementary variable method, combined with the matrix-analytic formalism and the censoring technique, and suggest a new treatment for boundary conditions of the underlying system of differential equations.

References

1. J. Abate and W. Whitt (1995), Numerical inversion of Laplace transforms of probability distributions, *ORSA Journal on Computing* **7**, 36-43.
2. J. Abate, G.L. Choudhury and W. Whitt (2000), An introduction to numerical transform inversion and its application to probability models. In: *Computational Probability* (Ed. W.K. Grassmann), pp. 257-323, Kluwer Academic Publishers, Boston.
3. J. Abate and W. Whitt (2006), A unified framework for numerically inverting Laplace transforms, *INFORMS Journal on Computing* **18**, 408-421.
4. A. Aboul-Hassan, S.I. Rabia and A. Kadry (2005), Analytical study of a discrete time retrial queue with balking customers and early arrival scheme, *Alexandria Engineering Journal* **44**, 911-917.
5. A. Aboul-Hassan, S.I. Rabia and A. Kadry (2005), A recursive approach for analyzing a discrete time retrial queue with balking customers and early arrival scheme, *Alexandria Engineering Journal* **44**, 919-925.
6. V.M. Abramov (2006), Analysis of multiserver retrial queueing system: A martingale approach and an algorithm of solution, *Annals of Operations Research* **141**, 19-50.
7. V.M. Abramov (2007), Multiserver queueing systems with retrials and losses, *ANZIAM Journal of Australian Mathematical Society* **48**, 297-314.
8. L.G. Afanas'eva (1994), Ergodicity conditions for queueing systems with repeated calls, *Journal of Mathematical Sciences* **72**, 2835-2838.
9. N. Agmon, Y. Alhassid and R.D. Levine (1979), An algorithm for finding the distribution of maximal entropy, *Journal of Computational Physics* **30**, 250-258.
10. M.S. Aguir, F. Karaesmen, O.Z. Akşin and F. Chauvet (2004), The impact of retrials on call center performance, *OR Spectrum* **26**, 353-376.
11. M.S. Aguir, O.Z. Akşin, F. Karaesmen and Y. Dallery (2008), On the interaction between retrials and sizing of call centers, *European Journal of Operational Research*, in press.
12. M. Aida, C. Takano, M. Murata and M. Imase (2008), A study of control plane stability with retry traffic: Comparison of hard- and soft-state protocols, *IEICE Transactions on Communications* **E91-B**, 437-445.
13. A. Aissani (1988), On the $M/G/1/1$ queueing system with repeated orders and unreliable server, *J. Technology* **6**, 98-123 (in French).

14. A. Aissani (1992), Heavy loading approximation of the unreliable queue with repeated orders. In: *Actes du Colloque "Méthodes et Outils d'Aide à la Décision", MOAD'92*, pp. 97-102 (Vol. 1), Béjaïa.
15. A. Aissani (1993), Unreliable queuing with repeated orders, *Microelectronics and Reliability* **33**, 2093-2106.
16. A. Aissani (1994), A retrial queue with redundancy and unreliable server, *Queueing Systems* **17**, 431-449.
17. A. Aissani and N. Oukid (1994), Bounds for the mean busy and idle periods of some unreliable queues. In: *Actes des Journees de Statistiques Appliquées, JSA 94*, pp. 78-84, Alger.
18. A. Aissani and J.R. Artalejo (1998), On the single server retrial queue subject to breakdowns, *Queueing Systems* **30**, 309-321.
19. A. Aissani (2000), An $M^X/G/1$ retrial queue with exhaustive vacations, *Journal of Statistics & Management Systems* **3**, 269-286.
20. A. Aissani (2001), On retrial queues with vacations or with breakdowns. In: *Queues, Flows, Systems, Networks. Proceedings of the International Conference "Modern Mathematical Methods of Investigating of the Information Networks"*, pp. 7-10, Minsk.
21. A. Aissani (2003), An $M^X/G/1$ retrial queue with unreliable server and vacations. In: *Proceedings of the 10th International Conference on Analytical and Stochastic Modelling Techniques and Applications, ASMTA'03* (Ed. D. Al-Dabass), pp. 175-180, SCS-European Publishing House, Nothingham.
22. M. Ajmone Marsan, G. De Carolis, E. Leonardi, R. Lo Cigno and M. Meo (2000), An approximate model for the computation of blocking probabilities in cellular networks with repeated calls, *Telecommunication Systems* **15**, 53-62.
23. M. Ajmone Marsan, G. De Carolis, E. Leonardi, R. Lo Cigno and M. Meo (2000), Approximate Markovian models of cellular mobile telephone networks with customer retrials. In: *IEEE International Conference on Communications, ICC 2000*, pp. 356-361 (Vol. 1).
24. M. Ajmone Marsan, G. De Carolis, E. Leonardi, R. Lo Cigno and M. Meo (2001), Efficient estimation of call blocking probabilities in cellular mobile telephony networks with customer retrials, *IEEE Journal on Selected Areas in Communications* **19**, 332-346.
25. N. Akar and K. Sohraby (1997), An invariant subspace approach in $M/G/1$ and $G/M/1$ type Markov chains, *Stochastic Models* **13**, 381-416.
26. N. Akar and K. Sohraby (1997), Finite and infinite QBD chains: A simple and unifying algorithmic approach. In: *Proceedings IEEE Infocom'97*, pp. 1105-1113, IEEE Computer Society Press.
27. A.M. Aleksandrov (1974), A queueing system with repeated orders, *Engineering Cybernetics* **12**, No. 3, 1-4.
28. A.S. Alfa and W. Li (2002), PCS networks with correlated arrival process and retrial phenomenon, *IEEE Transactions on Wireless Communications* **1**, 630-637.
29. A.S. Alfa and K.P. Sapna Isotupa (2004), An $M/PH/k$ retrial queue with finite number of sources, *Computers and Operations Research* **31**, 1455-1464.
30. A.S. Alfa (2006), Discrete-time analysis of the $GI/G/1$ system with Bernoulli retrials: An algorithmic approach, *Annals of Operations Research* **141**, 51-66.
31. B. Almási, G. Bolch and J. Sztrik (2004), Heterogeneous finite-source retrial queues, *Journal of Mathematical Sciences* **121**, 2590-2596.

32. B. Almási, J. Roszik and J. Sztrik (2005), Homogeneous finite-source retrial queues with server subject to breakdowns and repairs, *Mathematical and Computer Modelling* **42**, 673-682.
33. H. Alshaer and E. Horlait (2005), The joint distribution of server state and queue length of $M/M/1/1$ retrial queue with abandonment and feedback. In: *8th International Symposium on DSP and Communication Systems, DSPCS'2005*, Sunshine Coast.
34. E. Altman and A.A. Borovkov (1997), On the stability of retrial queues, *Queueing Systems* **26**, 343-363.
35. J. Amador and J.R. Artalejo (2007), On the distribution of the successful and blocked events in the $M/M/c$ retrial queue: A computational approach, *Applied Mathematics and Computation* **190**, 1612-1626.
36. J. Amador and J.R. Artalejo (2008), The $M/G/1$ retrial queue: New descriptors of the customer's behavior, *Journal of Computational and Applied Mathematics*, in press.
37. V.V. Anisimov and Kh.L. Atadzhanov (1992), The diffusion approximation of systems with repeated calls, *Theory of Probability and Mathematical Statistics* **44**, 1-5.
38. V.V. Anisimov and Kh.L. Atadzhanov (1994), Diffusion approximation of systems with repeated calls and an unreliable server, *Journal of Mathematical Sciences* **72**, 3032-3034.
39. V.V. Anisimov (1999), Averaging methods for transient regimes in overloading retrial queueing systems, *Mathematical and Computer Modelling* **30**, No. 3-4, 65-78.
40. V.V. Anisimov (1999), Switching stochastic models and applications in retrial queues, *Top* **7**, 169-186.
41. V.V. Anisimov (1999), Asymptotic analysis of highly reliable retrial systems with finite capacity. In: *Queues, Flows, Systems, Networks. Proceedings of the International Conference "Modern Mathematical Methods of Investigating the Telecommunicational Networks"*, pp. 7-12, Minsk.
42. V.V. Anisimov and J.R. Artalejo (2001), Analysis of Markov multiserver retrial queues with negative arrivals, *Queueing Systems* **39**, 157-182.
43. V.V. Anisimov and M. Kurtulush (2001), Some Markovian queuing retrial systems under light-traffic conditions, *Cybernetics and Systems Analysis* **37**, 876-887.
44. V.V. Anisimov and J.R. Artalejo (2002), Approximation of multiserver retrial queues by means of generalized truncated models, *Top* **10**, 51-66.
45. N.M. Apaolaza and J.R. Artalejo (2005), On the time to reach a certain orbit level in multi-server retrial queues, *Applied Mathematics and Computation* **168**, 686-703.
46. J.R. Artalejo (1992), Information theoretic approximations for retrial queueing systems. In: *Transactions of the 11th Prague Conference on Information Theory, Statistical Decision Functions and Random Processes* (Eds. S. Kubik and J.A. Vísek), pp. 263-270, Kluwer Academic Publishers, Dordrecht.
47. J.R. Artalejo (1993), Explicit formulae for the characteristics of the $M/H_2/1$ retrial queue, *Journal of the Operational Research Society* **44**, 309-313.
48. J.R. Artalejo (1994), New results in retrial queueing systems with breakdown of the servers, *Statistica Neerlandica* **48**, 23-36.
49. J.R. Artalejo and G.I. Falin (1994), Stochastic decomposition for retrial queues, *Top* **2**, 329–342.

50. J.R. Artalejo and A. Gómez-Corral (1994), Analysis of the modes of the stationary distribution in single server retrial queues with quasi-random input. In: *Transactions of the 12th Prague Conference on Information Theory, Statistical Decision Functions and Random Processes* (Eds. P. Lachout and J.A. Vísek), pp. 24-27, Academy of Sciences of the Czech Republic, Prague.
51. J.R. Artalejo and M. Martin (1994), A maximum entropy analysis of the $M/G/1$ queue with constant repeated attempts. In: *Selected Topics on Stochastic Modelling* (Eds. R. Gutiérrez and M.J. Valderrama), pp. 181-190, World Scientific, Singapore.
52. J.R. Artalejo (1995), A queueing system with returning customers and waiting line, *Operations Research Letters* **17**, 191-199.
53. J.R. Artalejo and A. Gómez-Corral (1995), Information theoretic analysis for queueing systems with quasi-random input, *Mathematical and Computer Modelling* **22**, No. 3, 65-76.
54. J.R. Artalejo (1996), Stationary analysis of the characteristics of the $M/M/2$ queue with constant repeated attempts, *Opsearch* **33**, 83-95.
55. J.R. Artalejo and G.I. Falin (1996), On the orbit characteristics of the $M/G/1$ retrial queue, *Naval Research Logistics* **43**, 1147-1161.
56. J.R. Artalejo and A. Gómez-Corral (1996), Stochastic analysis of the departure and quasi-input processes in a versatile single-server queue, *Journal of Applied Mathematics and Stochastic Analysis* **9**, 171-183.
57. J.R. Artalejo (1997), Analysis of an $M/G/1$ queue with constant repeated attempts and server vacations, *Computers & Operations Research* **24**, 493-504.
58. J.R. Artalejo and A. Gómez-Corral (1997), Steady state solution of a single-server queue with linear repeated requests, *Journal of Applied Probability* **34**, 223-233.
59. J.R. Artalejo (1998), Retrial queues with a finite number of sources, *Journal of the Korean Mathematical Society* **35**, 503-525.
60. J.R. Artalejo (1998), Retrial queues with negative arrivals. In: *Proceedings of the International Conference on Stochastic Processes* (Ed. A. Krishnamoorthy), pp. 159-167, Cochin University of Science and Technology, Cochin.
61. J.R. Artalejo and A. Gómez-Corral (1998), Unreliable retrial queues due to service interruptions arising from facsimile networks, *Belgian Journal of Operations Research, Statistics and Computer Science* **38**, 31-41.
62. J.R. Artalejo and A. Gómez-Corral (1998), Generalized birth and death processes with applications to queues with repeated attempts and negative arrivals, *OR Spektrum* **20**, 5-14.
63. J.R. Artalejo and A. Gómez-Corral (1998), Analysis of a stochastic clearing system with repeated attempts, *Stochastic Models* **14**, 623-645.
64. J.R. Artalejo (1999), Accessible bibliography on retrial queues, *Mathematical and Computer Modelling* **30**, No. 3-4, 1-6.
65. J.R. Artalejo (1999), A classified bibliography of research on retrial queues: Progress in 1990-1999, *Top* **7**, 187-211.
66. J.R. Artalejo (Ed.) (1999), *Retrial Queuing Systems, Mathematical and Computer Modelling* **30**, No. 3-4, 1-228.
67. J.R. Artalejo (Ed.) (1999), 1st *International Workshop on Retrial Queues, Top* **7**, No. 2, 169-353.
68. J.R. Artalejo (1999), Numerical investigation of multiserver retrial queues operating under the constant retrial policy. In: *Queues, Flows, Systems, Networks*.

Proceedings of the International Conference "Modern Mathematical Methods of Investigating the Telecommunicational Networks", pp. 13-16, Minsk.

69. J.R. Artalejo and A. Gómez-Corral (1999), On a single server queue with negative arrivals and request repeated, *Journal of Applied Probability* **36**, 907-918.
70. J.R. Artalejo and A. Gómez-Corral (1999), Performance analysis of a single-server queue with repeated attempts, *Mathematical and Computer Modelling* **30**, No. 3-4, 79-88.
71. J.R. Artalejo and A. Gómez-Corral (1999), Computation of the limiting distribution in queueing systems with repeated attempts and disasters, *RAIRO-Operations Research* **33**, 371-382.
72. J.R. Artalejo and A. Rodrigo (1999), On the single server queue with linear retrial policy and exhaustive vacations. In: *Stochastic Processes and Their Applications* (Eds. A. Vijayakumar and M. Sreenivasan), pp. 196-205, Narosa Publishing House, New Delhi.
73. J.R. Artalejo, A. Gómez-Corral and M.F. Neuts (2000), Numerical analysis of multiserver retrial queues operating under a full access policy. In: *Advances in Algorithmic Methods for Stochastic Models* (Eds. G. Latouche and P.G. Taylor), pp. 1-19, Notable Publications, Inc., New Jersey.
74. J.R. Artalejo and M.J. Lopez-Herrero (2000), On the single server retrial queue with balking, *Infor* **38**, 33-50.
75. J.R. Artalejo and M.J. Lopez-Herrero (2000), On the busy period of the $M/G/1$ retrial queue, *Naval Research Logistics* **47**, 115-127.
76. J.R. Artalejo and M.J. Lopez-Herrero (2000), Low retrial analysis of multiserver queues with repeated attempts due to impatience. In: *Proceedings of the 12th European Simulation Symposium* (Ed. D.P.F. Möller), pp. 535-538, A Publication of the Society for Computer Simulation International, Delft.
77. J.R. Artalejo, V. Rajagopalan and R. Sivasamy (2000), On finite Markovian queues with repeated attempts, *Investigacion Operativa* **9**, 83-94.
78. J.R. Artalejo, A.N. Dudin and V.I. Klimenok (2001), Stationary analysis of a retrial queue with preemptive repeated attempts, *Operations Research Letters* **28**, 173-180.
79. J.R. Artalejo, A. Gómez-Corral and M.F. Neuts (2001), Analysis of multiserver queues with constant retrial rate, *European Journal of Operational Research* **135**, 569-581.
80. J.R. Artalejo and M.J. Lopez-Herrero (2001), Analysis of the busy period for the $M/M/c$ queue: An algorithmic approach, *Journal of Applied Probability* **38**, 209-222.
81. J.R. Artalejo and M.J. Lopez-Herrero (2001), On the $M/G/1$ queue with quadratic repeated attempts, *Statistical Methods* **3**, 60-78.
82. J.R. Artalejo and G.I. Falin (2002), Standard and retrial queueing systems: A comparative analysis, *Revista Matemática Complutense* **15**, 101-129.
83. J.R. Artalejo, G.I. Falin and M.J. Lopez-Herrero (2002), A second order analysis of the waiting time in the $M/G/1$ retrial queue, *Asia-Pacific Journal of Operational Research* **19**, 131-148.
84. J.R. Artalejo, V.C. Joshua and A. Krishnamoorthy (2002), An $M/G/1$ retrial queue with orbital search by the server. In: *Advances in Stochastic Modelling* (Eds. J.R. Artalejo and A. Krishnamoorthy), pp. 41-54, Notable Publications, Inc., New Jersey.

85. J.R. Artalejo and M. Pozo (2002), Numerical calculation of the stationary distribution of the main multiserver retrial queue, *Annals of Operations Research* **116**, 41-56.
86. J.R. Artalejo and A. Gómez-Corral (2003), Channel idle periods in computer and telecommunication systems with customer retrials, *Telecommunication Systems* **24**, 29-46.
87. J.R. Artalejo and I. Atencia (2004), On the single server retrial queue with batch arrivals, *Sankhyā* **66**, 140-158.
88. J.R. Artalejo and G. Choudhury (2004), Steady state analysis of an $M/G/1$ queue with repeated attempts and two-phase service, *Quality Technology & Quantitative Management* **1**, 189-199.
89. J.R. Artalejo and M.J. Lopez-Herrero (2004), The $M/G/1$ retrial queue: An information theoretic approach. In: *Proceedings of the Fifth International Workshop on Retrial Queues* (Ed. B.D. Choi), pp. 1-16, Seoul.
90. J.R. Artalejo, I. Atencia and P. Moreno (2005), A discrete-time $Geo^{[X]}/G/1$ retrial queue with control of admission, *Applied Mathematical Modelling* **29**, 1100-1120.
91. J.R. Artalejo and A. Economou (2005), On the non-existence of product-form solutions for queueing networks with retrials, *Journal of Simulation Engineering* **27**, No. 1, 13-19.
92. J.R. Artalejo, A. Economou and M.J. Lopez-Herrero (2005), Algorithmic analysis for the number of customers served in a busy period of the $M/M/c$ retrial queue. In: *Proceedings of the Third National Conference on Mathematical and Computational Models* (Eds. R. Arumuganathan and R. Nadarajan), pp. 3-15, Allied Publishers, New Delhi.
93. J.R. Artalejo and A. Gómez-Corral (2005), Waiting time in the $M/M/c$ queue with finite retrial group, *Bulletin of Kerala Mathematics Association* **2**, 1-17.
94. J.R. Artalejo and M.J. Lopez-Herrero (2005), The $M/G/1$ retrial queue: An information theoretic approach, *Statistics and Operations Research Transactions* **29**, 119-137.
95. J.R. Artalejo and M.J. Lopez-Herrero (2005), A discrete-time multiserver retrial queue: Performance analysis and simulation. In: *Proceedings of the Fifth Workshop on Simulation* (Eds. S.M. Ermakov, V.B. Melas and A.N. Pepelyshev), pp. 85-90, NII Chemistry Saint Petersburg University Publishers, Saint Petersburg.
96. J.R. Artalejo, D.S. Orlovsky and A.N. Dudin (2005), Multi-server retrial model with variable number of active servers, *Computers & Industrial Engineering* **48**, 273-288.
97. J.R. Artalejo (Ed.) (2006), *Algorithmic Methods in Retrial Queues*, Annals of Operations Research **141**, 1-301.
98. J.R. Artalejo and S.R. Chakravarthy (2006), Computational analysis of the maximal queue length in the $MAP/M/c$ retrial queue, *Applied Mathematics and Computation* **183**, 1399-1409.
99. J.R. Artalejo and S.R. Chakravarthy (2006), Algorithmic analysis of the maximum level length in general-block two-dimensional Markov processes, *Mathematical Problems in Engineering*, Article ID 53570, pp. 1-15.
100. J.R. Artalejo and S.R. Chakravarthy (2006), Algorithmic analysis of the $MAP/PH/1$ retrial queue, *Top* **14**, 293-332.

101. J.R. Artalejo, A. Krishnamoorthy and M.J. Lopez-Herrero (2006), Numerical analysis of (s, S) inventory systems with repeated attempts, *Annals of Operations Research* **141**, 67-83.
102. J.R. Artalejo and J.A.C. Resing (2006), Mean value analysis of single server retrial queues, *SPOR-Report 2006-13*, Eindhoven University of Technology.
103. J.R. Artalejo, S.R. Chakravarthy and M.J. Lopez-Herrero (2007), The busy period and the waiting time analysis of a $MAP/M/c$ queue with finite retrial group, *Stochastic Analysis and Applications* **25**, 445-469.
104. J.R. Artalejo, A. Economou and A. Gómez-Corral (2007), Applications of maximum queue lengths to call center management, *Computers & Operations Research* **34**, 983-996.
105. J.R. Artalejo, A. Economou and M.J. Lopez-Herrero (2007), Algorithmic approximations for the busy period distribution of the $M/M/c$ retrial queue, *European Journal of Operational Research* **176**, 1687-1702.
106. J.R. Artalejo, A. Economou and M.J. Lopez-Herrero (2007), Algorithmic analysis of the maximum queue length in a busy period for the $M/M/c$ retrial queue, *INFORMS Journal on Computing* **19**, 121-126.
107. J.R. Artalejo and A. Gómez-Corral (2007), Modelling communication systems with phase type service and retrial times, *IEEE Communications Letters* **11**, 955-957.
108. J.R. Artalejo and A. Gómez-Corral (2007), Waiting time analysis of the $M/G/1$ queue with finite retrial group, *Naval Research Logistics* **54**, 524-529.
109. J.R. Artalejo and A. Gómez-Corral (2007), A note on the busy period of the $M/G/1$ queue with finite retrial group, *Probability in the Engineering and Informational Sciences* **21**, 77-82.
110. J.R. Artalejo and M.J. Lopez-Herrero (2007), On the distribution of the number of retrials, *Applied Mathematical Modelling* **31**, 478-489.
111. J.R. Artalejo and M.J. Lopez-Herrero (2007), A simulation study of a discrete-time multiserver retrial queue with finite population, *Journal of Statistical Planning and Inference* **137**, 2536-2542.
112. J.R. Artalejo, A. Economou and A. Gómez-Corral (2008), Algorithmic analysis of the $Geo/Geo/c$ retrial queue, *European Journal of Operational Research*, in press.
113. J.R. Artalejo and A. Gómez-Corral (Eds.) (2008), *Advances in Retrial Queues*, European Journal of Operational Research, in press.
114. J.R. Artalejo and V. Pla (2008), On the impact of customer balking, impatience and retrials in telecommunication systems, submitted.
115. S. Asmussen and G. Koole (1993), Marked point processes as limits of Markovian arrival streams, *Journal of Applied Probability* **30**, 365-372.
116. S. Asmussen (2000), Matrix-analytic models and their analysis, *Scandinavian Journal of Statistics* **27**, 193-226.
117. I. Atencia, P.P. Bocharov and N.H. Phong (1999), Some results for the $MAP/PH/1/0$ system with finite retrial queue and constant retrial rate. In: *Queues, Flows, Systems, Networks. Proceedings of the International Conference "Modern Mathematical Methods of Investigating the Telecommunicational Networks"*, pp. 17-21, Minsk.
118. I. Atencia, P.P. Bocharov and D.A. Puzikova (1999), A matrix-multiplicative solution of a single-server system with service interruption, a finite queue of repeat customers, and phase-type distributions, *Automation and Remote Control* **60**, 1273-1289.

119. I. Atencia, P.P. Bocharov and D.A. Puzikova (1999), Matrix-multiplicative solution for the PH/PH/1/0/s queueing system with limited retrial queue, *Vestnik Rossijskogo Universiteta Druzhby Narodov. Seriya Prikladnaya Matematika i Informatika*, No. 1, 52-65.
120. I. Atencia, P.P. Bocharov, C. D'Apice and N.H. Phong (2000), A single-server retrial queue system with multidimensional Poisson flow, *Automation and Remote Control* **61**, 1871-1884.
121. I. Atencia, C. D'Apice, R. Manzo and S. Salerno (2000), Retrial queueing system with several input flows, negative customers and LCFS PR discipline. In: *Technical Proceedings of the Fourth International Workshop on Queueing Networks with Finite Capacity, QNETs 2000* (Ed. D. Kouvatsos), pp. 02/1-9, Bradford.
122. I. Atencia (2001), A queueing system under LCFS PR discipline with general retrial times. In: *Queues, Flows, Systems, Networks. Proceedings of the International Conference "Modern Mathematical Methods of Investigating of the Information Networks"*, pp. 30-34, Minsk.
123. I. Atencia, S. Sánchez, I. Fortes and P. Moreno (2002), A queueing system with non-persistent customers, several input flows and constant repeated attempts. In: *Computational and Mathematical Methods on Science and Engineering* (Eds. J. Vigo-Aguiar and B.A. Wade), pp. 25-33 (Vol. III), Alicante.
124. I. Atencia, I. Fortes, P. Moreno and S. Sánchez (2003), The $M/G/1$ retrial queue with starting failures, feedback and linear retrial policy. In: *Proceedings of the Sixth Hellenic European Conference on Computer Mathematics and Its Applications* (Ed. E.A. Lipitakis), pp. 669-678, LEA Publishers, Athens.
125. I. Atencia and P. Moreno (2003), A queueing system with linear repeated attempts, Bernoulli schedule and feedback, *Top* **11**, 285-310.
126. I. Atencia, I. Fortes, P. Moreno and S. Sánchez (2004), A retrial queue with priority customers, several input flows and classical retrial policy. In: *Transactions of XXIV International Seminar on Stability Problems for Stochastic Models* (Eds. A. Andronov, P.P. Bocharov and V. Korolev), pp. 95-102, Transport and Telecommunication Institute, Jurmala.
127. I. Atencia and P. Moreno (2004), Discrete-time $Geo^{[X]}/G_H/1$ retrial queue with Bernoulli feedback, *Computers and Mathematics with Applications* **47**, 1273-1294.
128. I. Atencia and P. Moreno (2004), A discrete-time $Geo/G/1$ retrial queue with general retrial times, *Queueing Systems* **48**, 5-21.
129. I. Atencia and P. Moreno (2004), A discrete-time retrial queue with feedback. In: *Proceedings of the Fifth International Workshop on Retrial Queues* (Ed. B.D. Choi), pp. 61-65, Seoul.
130. I. Atencia and P. Moreno (2004), A discrete-time retrial queue with 2nd optional service. In: *Proceedings of the Fifth International Workshop on Retrial Queues* (Ed. B.D. Choi), pp. 117-121, Seoul.
131. I. Atencia and P. Moreno (2005), A single-server retrial queue with general retrial times and Bernoulli schedule, *Applied Mathematics and Computation* **162**, 855-880.
132. I. Atencia and P. Moreno (2005), A discrete-time retrial queue with exhaustive vacations. In: *Proceedings of the Fifth Workshop on Simulation* (Eds. S.M. Ermakov, V.B. Melas and A.N. Pepelyshev), pp. 95-100, NII Chemistry Saint Petersburg University Publishers, Saint Petersburg.

133. I. Atencia, P.P. Bocharov and P. Moreno (2006), A discrete-time $Geo/PH/1$ queueing system with repeated attempts, *Information Processes* **6**, 272-280.
134. I. Atencia, I. Fortes, P. Moreno and S. Sánchez (2006), An $M/G/1$ retrial queue with active breakdowns and Bernoulli schedule in the server, *International Journal of Information and Management Sciences* **17**, 1-17.
135. I. Atencia and P. Moreno (2006), A discrete-time $Geo/G/1$ retrial queue with the server subject to starting failures, *Annals of Operations Research* **141**, 85-107.
136. I. Atencia and P. Moreno (2006), A discrete-time $Geo/G/1$ retrial queue with server breakdowns, *Asia-Pacific Journal of Operational Research* **23**, 247-271.
137. I. Atencia and P. Moreno (2006), $Geo/G/1$ retrial queue with 2nd optional service, *International Journal of Operational Research* **1**, 340-362.
138. I. Atencia, P. Moreno and G. Bouza (2006), An $M_2/G_2/1$ retrial queue with priority customers, 2nd optional service and linear retrial policy, *Revista de Investigación Operacional* **27**, 229-248.
139. I. Atencia, G. Bouza and P. Moreno (2008), An $M^{[X]}/G/1$ retrial queue with server breakdowns and constant rate of repeated attempts, *Annals of Operations Research* **157**, 225-243.
140. F. Avram and A. Gómez-Corral (2006), On bulk-service $MAP/PH^{L,N}/1/N$ G-queues with repeated attempts, *Annals of Operations Research* **141**, 109-137.
141. J. Barner and G. Bolch (2002), Performance modeling of retrial systems using the *MOSEL* language. In: *Proceedings of the 9th International Conference on Analytical and Stochastic Modelling Techniques* (Eds. K. Amborski and H. Neuth), pp. 567-571, Darmstadt.
142. G.P. Basharin and V.E. Merkulov (2003), Blocking probability analysis of new and handover calls in cellular mobile networks with repeated attempts. In: *Proceedings of the 7th International Conference on Telecommunications, ConTEL 2003* (Eds. D. Jevtić and M. Mikuc), pp. 273-278, Faculty of Electrical Engineering and Computing, University of Zagreb, Croatia.
143. L. Berdjoudj and D. Aissani (2003), Strong stability in retrial queues, *Theory of Probability and Mathematical Statistics* **68**, 11-17.
144. L. Berdjoudj and D. Aissani (2005), Martingale methods for analyzing the $M/M/1$ retrial queue with negative arrivals, *Journal of Mathematical Sciences* **131**, 5595-5599.
145. L.T.M. Berry (1987), A repeated calls model encompassing non-Poisson traffic streams in alternative routing networks, *Australian Telecommunication Research* **21**, 21-28.
146. D.A. Bini, G. Latouche and B. Meini (2005), *Numerical Methods for Structured Markov Chains*, Oxford University Press, Oxford.
147. P.P. Bocharov, O.I. Pavlova and D.A. Puzikova (1997), The $M/G/1/r$ retrial queuing system with non-preemptive priority for primary customers. In: *Proceedings of the International Conference Distributed Computer Communication Networks, DCCN'97*, pp. 29-41, Tel-Aviv.
148. P.P. Bocharov, O.I. Pavlova and D.A. Puzikova (1999), $M/G/1/r$ retrial queueing systems with priority of primary customers, *Mathematical and Computer Modelling* **30**, No. 3-4, 89-98.
149. P.P. Bocharov and A.V. Pechinkin (1999), Application of branching processes to investigate the $MAP/G/1/0/n$ queueing system with retrials. In: *Proceed-*

ings of the International Conference Distributed Computer Communication Networks, DCCN'99, pp. 20-26, Tel-Aviv.
150. P.P. Bocharov, C. D'Apice, B. D'Auria and S. Salerno (2000), A queueing system of finite capacity with the server requiring a priority search for customers, *Vestnik Rossijskogo Universiteta Druzhby Narodov. Seriya Prikladnaya Matematika i Informatika*, No. 1, 49-59.
151. P.P. Bocharov, A.V. Pechinkin and N.H. Phong (2000), Stationary probabilities of the states of the retrial system $MAP/G/1/r$ with priority servicing of primary customers, *Automation and Remote Control* **61**, 1300-1309.
152. P.P. Bocharov, C. D'Apice, R. Manzo and N.H. Phong (2001), On a multiclass arrival retrial $M_K/G_K/1/r$ queueing system with finite buffer. In: *Queues, Flows, Systems, Networks. Proceedings of the International Conference "Modern Mathematical Methods of Investigating of the Information Networks"*, pp. 57-60, Minsk.
153. P.P. Bocharov, C. D'Apice and N.H. Phong (2001), On a retrial single-server queueing system with finite buffer and Poisson flow, *Problems of Information Transmission* **37**, 248-261.
154. P.P. Bocharov, C. D'Apice and N.H. Phong (2001), The $M/G/1/r$ queueing system with finite buffer and retrials. In: *Queues, Flows, Systems, Networks. Proceedings of the International Conference "Modern Mathematical Methods of Investigating of the Information Networks"*, pp. 53-56, Minsk.
155. P.P. Bocharov, C. D'Apice, G. Rizelian and S. Salerno (2001), On the single-server queueing system with fixed number of retrials. In: *Proceedings of the 15th European Simulation Multiconference* (Eds. E.J.H. Kerckhoffs and M. Snorek), pp. 751-753, A Publication of the Society for Computer Simulation International, Delft.
156. P.P. Bocharov and N.H. Phong (2001), Retrial queues with several input flows, *Journal of Simulation Engineering* **23**, No. 2, 44-53.
157. P.P. Bocharov, N.H. Phong and I. Atencia (2001), Retrial queueing system with several input flows, *Revista de Investigación Operacional* **22**, 135-143.
158. P.P. Bocharov, C. D'Apice, A.V. Pechinkin and S. Salerno (2004), *Queueing Theory*, Brill Academic Publishers, Utrech.
159. G. Bolch, J. Roszik and J. Sztrik (2003), Heterogeneous finite-source retrial queues in the analysis of communication systems with CSMA/CD protocols. In: *Queues, Flows, Systems, Networks. Proceedings of the International Conference "Modern Mathematical Methods of Analysis and Optimization of Telecommunication Networks"*, pp. 39-45, Gomel.
160. G. Bolch, S. Greiner, M. De Meer and K.S. Trivedi (2006), *Queueing Networks and Markov Chains: Modeling and Performance Evaluation with Computer Science Applications*, Wiley-Interscience, New Jersey.
161. T. Bonald (2006), The Erlang model with non-Poisson call arrivals, *ACM Sigmetrics* **34**, 276-286.
162. V.V. Borokhovsky (2001), A retrial $M/G/1$ system with passive service. In: *Queues, Flows, Systems, Networks. Proceedings of the International Conference "Modern Mathematical Methods of Investigating of the Information Networks"*, pp. 61-64, Minsk.
163. S.C. Borst, R.J. Boucherie and O.J. Boxma (1998), ERMR: A generalised equivalent random method for overflow systems with repacking, *CWI Report PNA-R9817*, Centrum voor Wiskunde en Informatica.

164. P. Boyer, A. Dupuis and A. Khelladi (1988), A simple model for repeated calls due to time-outs. In: *Teletraffic Science for New Cost-Effective Systems, Networks and Services, ITC-12* (Ed. M. Bonatti), pp. 356-363, Elsevier Science Publishers B.V., Amsterdam.
165. A. Brandt and T. Schiemann (2003), A multi-server batch arrival retrial system, *Technical Report 2003*, Institute for Operations Research, Humboldt Universität zu Berlin.
166. G. Bretschneider (1970), Repeated calls with limited repetition probability. In: *Proceedings of the 6th International Teletraffic Congress, ITC-6*, pp. 434/1-5, Munich.
167. L. Breuer, A.N. Dudin and V.I. Klimenok (2002), A retrial $BMAP/PH/N$ system, *Queueing Systems* **40**, 433-457.
168. L. Breuer, V.I. Klimenok, A. Birukov, A.N. Dudin and U.R. Krieger (2005), Mobile networks modeling the access to a wireless network at hot spots, *European Transactions on Telecommunications* **16**, 309-316.
169. L. Bright and P.G. Taylor (1995), Calculating the equilibrium distribution in level dependent quasi-birth-and-death processes, *Stochastic Models* **11**, 497-525.
170. A.C. Brooms (2000), Individual equilibrium dynamic routing in a multiple server retrial queue, *Probability in the Engineering and Informational Sciences* **14**, 9-26.
171. H. Bruneel and B.G. Kim (1993), *Discrete-Time Models for Communication Systems Including ATM*, Kluwer Academic Publishers, Boston.
172. C.D. Carothers, R.M. Fujimoto and Y.B. Lin (1995), A re-dial model for personal communications services networks. In: *IEEE 45th Vehicular Technology Conference*, pp. 135-139 (Vol. 1).
173. S.R. Chakravarthy (2001), The batch Markovian arrival process: A review and future work. In: *Advances in Probability and Stochastic Processes* (Eds. A. Krishnamoorthy, N. Raju and V. Ramaswami), pp. 21-49, Notable Publications, Inc., New Jersey.
174. S.R. Chakravarthy and A.N. Dudin (2002), A multi-server retrial queue with $BMAP$ arrivals and group services, *Queueing Systems* **42**, 5-31.
175. S.R. Chakravarthy and A.N. Dudin (2003), Analysis of a retrial queuing model with MAP arrivals and two types of customers, *Mathematical and Computer Modelling* **37**, 343-363.
176. S.R. Chakravarthy, A. Krishnamoorthy and V.C. Joshua (2006), Analysis of a multi-server retrial queue with search of customers from the orbit, *Performance Evaluation* **63**, 776-798.
177. M.L. Chaudhry (2000), On numerical computations of some discrete-time queues. In: *Computational Probability* (Ed. W.K. Grassmann), pp. 365-407, Kluwer Academic Publishers, Boston.
178. K.K. Cheng, K.T. Ko and S.W.C. Suen (1990), Optimization of telephone networks in developing nations with example. In: *IEEE Region 10 Conference on Computer and Communication Systems*, pp. 371-375, Hong-Kong.
179. B.D. Choi and K.K. Park (1990), The $M/G/1$ retrial queue with Bernoulli schedule, *Queueing Systems* **7**, 219-227.
180. B.D. Choi and V.G. Kulkarni (1992), Feedback retrial queueing systems. In: *Queueing and Related Models* (Eds. U.N. Bhat and I.V. Basawa), pp. 93-105, Oxford University Press, New York.

181. B.D. Choi, Y.W. Shin and W.C. Ahn (1992), Retrial queues with collision arising from unslotted CSMA/CD protocol, *Queueing Systems* **11**, 335-356.
182. B.D. Choi, D.H. Han and G.I. Falin (1993), On the virtual waiting time for an $M/G/1$ retrial queue with two types of calls, *Journal of Applied Mathematics and Stochastic Analysis* **6**, 11-23.
183. B.D. Choi, K.K. Park and C.E.M. Pearce (1993), An $M/M/1$ retrial queue with control policy and general retrial times, *Queueing Systems* **14**, 275-292.
184. B.D. Choi, K.H. Rhee and K.K. Park (1993), The $M/G/1$ retrial queue with retrial rate control policy, *Probability in the Engineering and Informational Sciences* **7**, 29-46.
185. B.D. Choi, K.B. Choi and Y.W. Lee (1995), $M/G/1$ retrial queueing systems with two types of calls and finite capacity, *Queueing Systems* **19**, 215-229.
186. B.D. Choi and K.H. Rhee (1996), An $M/G/1$ retrial queue with a threshold in the retrial group, *Kyungpook Mathematical Journal* **35**, 469-479.
187. B.D. Choi and J.W. Kim (1997), Discrete-time $Geo_1, Geo_2/G/1$ retrial queueing systems with two types of calls, *Computers and Mathematics with Applications* **33**, No. 10, 79-88.
188. B.D. Choi, Y.C. Kim and Y.W. Lee (1998), The $M/M/c$ retrial queue with geometric loss and feedback, *Computers and Mathematics with Applications* **36**, No. 6, 41-52.
189. B.D. Choi and D.B. Zhu (1998), The $M_1/M_2/G/1/K$ retrial queueing systems with priority, *Journal of the Korean Mathematical Society* **35**, 691-712.
190. B.D. Choi and Y. Chang (1999), Single server retrial queues with priority calls, *Mathematical and Computer Modelling* **30**, No. 3-4, 7-32.
191. B.D. Choi and Y. Chang (1999), $MAP_1, MAP_2/M/c$ retrial queue with the retrial group of finite capacity and geometric loss, *Mathematical and Computer Modelling* **30**, No. 3-4, 99-113.
192. B.D. Choi, Y. Chang and B. Kim (1999), $MAP_1, MAP_2/M/c$ retrial queue with guard channels and its application to cellular networks, *Top* **7**, 231-248.
193. B.D. Choi, Y.H. Chung and A.N. Dudin (2001), The $BMAP/SM/1$ retrial queue with controllable operation modes, *European Journal of Operational Research* **131**, 16-30.
194. Y.J. Choi, S. Park and S. Bahk (2006), Multichannel random access in OFDMA wireless networks, *IEEE Journal on Selected Areas in Communications* **24**, 603-613.
195. Q.H. Choo and B. Conolly (1979), New results in the theory of repeated orders queueing systems, *Journal of Applied Probability* **16**, 631-640.
196. G. Choudhury (2007), A two phase batch arrival retrial queueing system with Bernoulli vacation schedule, *Applied Mathematics and Computation* **188**, 1455-1466.
197. E. Çinlar (1975), *Introduction to Stochastic Processes*, Prentice-Hall, Inc., Englewood Cliffs, New Jersey.
198. C. Clos (1948), An aspect of the dialing behavior of subscribers and its effect on the trunk plant, *The Bell System Technical Journal* **27**, 424-445.
199. E.G. Coffman Jr., E.N. Gilbert and Y.A. Kogan (1999), Redialing policies: Optimality and success probabilities, *Probability in the Engineering and Informational Sciences* **13**, 37-53.
200. J.W. Cohen (1957), Basic problems of telephone traffic theory and the influence of repeated calls, *Philips Telecommunication Review* **18**, 49-100.

201. B. Conolly (1982), Letter to the editor, *Journal of Applied Probability* **19**, 904-905.
202. R.B. Cooper (1981), *Introduction to Queueing Theory*, Edward Arnold, London.
203. C. D'Apice, R. Manzo, N.H. Phong and S. Salerno (2000), Retrial queueing system with several input flows and with server vacations, *Vestnik Rossijskogo Universiteta Druzhby Narodov. Seriya Prikladnaya Matematika i Informatika*, No. 1, 60-71.
204. A.G. De Kok (1984), Algorithmic methods for single server systems with repeated attempts, *Statistica Neerlandica* **38**, 23-32.
205. J.R. De Los Mozos and A. Buchheister (1983), Blocking calculation method for digital switching networks with step-by-step hunting and retrials. In: *Proceedings of the 10th International Teletraffic Congress, ITC-10*, Montreal.
206. J.E. Dennis and R.B. Schnabel (1983), *Numerical Methods for Unconstrained Optimization and Nonlinear Equations*, Prentice Hall, Inc., Englewood Cliffs, New Jersey.
207. N. Deul (1980), Stationary conditions for multi-server queueing systems with repeated calls, *Journal of Information Processing and Cybernetics* **16**, 607-613.
208. N. Deul (1982), The influence of the perseverance function in queueing systems with repeated calls, *Journal of Information Processing and Cybernetics* **18**, 587-594.
209. J.E. Diamond and A.S. Alfa (1995), Matrix analytical methods for $M/PH/1$ retrial queues, *Stochastic Models* **11**, 447-470.
210. J.E. Diamond and A.S. Alfa (1998), The $MAP/PH/1$ retrial queue, *Stochastic Models* **14**, 1151-1177.
211. J.E. Diamond and A.S. Alfa (1999), Approximation method for $M/PH/1$ retrial queues with phase type inter-retrial times, *European Journal of Operational Research* **113**, 620-631.
212. J.E. Diamond and A.S. Alfa (1999), Matrix analytic methods for a multi-server retrial queue with buffer, *Top* **7**, 249-266.
213. N.V. Djellab (2001), $M/G/1$ retrial queue with random breakdowns and general retrial times. In: *Queues, Flows, Systems, Networks. Proceedings of the International Conference "Modern Mathematical Methods of Investigating of the Information Networks"*, pp. 73-78, Minsk.
214. N.V. Djellab (2002), On the $M/G/1$ retrial queue subjected to breakdowns, *RAIRO-Operations Research* **36**, 299-310.
215. N.V. Djellab (2003), On the decomposition property of $M/G/1$ retrial queues. In: *Queues, Flows, Systems, Networks. Proceedings of the International Conference "Modern Mathematical Methods of Analysis and Optimization of Telecommunication Networks"*, pp. 75-79, Gomel.
216. N.V. Djellab (2005), On the $M/G/1$ retrial queue with feedback. In: *Queues, Flows, Systems, Networks. Proceedings of the International Conference "Mathematical Methods of Optimization of Telecommunication Networks"*, pp. 32-35, Minsk.
217. N.V. Djellab (2006), On the single-server retrial queue, *Yugoslav Journal of Operations Research* **16**, 45-53.
218. M.J. Doménech-Benlloch, J.M. Giménez-Guzmán, V.C. Casares-Giner and J. Martínez-Bauset (2005), Solving retrial systems by trading accuracy and simplicity. In: *Proceedings of the First Conference on Next Generation Internet Networks Traffic Engineering, NGI 2005*, Rome.

219. M.J. Doménech-Benlloch, J.M. Giménez-Guzmán, J. Martínez-Bauset and V. Casares-Giner (2005), Efficient and accurate methodology for solving multi-server retrial systems, *IEE Electronics Letters* **41**, 967-969.
220. M.J. Doménech-Benlloch, J.M. Giménez-Guzmán, J. Martínez-Bauset and V. Casares-Giner (2006), A low computation cost algorithm to solve cellular systems with retrials accurately. In: *Wireless Systems and Network Architectures in Next Generation Internet. Lecture Notes in Computer Science Vol. 3883* (Eds. M. Cesana and L. Fratta), pp. 103-114, Springer-Verlag, Berlin.
221. B.T. Doshi (1986), Queueing systems with vacations – a survey, *Queueing Systems* **1**, 29-66.
222. V.I. Douz, A.M. Zelinskiy and Y.N. Kornyshev (1984), A study of traffic handling processes with repeated calls and variable parameters, *Telecommunications and Radio Engineering* **38**, No. 8, 32-37.
223. V.I. Dragieva (1994), A single-server queueing system with a finite source and repeated calls, *Problems of Information Transmission* **30**, 283-289.
224. A.N. Dudin and A.V. Karolik (1999), A retrial $BMAP/G/1$ system with linear repeated requests and with MAP-input of disasters. In: *Queues, Flows, Systems, Networks. Proceedings of the International Conference "Modern Mathematical Methods of Investigating the Telecommunicational Networks"*, pp. 26-30, Minsk.
225. A.N. Dudin and V.I. Klimenok (1999), Queueing system $BMAP/G/1$ with repeated calls, *Mathematical and Computer Modelling* **30**, No. 3-4, 115-128.
226. A.N. Dudin and V.I. Klimenok (1999), $BMAP/SM/1$ model with Markov modulated retrials, *Top* **7**, 267-278.
227. A.N. Dudin and V.I. Klimenok (1999), Retrial $BMAP/SM/1$ system operating in a synchronous random environment. In: *Probabilistic Analysis of Rare Events: Theory and Problems of Safety, Insurance and Ruin* (Eds. V.V. Kalashnikov and A.M. Andronov), pp. 143-148, Riga.
228. A.N. Dudin and V.I. Klimenok (1999), Investigation of the $BMAP/SM/1$ queueing systems with different strategies of retrials. In: *Proceedings of the International Conference Distributed Computer Communication Networks, DCCN'99*, pp. 29-34, Tel-Aviv.
229. A.N. Dudin and V.I. Klimenok (2000), A retrial $BMAP/SM/1$ system with linear repeated requests, *Queueing Systems* **34**, 47-66.
230. A.N. Dudin and V.I. Klimenok (2000), The $M_1; M_2/G_1^{(1)}, G_1^{(2)}; G_2/1$ model with the controlled service of the waiting flow and the low-priority retrying flow. In: *Advances in Algorithmic Methods for Stochastic Models* (Eds. G. Latouche and P.G. Taylor), pp. 99-114, Notable Publications, Inc., New Jersey.
231. A.N. Dudin, V.I. Klimenok, I.A. Klimenok, V.V. Borokhovsky, A.V. Karolik and G.V. Tsarenkov (2000), Software "SIRIUS+" for evaluation and optimization of queues with the $BMAP$-input. In: *Advances in Algorithmic Methods for Stochastic Models* (Eds. G. Latouche and P.G. Taylor), pp. 115-133, Notable Publications, Inc., New Jersey.
232. A.N. Dudin (2001), Further results for random multiple access protocols via retrial queueing models with unreliable two-phase service, *Journal of Simulation Engineering* **23**, No. 2, 64-73.
233. A.N. Dudin and S.R. Chakravarthy (2002), A single server retrial queueing model with batch arrivals and group services. In: *Advances in Stochastic Modelling* (Eds. J.R. Artalejo and A. Krishnamoorthy), pp. 1-21, Notable Publications, Inc., New Jersey.

234. A.N. Dudin, V.I. Klimenok and G.V. Tsarenkov (2002), Software "SIRIUS++" for performance evaluation of modern communication networks. In: *Proceedings of the the 16th European Simulation Multiconference*, pp. 489-493, Darmstadt.
235. A.N. Dudin and V.I. Klimenok (2004), The state dependent $M/M/1$ retrial system. In: *Proceedings of the Fifth International Workshop on Retrial Queues* (Ed. B.D. Choi), pp. 81-88, Seoul.
236. A.N. Dudin, V.I. Klimenok, C.S. Kim and G.V. Tsarenkov (2004), The $BMAP/G/1 \to \cdot/PH/1/M-1$ tandem queue with retrials and losses. In: *Proceedings of the Fifth International Workshop on Retrial Queues* (Ed. B.D. Choi), pp. 17-29, Seoul.
237. A.N. Dudin, A. Krishnamoorthy, V.C. Joshua and G.V. Tsarenkov (2004), Analysis of the $BMAP/G/1$ retrial system with search of customers from the orbit, *European Journal of Operational Research* **157**, 169-179.
238. F.P. Duffy and R.A. Mercer (1978), A study of network performance and customer behavior during direct-distance-dialing call attempts in the U.S.A., *The Bell System Technical Journal* **57**, 1-33.
239. D. Efrosinin and L. Breuer (2006), Threshold policies for controlled retrial queues with heterogeneous servers, *Annals of Operations Research* **141**, 139-162.
240. A. Elcan (1994), Optimal customer return rate for an $M/M/1$ queueing system with retrials, *Probability in the Engineering and Informational Sciences* **8**, 521-539.
241. A. Elcan (1999), Asymptotic bounds for an optimal state-dependent retrial rate of the $M/M/1$ queue with returning customers, *Mathematical and Computer Modelling* **30**, No. 3-4, 129-140.
242. A. Elldin (1967), Approach to the theoretical description of repeated call attempts, *Ericsson Technics* **23**, 345-407.
243. A. Elldin and G. Lind (1971), *Elementary Telephone Traffic Theory*, Ericsson Public Telecommunications.
244. A. Elldin (1977), Traffic engineering in developing countries. Some observations from the ESCAP region, *Telecommunication Journal* **44**, 427-436.
245. R.V. Evans (1967), Geometric distribution in some two-dimensional queuing systems, *Operations Research* **15**, 830-846.
246. R. Evers (1973), Measurement of subscriber reaction to unsuccessful call attempts and the influence of reasons of failure. In: *Proceedings of the 7th International Teletraffic Congress, ITC-7*, Stockholm.
247. R. Evers (1976), Analysis of traffic flows on subscriber-lines dependent of time and subscriber class. In: *Proceedings of the 8th International Teletraffic Congress, ITC-8*, Melbourne.
248. G.I. Falin (1976), Aggregate arrival of customers in a one-line system with repeated calls, *Ukrainian Mathematical Journal* **28**, 437-440.
249. G.I. Falin (1977), Waiting time in a single-channel queuing system with repeated calls, *Moscow University Computational Mathematics and Cybernetics*, No. 4, 66-69.
250. G.I. Falin (1978), The exit flow of a single-line queueing system when there are secondary orders, *Engineering Cybernetics* **16**, No. 5, 64-67.
251. G.I. Falin (1979), Model of coupled switching in presence of recurrent calls, *Engineering Cybernetics* **17**, No. 1, 53-59.
252. G.I. Falin (1979), A single-line system with secondary orders, *Engineering Cybernetics* **17**, No. 2, 76-83.

253. G.I. Falin (1979), Effect of the recurrent calls on output flow of a single channel system of mass service, *Engineering Cybernetics* **17**, No. 4, 99-102.
254. G.I. Falin (1980), Not completely accessible schemes with allowance for repeated calls, *Engineering Cybernetics* **18**, No. 5, 56-63.
255. G.I. Falin (1980), Repeated calls in structurally complex systems, *Engineering Cybernetics* **18**, No. 6, 46-51.
256. G.I. Falin (1980), $M/G/1$ queue with repeated calls in heavy traffic, *Moscow University Mathematics Bulletin* **35**, No. 6, 48-51.
257. G.I. Falin (1980), Switching systems with allowance for repeated calls, *Problems of Information Transmission* **16**, 145-151.
258. G.I. Falin (1981), Investigation of weakly loaded switching systems with repeated calls, *Engineering Cybernetics* **19**, No. 3, 69-73.
259. G.I. Falin (1981), Calculation of the load on a shared-use telephone instrument, *Moscow University Computational Mathematics and Cybernetics*, No. 2, 76-80.
260. G.I. Falin (1981), Functioning under nonsteady conditions of a single-channel system with group arrival of requests and repeated calls, *Ukrainian Mathematical Journal* **33**, 429-432.
261. G.I. Falin (1983), The influence of inhomogeneity of the composition of subscribers on the functioning of telephone systems with repeated calls, *Engineering Cybernetics* **21**, No. 6, 78-82.
262. G.I. Falin (1983), Calculation of probability characteristics of a multiline system with repeat calls, *Moscow University Computational Mathematics and Cybernetics*, No. 1, 43-49.
263. G.I. Falin (1984), On sufficient conditions for ergodicity of multichannel queueing systems with repeated calls, *Advances in Applied Probability* **16**, 447-448.
264. G.I. Falin (1984), Continuous approximation for a single server system with an arbitrary service time under repeated calls, *Engineering Cybernetics* **22**, No. 2, 66-71.
265. G.I. Falin (1984), Multiserver fully-available systems with repeat calls under conditions of heavy traffic, *Moscow University Computational Mathematics and Cybernetics*, No. 3, 82-85.
266. G.I. Falin (1984), Asymptotic investigation of fully available switching systems with high repetition intensity of blocked calls, *Moscow University Mathematics Bulletin* **39**, No. 6, 72-77.
267. G.I. Falin (1986), On the waiting-time process in a single-line queue with repeated calls, *Journal of Applied Probability* **23**, 185-192.
268. G.I. Falin (1986), Single-line repeated orders queueing systems, *Optimization* **17**, 649-667.
269. G.I. Falin (1986), On heavily loaded systems with repeated calls, *Soviet Journal of Computer and Systems Sciences* **24**, No. 4, 124-128.
270. G.I. Falin (1987), Multichannel queueing systems with repeated calls under high intensity of repetition, *Journal of Information Processing and Cybernetics* **23**, 37-47.
271. G.I. Falin (1987), Error estimates for approximations of countable Markov chains associated with repeated calls models, *Moscow University Mathematics Bulletin* **42**, No. 2, 12-15.
272. G.I. Falin (1987), The ergodicity of multilinear queueing systems with repeated calls, *Soviet Journal of Computer and Systems Sciences* **25**, No. 1, 60-65.
273. G.I. Falin (1988), On a multiclass batch arrival retrial queue, *Advances in Applied Probability* **20**, 483-487.

274. G.I. Falin (1988), Virtual waiting time in systems with repeated calls, *Moscow University Mathematics Bulletin* **43**, No. 6, 6-10.
275. G.I. Falin (1988), Comparability of migration processes, *Theory of Probability and Its Applications* **33**, 370-372.
276. G.I. Falin and Y.I. Sukharev (1988), Singularly disturbed equations and asymptotic analysis of stationary characteristics of repeat-call-queuing systems, *Moscow University Mathematics Bulletin* **43**, No. 5, 7-11.
277. G.I. Falin (1989), Ergodicity and stability of systems with repeated calls, *Ukrainian Mathematical Journal* **41**, 559-562.
278. G.I. Falin (1989), Phase transitions in queueing systems connected with a virtual expectation time, *Ukrainian Mathematical Journal* **41**, 813-817.
279. G.I. Falin (1990), A survey of retrial queues, *Queueing Systems* **7**, 127-167.
280. G.I. Falin (1991), A diffusion approximation for retrial queueing systems, *Theory of Probability and Its Applications* **36**, 149-152.
281. G.I. Falin (1991), Phase transitions in overloaded computer and communication systems. In: *Analysis and Geometry 1991. Proceedings of KAIST Mathematics Workshop* (Eds. B.D. Choi and J.W. Yim), pp. 21-34, Korea Advanced Institute of Science and Technology, Taejon.
282. G.I. Falin (1991), Space-heterogeneous random walks on semi-strips associated with retrial queues. In: *Analysis and Geometry 1991. Proceedings of KAIST Mathematics Workshop* (Eds. B.D. Choi and J.W. Yim), pp. 35-45, Korea Advanced Institute of Science and Technology, Taejon.
283. G.I. Falin and C. Fricker (1991), On the virtual waiting time in an $M/G/1$ retrial queue, *Journal of Applied Probability* **28**, 446-460.
284. G.I. Falin, J.R. Artalejo and M. Martin (1993), On the single server retrial queue with priority customers, *Queueing Systems* **14**, 439-455.
285. G.I. Falin, M. Martin and J.R. Artalejo (1994), Information theoretic approximations for the $M/G/1$ retrial queue, *Acta Informatica* **31**, 559-571.
286. G.I. Falin (1995), Estimation of retrial rate in a retrial queue, *Queueing Systems* **19**, 231-246.
287. G.I. Falin and J.R. Artalejo (1995), Approximations for multiserver queues with balking/retrial discipline, *OR Spektrum* **17**, 239-244.
288. G.I. Falin and J.G.C. Templeton (1997), *Retrial Queues*, Chapman and Hall, London.
289. G.I. Falin and J.R. Artalejo (1998), A finite source retrial queue, *European Journal of Operational Research* **108**, 409-424.
290. G.I. Falin (1999), A multiserver retrial queue with a finite number of sources of primary calls, *Mathematical and Computer Modelling* **30**, No. 3-4, 33-49.
291. G.I. Falin and A.I. Falin (1999), Heavy traffic analysis of $M/G/1$ type queueing systems with Markov-modulated arrivals, *Top* **7**, 279-291.
292. G.I. Falin and A. Gómez-Corral (2000), On a bivariate Markov process arising in the theory of single-server retrial queues, *Statistica Neerlandica* **54**, 67-78.
293. K. Farahmand (1990), Single line queue with repeated demands, *Queueing Systems* **6**, 223-228.
294. K. Farahmand (1996), Single line queue with recurrent repeated demands, *Queueing Systems* **22**, 425-435.
295. K. Farahmand and N.H. Smith (1996), Retrial queues with recurrent demand option, *Journal of Applied Mathematics and Stochastic Analysis* **9**, 221-228.
296. K. Farahmand and N. Cooke (1997), Single retrial queues with service option on arrival, *Journal of Applied Mathematics and Decision Sciences* **1**, 5-12.

297. K. Farahmand and N. Livingstone (2001), Recurrent retrial queues with service option on arrival, *European Journal of Operational Research* **131**, 530-535.
298. G. Fayolle (1986), A simple telephone exchange with delayed feedbacks. In: *Teletraffic Analysis and Computer Performance Evaluation* (Eds. O.J. Boxma, J.W. Cohen and H.C. Tijms), pp. 245-253, Elsevier Science Publishers B.V., Amsterdam.
299. G. Fayolle and M.A. Brun (1988), On a system with impatience and repeated calls. In: *Queueing Theory and Its Applications. Liber Amicorum for J.W. Cohen* (Eds. O.J. Boxma and R. Syski), pp. 283-305, North-Holland, Amsterdam.
300. M.A. Feinberg (1991), Analytical model of automated call distribution system. In: *Queueing, Performance and Control in ATM, ITC-13* (Eds. J.W. Cohen and C.D. Pack), pp. 193-197, Elsevier Science Publishers B.V., Amsterdam.
301. A.A. Fredericks and G.A. Reisner (1979), Approximations to stochastic service systems, with an application to a retrial model, *The Bell System Technical Journal* **58**, 557-576.
302. S.W. Fuhrmann and R.B. Cooper (1985), Stochastic decomposition in the $M/G/1$ queue with generalized vacations, *Operations Research* **33**, 1117-1129.
303. R.A. Gable (1993), *Inbound Call Centers: Design, Implementation and Management*, Artech House Telecommunications Library.
304. H.R. Gail, S.L. Hantler and B.A. Taylor (1996), Spectral analysis of $M/G/1$ and $G/M/1$ type Markov chains, *Advances in Applied Probability* **28**, 114-165.
305. N. Gans, G. Koole and A. Mandelbaum (2003), Telephone call centers: Tutorial, review and research prospects, *Manufacturing and Service Operations Management* **5**, 79-141.
306. D.P. Gaver and J.P. Lehoczky (1976), Gaussian approximations to service problems: A communication system example, *Journal of Applied Probability* **13**, 768-780.
307. D.P. Gaver, P.A. Jacobs and G. Latouche (1984), Finite birth-and-death models in randomly changing environments, *Advances in Applied Probability* **16**, 715-731.
308. E. Gelenbe and R. Iasnogorodski (1980), A queue with server of walking type (autonomous service), *Annales de l'Institut Henry Poincaré, Series* **B16**, 63-73.
309. N. Gharbi and M. Ioualalen (2002), Performance analysis of retrial queueing systems using generalized stochastic Petri nets, *Electronic Notes in Theoretical Computer Science* **65**, No. 6, 1-15.
310. N. Gharbi and M. Ioualalen (2006), GSPN analysis of retrial systems with servers breakdowns and repairs, *Applied Mathematics and Computation* **174**, 1151-1168.
311. G. Giambene (2005), *Queueing Theory and Telecommunications: Networks and Applications*, Springer, New York.
312. E.N. Gilbert (1988), Retrials and balks, *IEEE Transactions on Information Theory* **34**, 1502-1508.
313. J.M. Giménez-Guzmán, M.J. Doménech-Benlloch, J. Martínez-Bauset, V. Pla and V. Casares-Giner (2005), Analysis of a handover procedure with queueing, retrials and impatient customers. In: *Proceedings HET-NETs'05 - Performance Modelling and Evaluation of Heterogeneous Networks*, pp. P27/1-10, Ilkley.
314. A. Gómez-Corral (1998), On single-server queues governed by a clearing mechanism and a secondary input of repeated attempts. In: *Proceedings of the In-*

ternational Conference on Stochastic Processes (Ed. A. Krishnamoorthy), pp. 169-179, Cochin University of Science and Technology, Cochin.
315. A. Gómez-Corral (1999), Stochastic analysis of a single server retrial queue with general retrial times, *Naval Research Logistics* **46**, 561-581.
316. A. Gómez-Corral and M.F. Ramalhoto (1999), The stationary distribution of a Markovian process arising in the theory of multiserver retrial queueing systems, *Mathematical and Computer Modelling* **30**, No. 3-4, 141-158.
317. A. Gómez-Corral and M.F. Ramalhoto (2000), On the waiting time distribution and the busy period of a retrial queue with constant retrial rate, *Stochastic Modelling and Applications* **3**, 37-47.
318. A. Gómez-Corral (2001), On extreme values of orbit lengths in $M/G/1$ queues with constant retrial rate, *OR Spektrum* **23**, 395-409.
319. A. Gómez-Corral (2002), Analysis of a single-server retrial queue with quasi-random input and nonpreemptive priority, *Computers and Mathematics with Applications* **43**, 767-782.
320. A. Gómez-Corral (2002), A matrix-geometric approximation for tandem queues with blocking and repeated attempts, *Operations Research Letters* **30**, 360-374.
321. A. Gómez-Corral (2004), Bulk-service finite-buffer retrial queues. In: *Proceedings of the Fifth International Workshop on Retrial Queues* (Ed. B.D. Choi), pp. 67-80, Seoul.
322. A. Gómez-Corral (2006), A bibliographical guide to the analysis of retrial queues through matrix analytic techniques, *Annals of Operations Research* **141**, 163-191.
323. G. Gosztony (1976), Repeated call attempts and their effect on traffic engineering, *Budavox Telecommunication Review*, No. 2, 16-26.
324. G. Gosztony (1976), Stochastic service systems with two input processes and repeated calls (Traffic engineering of telex stations). In: *Progress in Operations Research*, pp. 467-492, North-Holland, Amsterdam.
325. G. Gosztony (1977), Comparison of calculated and simulated results for trunk groups with repeated attempts, *Budavox Telecommunication Review*, No. 1, 1-18.
326. G. Gosztony, K. Rahko and R. Chapuis (1979), The grade of service in the world-wide telephone network, Part I, *Telecommunication Journal* **46**, 556-565.
327. G. Gosztony (1985), A general (rHβ) formula of call repetition: Validity and constraints. In: *Teletraffic Issues in an Advanced Information Society, ITC-11* (Ed. M. Akiyama), pp. 1010-1016, Elsevier Science Publishers B.V., Amsterdam.
328. A. Graham (1981), *Kronecker Products and Matrix Calculus with Applications*, Ellis Horwood Ltd., Chichester.
329. W.K. Grassmann and D.A. Stanford (2000), Matrix analytic methods. In: *Computational Probability* (Ed. W.K. Grassmann), pp. 153-203, Kluwer Academic Publishers, Boston.
330. B.S. Greenberg and R.W. Wolff (1987), An upper bound on the performance of queues with returning customers, *Journal of Applied Probability* **24**, 466-475.
331. B.S. Greenberg (1989), $M/G/1$ queueing systems with returning customers, *Journal of Applied Probability* **26**, 152-163.
332. N. Grier, W.A. Massey, T. McKoy and W. Whitt (1997), The time-dependent Erlang loss model with retrials, *Telecommunication Systems* **7**, 253-265.

333. D. Grillo (1979), Telephone network behaviour in repeated attempts environment: A simulation analysis. In: *Proceedings of the 9th International Teletraffic Congress, ITC-9*, Torremolinos.
334. S.A. Grishechkin (1990), Branching processes and queueing systems with repeated orders or with random discipline, *Theory of Probability and Its Applications* **35**, 38-53.
335. S.A. Grishechkin (1991), On the use of branching processes for finding steady-state distributions in queueing theory, *Theory of Probability and Its Applications* **36**, 477-493.
336. S.A. Grishechkin (1992), Multiclass batch arrival retrial queues analyzed as branching processes with immigration, *Queueing Systems* **11**, 395-418.
337. G. Gupur (2005), Semigroup method for $M/G/1$ retrial queue with general retrial times, *International Journal of Pure and Applied Mathematics* **18**, 405-429.
338. J.L. Hammond and P.J.P. O'Reilly (1988), *Performance Analysis of Local Computer Networks*, Addison-Wesley, Massachusetts.
339. D.H. Han and Y.W. Lee (1996), $MMPP, M/G/1$ retrial queue with two classes of customers, *Communication of the Korean Mathematical Society* **11**, 481-493.
340. D.H. Han and C.G. Park (1998), $M_1, M_2/M/1$ retrial queueing systems with two classes of customers and smart machine, *Communication of the Korean Mathematical Society* **13**, 393-403.
341. T. Hanschke (1985), The $M/G/1/1$ queue with repeated attempts and different types of feedback effects, *OR Spektrum* **7**, 209-215.
342. T. Hanschke (1985), A model for planning switching networks. In: *Operations Research Proceedings 1984*, pp. 555-562, Springer-Verlag, Berlin.
343. T. Hanschke (1986), A computational procedure for the variance of the waiting time in the $M/M/1/1$ queue with repeated attempts. In: *Operations Research Proceedings 1985* (Eds. L. Streitferdt, H. Hauptmann, A.W. Marusev, D. Ohse and U. Pape), pp. 525-532, Springer-Verlag, Berlin.
344. T. Hanschke (1987), Explicit formulas for the characteristics of the $M/M/2/2$ queue with repeated attempts, *Journal of Applied Probability* **24**, 486-494.
345. T. Hanschke (1999), A matrix continued fraction algorithm for the multiserver repeated order queue, *Mathematical and Computer Modelling* **30**, No. 3-4, 159-170.
346. L. Haque and M.J. Amstrong (2007), A survey of the machine interference problem, *European Journal of Operational Research* **179**, 469-482.
347. C.M. Harris, K.L. Hoffman and P.B. Saunders (1987), Modeling the IRS telephone taxpayer information system, *Operations Research* **35**, 504-523.
348. O. Hashida and K. Kodaira (1975), Repeated calls resulting from data terminal busy and delayed delivery service, *Electronics and Communications in Japan* **58-A**, No. 3, 1-9.
349. O. Hashida and K. Kawashima (1979), Buffer behavior with repeated attempts, *Electronics and Communications in Japan* **62-B**, No. 3, 27-35.
350. R. Hassin and M. Haviv (1996), On optimal and equilibrium retrial rates in a queueing system, *Probability in the Engineering and Informational Sciences* **10**, 223-227.
351. W.S. Haywarder and R.I. Wilkinson (1970), Human factors in telephone systems and their influence on traffic theory, specially with regard to future facilities. In: *Proceedings of the 6th International Teletraffic Congress, ITC-6*, Munich.

352. Q.M. He, H. Li and Y.Q. Zhao (2000), Ergodicity of the $BMAP/PH/s/s+K$ retrial queue with PH-retrial times, *Queueing Systems* **35**, 323-347.
353. S.L. Hew and L.B. White (2005), Optimal integrated call admission control and dynamic pricing with handoffs and price-affected arrivals. In: *2005 Asia-Pacific Conference on Communications*, pp. 396-400, Perth.
354. K.L. Hoffman and C.M. Harris (1986), Estimation of a caller retrial rate for a telephone information system, *European Journal of Operational Research* **27**, 207-214.
355. G. Honi (1975), Some macro-models for discussing repeated call attempts, *Budavox Telecommunication Review*, No. 2, 21-39.
356. G. Honi and G. Gosztony (1976), Some practical problems of the traffic engineering of overloaded telephone networks. In: *Proceedings of the 8th International Teletraffic Congress, ITC-8*, Melbourne.
357. D.J. Houck and W.S. Lai (1998), Traffic modeling and analysis of hybrid fiber-coax systems, *Computer Networks and ISDN Systems* **30**, 821-834.
358. J.J. Hunter (1983), *Discrete Time Models: Techniques and Applications, Volume 2 of Mathematical Techniques of Applied Probability*, Academic Press, New York.
359. H. Inamori, M. Sawai, T. Endo and K. Tanabe (1985), An automatically repeated call model in NTT public facsimile communication systems. In: *Teletraffic Issues in an Advanced Information Society, ITC-11* (Ed. M. Akiyama), pp. 1017-1023, Elsevier Science Publishers B.V., Amsterdam.
360. W.B. Iversen (1973), Analysis of real teletraffic processes based on computerized measurements, *Ericsson Technics* **29**, 3-64.
361. V.B. Iversen, S.N. Stepanov and E.O. Naumova (1999), The approximate evaluation of stationary performance measures of multi-service systems with fixed number of retrials. In: *Queues, Flows, Systems, Networks. Proceedings of the International Conference "Modern Mathematical Methods of Investigating the Telecommunicational Networks"*, pp. 31-35, Minsk.
362. V.B. Iversen, S.N. Stepanov, E.O. Naumova and I.A. Ovseevich (1999), Performance measures of multi-service systems with fixed number of retrials. In: *Proceedings of the International Conference Distributed Computer Communication Networks, DCCN'99*, pp. 57-62, Tel-Aviv.
363. V.B. Iversen and S.N. Stepanov (2001), Estimation of the characteristics of multiflow models with fixed number of retrials, *Automation and Remote Control* **62**, 772-781.
364. G.K. Janssens (1994), Delay time computation for an active star topology local area network, *Microprocessing and Microprogramming* **40**, 241-248.
365. G.K. Janssens (1997), The quasi-random input queueing system with repeated attempts as a model for a collision-avoidance star local area network, *IEEE Transactions on Communications* **45**, 360-364.
366. G.L. Jonin and J.J. Sedol (1970), Telephone systems with repeated calls. In: *Proceedings of the 6th International Teletraffic Congress, ITC-6*, pp. 435/1-5, Munich.
367. G.L. Jonin and J.J. Sedol (1973), Fully-availability groups with repeated calls and time of advanced service. In: *Proceedings of the 7th International Teletraffic Congress, ITC-7*, pp. 137/1-4, Stockholm.
368. G.L. Jonin (1984), The systems with repeated calls: Models, measurement, results. In: *Fundamentals of Teletraffic Theory. Proceedings of the Third International Seminar on Teletraffic Theory*, pp. 197-208, Moscow.

369. G.L. Jonin and J.J. Sedol (1988), Investigation of a system with repeated calls taking subscriber busyness into account, *Telecommunications and Radio Engineering*, No. 11, 1-6.
370. G.L. Jonin (1990), An approximate method for calculations of the probability characteristics of multilevel switching systems with repeated calls, *Telecommunications and Radio Engineering*, No. 2, 26-29.
371. A. Jrad, G. O'Reilly, S.H. Richman, S. Conrad and A. Kelic (2006), Dynamic changes in subscriber behavior and their impact on the telecom network in cases of emergency. In: *Proceedings - IEEE Military Communications Conference, MILCOM 2006*, Article DOI 10.1109/MILCOM.2006.302071, pp. 1-7.
372. A.I. Kalmychkov and G.A. Medvedev (1990), Probability characteristics of Markov local-area networks with random-access protocols, *Automatic Control and Computer Sciences* **24**, No. 5, 38-45.
373. R. Kalyanaraman and B. Srinivasan (2003), A single server retrial queue with two types of calls and recurrent repeated calls, *International Journal of Information and Management Sciences* **14**, No. 4, 49-62.
374. A. Kanechny (2003), Some results for a single server retrial queueing model with batch arrivals and group services employing threshold strategy. In: *Queues, Flows, Systems, Networks. Proceedings of the International Conference "Modern Mathematical Methods of Analysis and Optimization of Telecommunication Networks"*, pp. 110-114, Gomel.
375. V.A. Kapyrin (1977), A study of the stationary characteristics of a queueing system with recurring demands, *Cybernetics* **13**, 584-590.
376. N. Karasawa, R. Noto, K. Mase, K. Nakano, M. Sengoku and S. Shinoda (2000), PHS based ad hoc networks. In: *26th Annual Conference of the IEEE Industrial Electronics Society, IECON 2000*, pp. 1141-1146 (Vol. 2).
377. P. Kárász and G. Farkas (2005), Exact solution for a two-type customers retrial system, *Computers and Mathematics with Applications* **49**, 95-102.
378. J.S. Kaufman (1992), Blocking with retrials in a completely shared resource environment, *Performance Evaluation* **15**, 99-113.
379. J.C. Ke, H.I. Huang and C.H. Lin (2007), On retrial queueing model with fuzzy parameters, *Physica A* **374**, 272-280.
380. J. Keilson, J. Cozzolino and H. Young (1968), A service system with unfilled requests repeated, *Operations Research* **16**, 1126-1137.
381. J. Keilson and L.D. Servi (1993), The matrix $M/M/\infty$ system: Retrial models and Markov modulated sources, *Advances in Applied Probability* **25**, 453-471.
382. F.P. Kelly (1986), On auto-repeat facilities and telephone network performance, *Journal of the Royal Statistical Society* **B 48**, 123-132.
383. T. Kernane and A. Aissani (2004), Strong coupling convergence for retrial queues. In: *Proceedings of the Fifth International Workshop on Retrial Queues* (Ed. B.D. Choi), pp. 123-128, Seoul.
384. T. Kernane and A. Aissani (2006), Stability of retrial queues with versatile retrial policy, *Journal of Applied Mathematics and Stochastic Analysis*, Article ID 54359, pp. 1-16.
385. Z. Khalil, G.I. Falin and T. Yang (1992), Some analytical results for congestion in subscriber line modules, *Queueing Systems* **10**, 381-402.
386. Z. Khalil and G.I. Falin (1994), Stochastic inequalities for $M/G/1$ retrial queues, *Operations Research Letters* **16**, 285-290.

387. A. Kharkevich, I. Endaltsev, E. Melik-Gaikazova, N. Pevtsov and S. Stepanov (1985), Approximate analysis of systems with repeated calls and multiphase service. In: *Teletraffic Issues in an Advanced Information Society, ITC-11* (Ed. M. Akiyama), pp. 1029-1035, Elsevier Science Publishers B.V., Amsterdam.
388. I.I. Khomichkov (1988), Calculation of the characteristics of a queueing system with repeated customers and with paired connections, *Automation and Remote Control* **49**, 458-463.
389. I.I. Khomichkov (1993), Study of models of local networks with multiple-access protocols, *Automation and Remote Control* **54**, 1801-1811.
390. I.I. Khomichkov (1995), Calculation of the characteristics of local area network with p-persistent protocol of multiple random access, *Automation and Remote Control* **56**, 208-218.
391. C.S. Kim, V.I. Klimenok and D.S. Orlovsky (2004), $BMAP/PH/N/R$ retrial queuing system. In: *Proceedings of the VII-th International Conference "Computer Data Analysis and Modelling: Robustness and Computer Intensive Methods"*, pp. 76-79 (Vol. 2), Minsk.
392. C.S. Kim, A. Kanechny and A.N. Dudin (2006), Threshold control by a single-server retrial queue with batch arrivals and group services, *Operations Research Letters* **34**, 548-556.
393. C.S. Kim, V.I. Klimenok, A. Birukov and A.N. Dudin (2006), Optimal multi-threshold control by the $BMAP/SM/1$ retrial system, *Annals of Operations Research* **141**, 193-210.
394. C.S. Kim, V.I. Klimenok, S.C. Lee and A.N. Dudin (2007), The $BMAP/PH/1$ retrial queueing system operating in random environment, *Journal of Statistical Planning and Inference* **137**, 3904-3916.
395. C.S. Kim, V.I. Klimenok and D.S. Orlovsky (2008), The $BMAP/PH/N$ retrial queue with Markovian flow of breakdowns, *European Journal of Operational Research*, in press.
396. Y.C. Kim (1995), On $M/M/3/3$ retrial queueing system, *Honam Mathematical Journal* **17**, 141-147.
397. K. Kinoshita, M. Fujisawa, T. Takine and K. Murakami (1999), Performance evaluation method of cellular networks with retrial. In: *IEEE International Conference on Personal Wireless Communication, ICPWC'99*, pp. 389-393, Jaipur.
398. L. Kleinrock (1975), *Queueing Systems, Volume 1: Theory*, Wiley, New York.
399. V.I. Klimenok (1990), Optimization of dynamic management of the operating mode of data systems with repeat calls, *Automatic Control and Computer Sciences* **24**, No. 1, 23-28.
400. V.I. Klimenok (1999), The $BMAP/SM/1$ queue with heterogeneous retrial politics. In: *Queues, Flows, Systems, Networks. Proceedings of the International Conference "Modern Mathematical Methods of Investigating the Telecommunicational Networks"*, pp. 37-41, Minsk.
401. V.I. Klimenok (2001), A multiserver retrial queueing system with batch Markov arrival process, *Automation and Remote Control* **62**, 1312-1322.
402. V.I. Klimenok (2001), The $BMAP/PH/N$ system with threshold strategy of retrials. In: *Queues, Flows, Systems, Networks. Proceedings of the International Conference "Modern Mathematical Methods of Investigating of the Information Networks"*, pp. 108-113, Minsk.

403. V.I. Klimenok (2003), The controlled $BMAP/SM/1$ retrial queue with Markov modulated retrials. In: *Proceedings of the International Conference "Distributed Computer and Communication Networks. Stochastic Modelling and Optimization"*, pp. 81-88, Moscow.
404. V.I. Klimenok, A.N. Dudin and S. Kim (2003), A loss-retrial $BMAP/PH/N$ system. In: *Queues, Flows, Systems, Networks. Proceedings of the International Conference "Modern Mathematical Methods of Analysis and Optimization of Telecommunication Networks"*, pp. 129-135, Gomel.
405. V.I. Klimenok, D.S. Orlovsky and A.N. Dudin (2004), A retrial $BMAP/PH/N$ system with impatient calls. In: *Proceedings of the Fifth International Workshop on Retrial Queues* (Ed. B.D. Choi), pp. 89-102, Seoul.
406. V.I. Klimenok, D.S. Orlovsky and C.S. Kim (2004), The $BMAP/PH/N/N+R$ retrial queuing system with different disciplines of retrials. In: *Proceedings of the 11th International Conference on Analytical and Stochastic Modelling Techniques, ASMTA'04* (Eds. K. Al-Begain and G. Bolch), pp. 93-98, Magdeburg.
407. V.I. Klimenok (2005), A $BMAP/SM/1$ queueing system with hybrid operation mechanism, *Automation and Remote Control* **66**, 779-790.
408. V.I. Klimenok, S.R. Chakravarthy and A.N. Dudin (2005), Algorithmic analysis of a multiserver Markovian queue with primary and secondary services, *Computers and Mathematics with Applications* **50**, 1251-1270.
409. V.I. Klimenok, S.R. Chakravarthy and A.N. Dudin (2005), A multi-server Markovian queueing model with primary and secondary services. In: *Queues, Flows, Systems, Networks. Proceedings of the International Conference "Mathematical Methods of Optimization of Telecommunication Networks"*, pp. 67-75, Minsk.
410. V.I. Klimenok and C.S. Kim (2005), $BMAP/PH/1$ retrial system operating in random environment. In: *Proceedings of the Fifth Workshop on Simulation* (Eds. S.M. Ermakov, V.B. Melas and A.N. Pepelyshev), pp. 367-372, NII Chemistry Saint Petersburg University Publishers, Saint Petersburg.
411. V.I. Klimenok and A.N. Dudin (2006), Multi-dimensional asymptotically quasi-Toeplitz Markov chains and their application in queueing theory, *Queueing Systems* **54**, 245-259.
412. V.I. Klimenok, D.S. Orlovsky and A.N. Dudin (2007), A $BMAP/PH/N$ system with impatient repeated calls, *Asia-Pacific Journal of Operational Research* **24**, 293-312.
413. E.V. Koba (2000), On a condition for the stability of the $M/D/1$ retrial queueing system with limited waiting time, *Cybernetics and Systems Analysis* **36**, 312-314.
414. E.V. Koba (2000), An $M/D/1$ queueing system with partial synchronization of its incoming flow and demands repeating at constant intervals, *Cybernetics and Systems Analysis* **36**, 946-948.
415. E.V. Koba (2002), On a $GI/G/1$ retrial queueing system with a FIFO queueing discipline, *Theory of Stochastic Processes* **8**, 201-208.
416. E.V. Koba (2005), Stability conditions for some typical retrial queues, *Cybernetics and Systems Analysis* **41**, 100-103.
417. E.V. Koba and S.V. Pustovaya (2007), Call center as retrial queueing system, *Journal of Automation and Information Sciences* **39**, 37-47.
418. D.V. Kolousov, A.A. Nazarov and S.A. Tsoi (2006), Probabilistic-time characteristics of bistable random access networks, *Automation and Remote Control* **67**, 251-264.

419. G. Koole and A. Mandelbaum (2002), Queueing models of call centers: An introduction, *Annals of Operations Research* **113**, 41-59.
420. Y.N. Kornyshev (1969), Design of a fully accessible switching system with repeated calls, *Telecommunications* **23**, No. 11, 46-52.
421. Y.N. Kornyshev (1977), A single-line system with repeated orders and preliminary servicing, *Engineering Cybernetics* **15**, No. 2, 63-68.
422. Y.N. Kornyshev and V.I. Duz (1990), The accuracy of evaluating call servicing quality characteristics in communication networks, *Telecommunications and Radio Engineering* **44**, No. 4, 1-8.
423. L. Kosten (1973), *Stochastic Theory of Service Systems*, Pergamon Press, Oxford.
424. D.D. Kouvatsos (1994), Entropy maximization and queueing networks models, *Annals of Operations Research* **48**, 63-126.
425. B. Krishna Kumar and D. Arivudainambi (2002), The $M/G/1$ retrial queue with Bernoulli schedules and general retrial times, *Computers and Mathematics with Applications* **43**, 15-30.
426. B. Krishna Kumar and D. Arivudainambi (2002), On the busy period of an $M/G/1$ Bernoulli feedback retrial queue with negative customers and preemptive resume. In: *Advances in Stochastic Modelling* (Eds. J.R. Artalejo and A. Krishnamoorthy), pp. 205-218, Notable Publications, Inc., New Jersey.
427. B. Krishna Kumar, D. Arivudainambi and A. Vijayakumar (2002), On the $M^X/G/1$ retrial queue with Bernoulli schedules and general retrial times, *Asia-Pacific Journal of Operational Research* **19**, 177-194.
428. B. Krishna Kumar, S. Pavai Madheswari and A. Vijayakumar (2002), The $M/G/1$ retrial queue with feedback and starting failures, *Applied Mathematical Modelling* **26**, 1057-1075.
429. B. Krishna Kumar, A. Vijayakumar and D. Arivudainambi (2002), An $M/G/1$ retrial queueing system with two-phase service and preemptive resume, *Annals of Operations Research* **113**, 61-79.
430. B. Krishna Kumar and S. Pavai Madheswari (2003), $M^X/G/1$ retrial queue with multiple vacations and starting failures, *Opsearch* **40**, 115-137.
431. B. Krishna Kumar and S. Pavai Madheswari (2003), Mixed loss and delay retrial queueing system with two classes of customers. In: *Stochastic Point Processes* (Eds. S.K. Srinivasan and A. Vijayakumar), pp. 196-211, Narosa Publishing House, New Delhi.
432. B. Krishna Kumar and S. Pavai Madheswari (2004), Mixed loss and delay retrial queueing system with two classes of customers, *Statistica* **LXIV**, 57-73.
433. B. Krishna Kumar and J. Raja (2006), On multiserver feedback retrial queues with balking and control retrial rate, *Annals of Operations Research* **141**, 211-232.
434. A. Krishnamoorthy and P.V. Ushakumari (1999), Reliability of a k-out-of-n system with repair and retrial of failed units, *Top* **7**, 293-304.
435. A. Krishnamoorthy and P.V. Ushakumari (2002), $GI/M/1/1$ queue with finite retrials and finite orbits, *Stochastic Analysis and Applications* **20**, 357-374.
436. A. Krishnamoorthy and M.E. Islam (2003), Production inventory with retrial of customers in an (s, S) policy, *Stochastic Modelling and Applications* **6**, No. 2, 1-11.
437. A. Krishnamoorthy, V.C. Narayanan and M.E. Islam (2003), Retrial production inventory with MAP and service time. In: *Queues, Flows, Systems,*

Networks. Proceedings of the International Conference "Modern Mathematical Methods of Analysis and Optimization of Telecommunication Networks", pp. 148-156, Gomel.
438. A. Krishnamoorthy and M.E. Islam (2004), (s, S) Inventory system with postponed demands, *Stochastic Analysis and Applications* **22**, 827-842.
439. A. Krishnamoorthy, V.C. Narayanan and T.G. Deepak (2004), Maximization of reliability of a k-out-of-n system with repair by a facility attending external customers in a retrial queue. In: *Proceedings of the Fifth International Workshop on Retrial Queues* (Ed. B.D. Choi), pp. 31-38, Seoul.
440. A. Krishnamoorthy, T.G. Deepak and V.C. Joshua (2005), An $M/G/1$ retrial queue with nonpersistent customers and orbital search, *Stochastic Analysis and Applications* **23**, 975-997.
441. A. Krishnamoorthy, V.C. Narayanan and T.G. Deepak (2005), Retrial queues with self generation of priority of orbital customers. In: *Proceedings of the Third National Conference on Mathematical and Computational Models* (Eds. R. Arumuganathan and R. Nadarajan), pp. 289-294, Allied Publishers, New Delhi.
442. A. Krishnamoorthy, V.C. Narayanan and T.G. Deepak (2007), Optimal utilization of service facility for a k-out-of-n system with repair by extending service to external customers in a retrial queue, *Journal of Applied Mathematics and Computing* **25**, 389-405.
443. V.G. Kulkarni (1982), Letter to the editor, *Journal of Applied Probability* **19**, 901-904.
444. V.G. Kulkarni (1983), On queueing systems with retrials, *Journal of Applied Probability* **20**, 380-389.
445. V.G. Kulkarni (1983), A game theoretic model for two types of customers competing for service, *Operations Research Letters* **2**, 119-122.
446. V.G. Kulkarni (1986), Expected waiting times in a multiclass batch arrival retrial queue, *Journal of Applied Probability* **23**, 144-154.
447. V.G. Kulkarni and S.P. Sethi (1988), Deterministic retrial times are optimal in queues with forbidden states, *Infor* **27**, 374-386.
448. V.G. Kulkarni (1989), Optimal retrial policies for restrained Markov chains, *Stochastic Models* **5**, 401-429.
449. V.G. Kulkarni and B.D. Choi (1990), Retrial queues with server subject to breakdowns and repairs, *Queueing Systems* **7**, 191-208.
450. V.G. Kulkarni (1995), *Modeling and Analysis of Stochastic Systems*, Chapman and Hall, London.
451. V.G. Kulkarni and H.M. Liang (1997), Retrial queues revisited. In: *Frontiers in Queueing* (Ed. J.H. Dshalalow), pp. 19-34, CRC Press, Boca Raton.
452. A. Kumar, D. Manjunath and J. Kuri (2004), *Communication Networking: An Analytical Approach*, Morgan Kaufmann Publishers, San Francisco.
453. L. Lakatos (1994), On a simple continuous cyclic-waiting problem, *Annales Universitatis Scientiarum Budapestinensis de Rolando Eötvös Nominatae. Sectio Computatorica* **14**, 105-113.
454. L. Lakatos (1998), On a simple discrete cyclic-waiting queueing problem, *Journal of Mathematical Sciences* **92**, 4031-4034.
455. L. Lakatos (2002), A retrial system with time-limited tasks, *Theory of Stochastic Processes* **8**, 249-256.

456. L. Lakatos (2002), Limit distributions for some cyclic-waiting queueing systems. In: *Proceedings of the Ukrainian Mathematical Congress - 2001. Probability Theory and Mathematical Statistics*, pp. 102-106, Kiev.
457. L. Lakatos (2006), A retrial queueing system with urgent customers, *Journal of Mathematical Sciences* **138**, 5405-5409.
458. C. Langaris and E. Moutzoukis (1995), A retrial queue with structured batch arrivals, priorities and server vacations, *Queueing Systems* **20**, 341-368.
459. C. Langaris (1997), A polling model with retrial customers, *Journal of the Operations Research Society of Japan* **40**, 489-508.
460. C. Langaris and E. Moutzoukis (1997), A batch arrival reader-writer queue with retrial writers, *Stochastic Models* **13**, 523-545.
461. C. Langaris (1999), Gated polling models with customers in orbit, *Mathematical and Computer Modelling* **30**, No. 3-4, 171-187.
462. C. Langaris (1999), Markovian polling systems with mixed service disciplines and retrial customers, *Top* **7**, 305-322.
463. G. Latouche and V. Ramaswami (1993), A logarithmic reduction algorithm for quasi-birth-death processes, *Journal of Applied Probability* **30**, 650-674.
464. G. Latouche and V. Ramaswami (1999), *Introduction to Matrix Analytic Methods in Stochastic Modeling*, ASA-SIAM, Philadelphia.
465. P. Le Gall (1984), The repeated call model and the queue with impatience. In: *Fundamentals of Teletraffic Theory. Proceedings of the Third International Seminar on Teletraffic Theory*, pp. 278-289, Moscow.
466. E.A. Lebedev (2002), On the first passage time of removing level for retrial queues, *Reports of the National Academy of Sciences of Ukraine*, No. 3, 47-50.
467. E.A. Lebedev and I. Usar (2005), On stationary distribution for migration processes of a special type. In: *Queues, Flows, Systems, Networks. Proceedings of the International Conference "Mathematical Methods of Optimization of Telecommunication Networks"*, pp. 110-113, Minsk.
468. S. Lederman (1985), Congestion model for subscriber line modules. In: *IEEE Global Telecommunications Conference, GLOBECOM'85*, pp. 395-401.
469. Y.W. Lee (2005), The $M/G/1$ feedback retrial queue with two types of customers, *Bulletin of the Korean Mathematical Society* **42**, 875-887.
470. A. Lewis and G. Leonard (1983), Measurements of repeat call attempts in the intercontinental telephone service. In: *Proceedings of the 10th International Teletraffic Congress, ITC-10*, Montreal.
471. H. Li and T. Yang (1995), A single-server retrial queue with server vacations and a finite number of input sources, *European Journal of Operational Research* **85**, 149-160.
472. H. Li and T. Yang (1998), $Geo/G/1$ discrete time retrial queue with Bernoulli schedule, *European Journal of Operational Research* **111**, 629-649.
473. H. Li and T. Yang (1998), The steady-state distribution of the $PH/M/1$ retrial queue. In: *Advances in Matrix Analytic Methods for Stochastic Models* (Eds. A.S. Alfa and S.R. Chakravarthy), pp. 135-149, Notable Publications, Inc., New Jersey.
474. H. Li and T. Yang (1999), Steady-state queue size distribution of discrete-time $PH/Geo/1$ retrial queues, *Mathematical and Computer Modelling* **30**, No. 3-4, 51-63.
475. H. Li and Y.Q. Zhao (2005), A retrial queue with a constant retrial rate, server break downs and impatient customers, *Stochastic Models* **21**, 531-550.

476. J. Li and J. Wang (2006), An $M/G/1$ retrial queue with second multi-optional service, feedback and unreliable server, *Applied Mathematics. A Journal of Chinese Universities, Series* **B 21**, 252-262.
477. Q.L. Li and J. Cao (2004), Two types of RG-factorizations of quasi-birth-and-death processes and their applications to stochastic integral functionals, *Stochastic Models* **20**, 299-340.
478. Q.L. Li and Y.Q. Zhao (2004), The RG-factorizations in block-structured Markov renewal processes with applications. In: *Observation, Theory and Modeling of Atmospheric Variability* (Ed. X. Zhu), pp. 545-568, World Scientific, New Jersey.
479. Q.L. Li, Y. Ying and Y.Q. Zhao (2006), A $BMAP/G/1$ retrial queue with a server subject to breakdowns and repairs, *Annals of Operations Research* **141**, 233-270.
480. Y.J. Li, W.H. Zhou and D.Y. Qi (2006), A new congestion prevention policy for the router, *Journal of Shangai Jiatong University (Science)* **E-11**, No.3, 286-289.
481. H.M. Liang and V.G. Kulkarni (1993), Stability condition for a single-server retrial queue, *Advances in Applied Probability* **25**, 690-701.
482. H.M. Liang and V.G. Kulkarni (1993), Monotonicity properties of single-server retrial queues, *Stochastic Models* **9**, 373-400.
483. H.M. Liang (1999), Service station factors in monotonicity of retrial queues, *Mathematical and Computer Modelling* **30**, No. 3-4, 189-196.
484. H.M. Liang and V.G. Kulkarni (1999), Optimal routing control in retrial queues. In: *Applied Probability and Stochastic Processes* (Eds. J.G. Shanthikumar and U. Sumita), pp. 203-218, Kluwer, Boston.
485. L. Libman and A. Orda (2002), Optimal retrial and timeout strategies for accessing network resources, *IEEE/ACM Transactions on Networking* **10**, 551-564.
486. R.E. Lillo (1996), A $G/M/1$-queue with exponential retrial, *Top* **4**, 99-120.
487. G. Lind (1976), Studies on the probability of a called subscriber being busy. In: *Proceedings of the 8th International Teletraffic Congress, ITC-8*, Melbourne.
488. K.S. Liu (1980), Direct distance dialing: Call completion and customer retrial behavior, *The Bell System Technical Journal* **59**, 295-311.
489. X. Liu and A.O. Fapojuwo (2006), Performance analysis of hierarchical cellular networks with queueing and user retrials, *International Journal of Communication Systems* **19**, 699-721.
490. M.J. Lopez-Herrero (2001), The $M/G/1$ retrial queue: A maximum entropy approach. In: *Proceedings of the 15th European Simulation Multiconference* (Eds. E.J.H. Kerckhoffs and M. Snorek), pp. 758-762, A Publication of the Society for Computer Simulation International, Delft.
491. M.J. Lopez-Herrero (2002), On the number of customers served in the $M/G/1$ retrial queue: First moments and maximum entropy approach, *Computers & Operations Research* **29**, 1739-1757.
492. M.J. Lopez-Herrero (2002), Distribution of the number of customers served in an $M/G/1$ retrial queue, *Journal of Applied Probability* **39**, 407-412.
493. M.J. Lopez-Herrero and M.F. Neuts (2002), The distribution of the maximum orbit size of an $M/G/1$ retrial queue during the busy period. In: *Advances in Stochastic Modelling* (Eds. J.R. Artalejo and A. Krishnamoorthy), pp. 219-231, Notable Publications, Inc., New Jersey.

494. M.J. Lopez-Herrero (2006), A maximum entropy approach for the busy period of the $M/G/1$ retrial queue, *Annals of Operations Research* **141**, 271-281.
495. J. Lubacz and J. Roberts (1984), A new approach to the single server repeated attempt system with balking. In: *Fundamentals of Teletraffic Theory. Proceedings of the Third International Seminar on Teletraffic Theory*, pp. 290-293, Moscow.
496. D.M. Lucantoni (1991), New results on the single server queue with a batch Markovian arrival process, *Stochastic Models* **7**, 1-46.
497. D.M. Lucantoni and M.F. Neuts (1994), Simpler proofs of some properties of the fundamental period of the $MAP/G/1$ queue, *Journal of Applied Probability* **31**, 235-243.
498. D.M. Lucantoni and M.F. Neuts (1994), Some steady-state distribution for the $MAP/SM/1$ queue, *Stochastic Models* **10**, 575-598.
499. L.I. Lukashuk (1990), Diffusion approximation and filtering of characteristics of a queueing system with repeated calls, *Cybernetics* **26**, No. 2, 253-264.
500. L.I. Lukashuk and Y.A. Semenchenko (1991), Filtering of a semi-Markov queueing system with repeated calls, *Cybernetics and Systems Analysis* **27**, No. 4, 627-631.
501. S.J. Lupker, G.J. Fleet and B.R. Shelton (1988), Caller's perceptions of post-dialing delays: The effects of a new signalling technology, *Behaviour and Information Technology* **7**, 263-274.
502. N.W. Macfadyen (1979), Statistical observation of repeated attempts in the arrival process. In: *Proceedings of the 9th International Teletraffic Congress, ITC-9*, Torremolinos.
503. F. Machihara and M. Saitoh (2008), Mobile customers model with retrials, *European Journal of Operational Research*, in press.
504. V.A. Malyshev (1972), Homogeneous random walks on the product of a finite set and a half-line. In: *Probabilistic Methods of Research*, Moscow State University (in Russian), pp. 5-13.
505. A. Mandelbaum, W.A. Massey, M.I. Reiman and B. Rider (1999), Time varying multiserver queues with abandonment and retrials. In: *Teletraffic Engineering in a Competitive World, ITC-16* (Eds. P. Key and D. Smith), pp. 355-364, Elsevier Science Publishers B.V., Amsterdam.
506. A. Mandelbaum, W.A. Massey, M.I. Reiman and A. Stolyar (1999), Waiting time asymptotics for time varying multiserver queues with abandonment and retrials. In: *Proceedings of the Thirty-Seventh Annual Allerton Conference on Communication, Control and Computing*, pp. 1095-1104.
507. A. Mandelbaum, W.A. Massey, M.I. Reiman, A. Stolyar and B. Rider (2002), Queue lengths and waiting times for multiserver queues with abandonment and retrials, *Telecommunication Systems* **21**, 149-171.
508. M. Mandjes and K. Tutschku (1996), Efficient call handling procedures in cellular mobile networks, *Research Report Series*, Institute of Computer Science, University of Würzburg.
509. P. Manuel, B. Sivakumar and G. Arivarignan (2007), Perishable inventory system with postponed demands and negative-customers, *Journal of Applied Mathematics and Decision Sciences*, Article ID 94850, pp. 1-12.
510. M. Martin and J.R. Artalejo (1995), Analysis of an $M/G/1$ queue with two types of impatient units, *Advances in Applied Probability* **27**, 840-861.
511. M. Martin and A. Gómez-Corral (1995), On the $M/G/1$ retrial queueing system with linear control policy, *Top* **3**, 285-305.

512. J. Medhi (2003), *Stochastic Models in Queueing Theory*, Academic Press, Amsterdam.
513. G. Medvedev (1994), Random process characteristic in *LAN* with random access and asymmetric load, *Automatic Control and Computer Sciences* **28**, No. 3, 34-41.
514. V.I. Meykshan and I.G. Fidel'man (1995), The design of communications networks with bypass routings when there are repeated calls and lines with undetected breakdowns, *Telecommunications and Radio Engineering* **49**, No. 7, 40-44.
515. G.S. Mokaddis, S.A. Metwally and B.M. Zaki (2007), A feedback retrial queuing system with starting failures and single vacation, *Tamkang Journal of Science and Engineering* **10**, 183-192.
516. P. Moreno (2004), An $M/G/1$ retrial queue with recurrent customers and general retrial times, *Applied Mathematics and Computation* **159**, 651-666.
517. P. Moreno (2006), A discrete-time retrial queue with unreliable server and general server lifetime, *Journal of Mathematical Sciences* **132**, 643-655.
518. E. Morozov (2007), A multiserver retrial queue: Regenerative stability analysis, *Queueing Systems* **56**, 157-168.
519. M. Morrison and P. Fleming Jr. (1976), Some blocking formulas for three-stage switching arrays, with multiple connection attempts. In: *Proceedings of the 8th International Teletraffic Congress, ITC-8*, Melbourne.
520. I.D. Moscholios and M.D. Logothetis (2001), Call-level QoS assessment in ATM networks supporting elastic traffic, *IEEE International Conference on Communications, ICC 2001*, pp. 11-14 (Vol. 6), Helsinki.
521. I.D. Moscholios, M.D. Logothetis and G.K. Kokkinakis (2002), Connection-dependent threshold model: A generalization of the Erlang multiple rate loss model, *Performance Evaluation* **48**, 177-200.
522. I.D. Moscholios, P.I. Nikolaropoulos and M.D. Logothetis (2003), Call level blocking of ON-OFF traffic sources with retrials under the complete sharing policy. In: *Providing Quality of Service in Heterogeneous Environments, ITC-18* (Eds. J. Charzinski, R. Lehnert and P. Tran-Gia), pp. 811-820, Elsevier Science Publishers B.V., Amsterdam.
523. I.D. Moscholios, M.D. Logothetis and G.K. Kokkinakis (2005), Call-burst blocking of ON-OFF traffic sources with retrials under the complete sharing policy, *Performance Evaluation* **59**, 279-312.
524. E. Moutzoukis and C. Langaris (1996), Non-preemptive priorities and vacations in a multiclass retrial queueing system, *Stochastic Models* **12**, 455-472.
525. E. Moutzoukis and C. Langaris (2001), Two queues in tandem with retrial customers, *Probability in the Engineering and Informational Sciences* **15**, 311-325.
526. V.V. Mushko, A.N. Dudin and C.S. Kim (2004), The $MMAP/M/c$ system with heterogenous servers and addressed retrials. In: *Proceedings of the VII-th International Conference "Computer Data Analysis and Modelling: Robustness and Computer Intensive Methods"*, pp. 100-104 (Vol. 2), Minsk.
527. V.V. Mushko (2006), $M/M/c$ system with address retrial strategy and back-up servers, *Automatic Control and Computer Sciences* **40**, No. 2, 58-65.
528. V.V. Mushko, M.J. Jacob, K.O. Ramakrishnan, A. Krishnamoorthy and A.N. Dudin (2006), Multiserver queue with addressed retrials, *Annals of Operations Research* **141**, 283-301.

529. K.V. Mykhalevych (2003), A comparison of a classical retrial $M/G/1$ queueing system and a Lakatos-type $M/G/1$ cyclic-waiting time queueing system, *Annales Universitatis Scientiarum Budapestinensis de Rolando Eötrös Nominatae. Sectio Computatorica* **23**, 229-238.
530. K.V. Mykhalevych (2005), On the ergodicity condition of a $GI/D/1$ retrial queueing system with constant retrial times and a dynamic service priority, *Annales Universitatis Scientiarum Budapestinensis de Rolando Eötrös Nominatae. Sectio Computatorica* **25**, 103-111.
531. A. Myskja (1971), A recording and processing system for accounting and traffic analysis on a large PABX, *IEEE Transactions on Communication Technology* **19**, 692-699.
532. A. Myskja and O.O. Walmann (1973), An investigation of telephone-user habits by means of computer techniques, *IEEE Transactions on Communications* **21**, 663-671.
533. A. Myskja and O.O. Walmann (1973), A statistical study of telephone traffic data with emphasis on subscriber behaviour. In: *Proceedings of the 7th International Teletraffic Congress, ITC-7*, Stockholm.
534. A. Myskja and F.A. Aagesen (1977), On the interaction between subscribers and a telephone system, *Telektronikk* **3**, 271-282.
535. A. Myskja (1979), Modelling of non-stationary traffic processes. In: *Proceedings of the 9th International Teletraffic Congress, ITC-9*, Torremolinos.
536. L. Nador (1988), The effects of traffic overloads in automatic telephone networks, *Budavox Telecommunication Review*, No. 1, 2-17.
537. E.O. Naumova (1983), The loss probability measurement accuracy for the subscriber line with repeated calls. In: *Proceedings of the 10th International Teletraffic Congress, ITC-10*, Montreal.
538. E.O. Naumova and E.I. Shkolny (1985), Fault lines' influence to the values of probabilistic characteristics and their measurement accuracy in system with expectation and repeated calls. In: *Teletraffic Issues in an Advanced Information Society, ITC-11* (Ed. M. Akiyama), pp. 1024-1028, Elsevier Science Publishers B.V., Amsterdam.
539. A.A. Nazarov and S.L. Shokhor (2000), Study of controlled asynchronous multiple access in satellite communication networks with conflict warning, *Problems of Information Transmission* **36**, No. 1, 71-83.
540. J.A. Nelder and R. Mead (1965), A simplex method for function minimization, *Computer Journal* **7**, 308-313.
541. M. Nesenbergs (1979), A hybrid of Erlang B and C formulas and its applications, *IEEE Transactions on Communications* **27**, 59-68.
542. M.F. Neuts (1964), The distribution of the maximum length of a Poisson queue during a busy period, *Operations Research* **12**, 281-285.
543. M.F. Neuts (1978), Markov chains with applications in queueing theory, which have a matrix-geometric invariant probability vector, *Advances in Applied Probability* **10**, 185-212.
544. M.F. Neuts (1981), *Matrix-Geometric Solutions in Stochastic Models: An Algorithmic Approach*, The Johns Hopkins University Press, Baltimore.
545. M.F. Neuts and M.F. Ramalhoto (1984), A service model in which the server is required to search for customers, *Journal of Applied Probability* **21**, 157-166.
546. M.F. Neuts (1989), *Structured Stochastic Matrices of $M/G/1$ Type and Their Applications*, Marcel Dekker, Inc., New York.

547. M.F. Neuts and B.M. Rao (1990), Numerical investigation of a multiserver retrial model, *Queueing Systems* **7**, 169-190.
548. M.F. Neuts (1995), *Algorithmic Probability: A Collection of Problems*, Chapman and Hall, London.
549. R.D. Nobel (2004), A discrete-time retrial queueing model with one server. In: *Proceedings of the Fifth International Workshop on Retrial Queues* (Ed. B.D. Choi), pp. 39-51, Seoul.
550. R.D. Nobel (2005), A discrete-time priority loss/retrial queueing model with two types of traffic. In: *Proceedings of the Korea-Netherlands Joint Conference on Queueing Theory and Its Applications to Telecommunication Systems* (Ed. B.D. Choi), pp. 189-207, Seoul.
551. R.D. Nobel and P. Moreno (2005), A discrete-time priority loss/retrial queueing model with two types of traffic. In: *Proceedings of the National Conference on Mathematical and Computational Models* (Eds. R. Arumuganathan and R. Nadarajan), pp. 41-54, Allied Publishers, New Delhi.
552. R.D. Nobel and H.C. Tijms (2006), Waiting-time probabilities in the $M/G/1$ retrial queue, *Statistica Neerlandica* **60**, 73-78.
553. R.D. Nobel and P. Moreno (2008), A discrete-time retrial queueing model with one server, *European Journal of Operational Research*, in press.
554. H. Ohmura and Y. Takahashi (1985), An analysis of repeated call model with a finite number of sources, *Electronics and Communications in Japan* **68**, No. 6, 112-121.
555. E. Onur, H. Delic, C. Ersoy and M.U. Caglayan (2000), On the retrial and redial phenomena in GSM networks. In: *IEEE Wireless Communications and Networking Conference, WCNC-2000*, pp. 885-889 (Vol. 2), Chicago.
556. E. Onur, H. Delic, C. Ersoy and M.U. Caglayan (2002), Measurement-based replanning of GSM cell capacities considering retrials, redials and hand-offs. In: *IEEE International Conference on Communications, ICC 2002*, pp. 3361-3365 (Vol. 5), New York.
557. P.R. Parthasarathy and R. Sudhesh (2007), Time-dependent analysis of a single-server retrial queue with state-dependent rates, *Operations Research Letters* **35**, 601-611.
558. M. Paterok, U. Herzog and C. Bleisteiner (1988), The influence of repeated calls on the performance measures of loss systems. In: *Teletraffic Science for New Cost-Effective Systems, Networks and Services, ITC-12* (Ed. M. Bonatti), pp. 1413-1419, Elsevier Science Publishers B.V., Amsterdam.
559. C.E.M. Pearce (1987), On the problem of the re-attempted calls in teletraffic, *Stochastic Models* **3**, 393-407.
560. C.E.M. Pearce (1989), Extended continued fractions, recurrence relations and two-dimensional Markov processes, *Advances in Applied Probability* **21**, 357-375.
561. C.E.M. Pearce and K.H. Rhee (1999), An algorithmic approach to the $M/PH/c$ retrial queue with homogeneous servers. In: *Proceedings of the Applied Mathematics Workshop, Centre for Applied Mathematics, KAIST* **6**, 135-155.
562. B. Pourbabai (1987), Markovian queueing systems with retrials and heterogeneous servers, *Computers and Mathematics with Applications* **13**, 917-923.
563. B. Pourbabai (1987), Analysis of a $G/M/K/0$ queueing loss system with heterogeneous servers and retrials, *International Journal of Systems Science* **18**, 985-992.

564. B. Pourbabai (1987), Approximation of the overflow process from a $G/M/N/K$ queueing system, *Management Science* **33**, 931-938.
565. B. Pourbabai (1988), Asymptotic analysis of $G/G/K$ queueing-loss system with retrials and heterogeneous servers, *International Journal of Systems Sciences* **19**, 1047-1052.
566. B. Pourbabai (1988), A random access telecommunication system, *Journal of Information Processing and Cybernetics* **24**, 613-625.
567. B. Pourbabai (1989), A finite capacity telecommunication system with repeated calls, *Journal of Information Processing and Cybernetics* **25**, 457-467.
568. B. Pourbabai (1989), Tandem behaviour of a telecommunication system with repeated calls: A Markovian case with buffers, *Journal of the Operational Research Society* **40**, 671-680.
569. B. Pourbabai (1990), A note on a $D/G/K$ loss system with retrials, *Journal of Applied Probability* **27**, 385-392.
570. B. Pourbabai (1990), Tandem behavior of a telecommunication system with finite buffers and repeated calls, *Queueing Systems* **6**, 89-108.
571. B. Pourbabai (1993), Tandem behavior of a telecommunication system with repeated calls: II, A general case without buffers, *European Journal of Operational Research* **65**, 247-258.
572. K. Prakash Rani and A. Srinivasan (2005), An $M/G/1$ retrial queue with server breakdown. In: *Proceedings of the National Conference on Mathematical and Computational Models* (Eds. R. Arumuganathan and R. Nadarajan), pp. 153-158, Allied Publishers, New Delhi.
573. W.H. Press, S.A. Teukolsky, W.T. Vetterling and B.P. Flannery (1992), *Numerical Recipes in Fortran. The Art of Scientific Computing*, Cambridge University Press, Cambrigde.
574. M.F. Ramalhoto and A. Gómez-Corral (1998), Some decomposition formulae for $M/M/r/r+d$ queues with constant retrial rate, *Stochastic Models* **14**, 123-145.
575. M.F. Ramalhoto (1999), The infinite server queue and heuristic approximations to the multi-server queue with and without retrials, *Top* **7**, 333-350.
576. K. Ramanath and P. Lakshmi (2005), Performance analysis of the $M/G/c$ retrial queueing systems using the theory of Markov regenerative stochastic Petri nets, *Opsearch* **42**, 134-151.
577. P.K. Reeser (1989), Simple approximation for blocking seen by peaked traffic with delayed, correlated reattempts. In: *Teletraffic Science for New Cost-Effective Systems, Networks and Services, ITC-12* (Ed. M. Bonatti), pp. 1420-1426, Elsevier Science Publishers B.V., Amsterdam.
578. M.A. Remiche (1998), Time to congestion in homogeneous quasi-birth-and-death processes, *Opsearch* **35**, 169-192.
579. N. Renganathan, R. Kalyanaraman and B. Srinivasan (2002), A finite capacity single server retrial queue with two types of calls, *International Journal of Information and Management Sciences* **13**, No. 3, 47-56.
580. A. Ridder (2000), Fast simulation of retrial queues. In: *Third Workshop on Rare Event Simulation and Related Combinatorial Optimization Problems*, pp. 1-5, Pisa.
581. J. Riordan (1962), *Stochastic Service Systems*, Wiley, New York.
582. J.W. Roberts (1979), Recent observations of subscriber behaviour. In: *Proceedings of the 9th International Teletraffic Congress, ITC-9*, Torremolinos.

583. R.S. Robeva (1991), A semi-Markov model of a non-homogeneous telephone subscribers system. In: *Teletraffic and Datatraffic in a Period of Change, ITC-13* (Eds. A. Jensen and V.B. Iversen), pp. 701-706, Elsevier Science Publishers B.V., Amsterdam.
584. A. Rodrigo, M. Vázquez and G.I. Falin (1998), A new Markovian description of the $M/G/1$ retrial queue, *European Journal of Operational Research* **104**, 231-240.
585. A. Rodrigo and M. Vázquez (1999), Large sample inference in retrial queues, *Mathematical and Computer Modelling* **30**, No. 3-4, 197-206.
586. A. Rodrigo (2006), Estimators of the retrial rate in $M/G/1$ retrial queues, *Asia-Pacific Journal of Operational Research* **23**, 193-213.
587. R. Rom and M. Sidi (1990), *Multiple Access Protocols: Performance and Analysis*, Springer-Verlag, New York.
588. J. Roszik and J. Sztrik (2004), The effect of server's breakdown on the performance of finite-source retrial queueing systems. In: *Proceedings of the 6th International Conference on Applied Informatics* (Eds. L. Csőke, P. Olajos, P. Szigetváry and T. Tómács), pp. 221-230 (Vol. II), Eger.
589. J. Roszik, J. Sztrik and C.S. Kim (2005), Retrial queues in the performance modeling of cellular mobile networks using *MOSEL*, *International Journal of Simulation* **6**, 38-47.
590. J. Roszik and J. Sztrik (2007), Performance analysis of finite-source retrial queues with nonreliable heterogeneous servers, *Journal of Mathematical Sciences* **146**, 6033-6038.
591. J. Roszik, J. Sztrik and J. Virtamo (2007), Performance analysis of finite-source retrial queues operating in random environments, *International Journal of Operational Research* **2**, 254-268.
592. G.G. Scavo and G. Miranda (1985), Traffic offered grade of service and call completion ratio in a toll network versus subscriber retrial behaviour. In: *Teletraffic Issues in an Advanced Information Society, ITC-11* (Ed. M. Akiyama), pp. 689-695, Elsevier Science Publishers B.V., Amsterdam.
593. M. Senthil Kumar and R. Arumuganathan (2008), On the single server batch arrival retrial queue with general vacation time under Bernoulli schedule and two phase of heterogeneous service, *Quality Technology & Quantitative Management* **5**, 145-160.
594. R.F. Serfozo (1988), Extreme values of queue lengths in $M/G/1$ and $GI/M/1$ systems, *Mathematics of Operations Research* **13**, 349-357.
595. R.F. Serfozo (1988), Extreme values of birth-and-death processes and queues, *Stochastic Processes and Their Applications* **27**, 291-306.
596. L.D. Servi (2002), Algorithmic solutions to two-dimensional birth-death processes with application to capacity planning, *Telecommunication Systems* **21**, 205-212.
597. W. Shang, L. Liu and Q.L. Li (2006), Tail asymptotics for the queue length in an $M/G/1$ retrial queue, *Queueing Systems* **52**, 193-198.
598. V. Sharma and E.A. Varvarigos (1997), Circuit switching with input queuing: An analysis for the d-dimensional wraparound mesh and the hypercube, *IEEE Transactions Parallel and Distributed Systems* **8**, 349-366.
599. N.P. Sherman and J.P. Kharoufeh (2006), An $M/M/1$ retrial queue with unreliable server, *Operations Research Letters* **34**, 697-705.

600. N.P. Sherman, J.P. Kharoufeh and M.A. Abramson (2008), An $M/G/1$ retrial queue with unreliable server for streaming multimedia applications, *Probability in the Engineering and Informational Sciences*, in press.
601. H. Shimonishi, T. Takine, M. Murata and H. Miyahara (1996), Performance analysis of fast reservation protocol in ATM networks with arbitrary topologies, *Performance Evaluation* **28**, 41-69.
602. Y.W. Shin and C.E.M. Pearce (1998), An algorithmic approach to the Markov chain with transition probability matrix of upper block-Hessenberg form, *Korean Journal of Computational and Applied Mathematics* **5**, 403-426.
603. Y.W. Shin (2000), Transient distributions of level dependent quasi-birth-death processes with linear transition rates, *Korean Journal of Computational and Applied Mathematics* **7**, 83-100.
604. Y.W. Shin and Y.C. Kim (2000), Stochastic comparisons of Markovian retrial queues, *Journal of the Korean Statistical Society* **29**, 473-488.
605. Y.W. Shin (2001), Approximations for Markov chains with upper Hessenberg transition matrices, *Journal of the Operations Research Society of Japan* **44**, 90-98.
606. Y.W. Shin (2004), Multi-server retrial queue with negative customers and disasters. In: *Proceedings of the Fifth International Workshop on Retrial Queues* (Ed. B.D. Choi), pp. 53-60, Seoul.
607. Y.W. Shin (2005), Convergence of generalized truncation method in retrial queues. In: *Queues, Flows, Systems, Networks. Proceedings of the International Conference "Mathematical Methods of Optimization of Telecommunication Networks"*, pp. 196-201, Minsk.
608. Y.W. Shin (2006), Monotonicity properties in various retrial queues and their applications, *Queueing Systems* **53**, 147-157.
609. Y.W. Shin and D.H. Moon (2006), An algorithmic solution for the stationary distribution of $M/M/c/K$ retrial queue. In: *Lecture Notes in Operations Research 6. The Sixth International Symposium on Operations Research and Its Applications, ISORA'06* (Eds. X.S. Zhang, D.G. Liu and L.Y. Wu), pp. 60-74, Xinjiang.
610. Y.W. Shin (2007), Multi-server retrial queue with negative customers and disasters, *Queueing Systems* **55**, 223-237.
611. Y.W. Shin (2008), Convergence of the stationary distributions of $M/M/s/K$ retrial queue as K tends to infinity, *European Journal of Operational Research*, in press.
612. N. Shinagawa, T. Kobayashi, K. Nakano and M. Sengoku (2000), Teletraffic characteristics in prioritized handoff control method considering reattempt calls, *IEICE Transactions on Communications*, Vol. E83-B, 1810-1818.
613. E.I. Shkol'nyi (1977), Estimates of the loss probabilities for a queueing system with repeated calls, *Engineering Cybernetics* **15**, No. 2, 74-78.
614. M.A. Shneps-Shneppe (1970), The effect of repeated calls on communication system. In: *Proceedings of the 6th International Teletraffic Congress, ITC-6*, Munich.
615. M.A. Shneps-Shneppe and A.F. Petrov (1984), Optimal distribution of overall grade of service in hierarchical telephone network with repeated attempts. In: *Fundamentals of Teletraffic Theory. Proceedings of the Third International Seminar on Teletraffic Theory*, pp. 377-386, Moscow.
616. J.E. Shore and R.W. Johnson (1981), Properties of cross-entropy minimization, *IEEE Transactions on Information Theory* **27**, 472-482.

617. B. Sivakumar (2008), Two-commodity inventory system with retrial demand, *European Journal of Operational Research* **187**, 70-83.
618. D. Sonderman and B. Pourbabai (1987), Single server stochastic recirculation systems, *Computers & Operations Research* **14**, 75-84.
619. D.J. Songhurst (1984), Subscriber repeat attempts, congestion and the quality of service: A study based on networking simulation, *British Telecommunication Technology Journal* **2**, 47-55.
620. F.M. Spieksma and R.L. Tweedie (1994), Strengthening ergodicity to geometric ergodicity for Markov chains, *Stochastic Models* **10**, 45-74.
621. M. Stastny (1984), On the substitution of the basic retrial model for a complex loss model with retrials. In: *Fundamentals of Teletraffic Theory. Proceedings of the Third International Seminar on Teletraffic Theory*, pp. 395-399, Moscow.
622. S.N. Stepanov (1980), Integral equilibrium relations of non-full-access systems with repeated calls, and their applications, *Problems of Information Transmission* **16**, No. 4, 323-327.
623. S.N. Stepanov (1981), Probabilistic characteristics of an incompletely accessible multi-phase service system with several types of repeated call, *Problems of Control and Information Theory* **10**, No. 6, 1-12.
624. S.N. Stepanov (1983), Algorithms of approximate design of systems with repeated calls, *Automation and Remote Control* **44**, No. 1, 63-71.
625. S.N. Stepanov (1983), Asymptotic formulae and estimations for probability characteristics of full-available group with absolutely persistent subscribers, *Problems of Control and Information Theory* **12**, No. 5, 361-369.
626. S.N. Stepanov (1983), Properties of probability characteristics of a communication network with repeated calls, *Problems of Information Transmission* **19**, No. 1, 69-76.
627. S.N. Stepanov (1983), The design of a trunk group based on repeated calls and waiting, *Telecommunications and Radio Engineering*, No. 6, 1-7.
628. S.N. Stepanov (1984), Probabilistic characteristics of an incompletely accessible service system with repeated calls for arbitrary values of subscriber persistent function, *Problems of Control and Information Theory* **13**, No. 2, 69-78.
629. S.N. Stepanov (1984), Numerical calculation accuracy of communication models with repeated calls, *Problems of Control and Information Theory* **13**, No. 6, 371-381.
630. S.N. Stepanov (1984), Estimation of characteristics of multilinear systems with repeated calls. In: *Fundamentals of Teletraffic Theory. Proceedings of the Third International Seminar on Teletraffic Theory*, pp. 400-409, Moscow.
631. S.N. Stepanov (1985), The construction of effective algorithms for numerical analysis of multilinear systems with repeated calls. In: *Teletraffic Issues in an Advanced Information Society, ITC-11* (Ed. M. Akiyama), pp. 1151, Elsevier Science Publishers B.V., Amsterdam.
632. S.N. Stepanov and I.I. Tsitovich (1985), The model of a full-available group with repeated calls and waiting positions in the case of extreme load, *Problems of Control and Information Theory* **14**, No. 1, 25-32.
633. S.N. Stepanov (1986), Increasing the efficiency of numerical methods for models with repeated calls, *Problems of Information Transmission* **22**, No. 4, 313-326.
634. S.N. Stepanov and I.I. Tsitovich (1987), Qualitative methods of analysis of systems with repeated calls, *Problems of Information Transmission* **23**, No. 2, 156-173.

635. S.N. Stepanov (1988), Optimal calculation of characteristics of models with repeated calls. In: *Teletraffic Science for New Cost-Effective Systems, Networks and Sciences, ITC-12* (Ed. M. Bonatti), pp. 1427-1433, Elsevier Science Publishers B.V., Amsterdam.
636. S.N. Stepanov (1989), Solution of simultaneous large-scale equilibrium equations, *Automation and Remote Control* **50**, No. 5, 647-655.
637. S.N. Stepanov (1989), Optimized algorithms for numerical calculation of the characteristics of multistream models with repeated calls, *Problems of Information Transmission* **25**, No. 2, 136-144.
638. S.N. Stepanov and I.I. Tsitovich (1989), Equivalent definitions of the probabilistic characteristics of models with repeated calls and their application, *Problems of Information Transmission* **25**, No. 2, 145-153.
639. S.N. Stepanov (1991), Asymptotic analysis models with repeated calls in case of extreme load. In: *Teletraffic and Datatraffic in a Period of Change, ITC-13* (Eds. A. Jensen and V.B. Iversen), pp. 21-26, Elsevier Science Publishers B.V., Amsterdam.
640. S.N. Stepanov (1993), Asymptotic analysis of models with repeated calls in case of extreme load, *Problems of Information Transmission* **29**, No. 3, 248-267.
641. S.N. Stepanov (1994), The analysis of the model with finite number of sources and taking into account the subcriber behavior, *Automation and Remote Control* **55**, No. 4, 541-551.
642. S.N. Stepanov (1997), Generalized model with repeated calls in case of extreme load, *Queueing Systems* **27**, 131-151.
643. S.N. Stepanov (1999), Markov models with retrials: The calculation of stationary performance measures based on the concept of truncation, *Mathematical and Computer Modelling* **30**, No. 3-4, 207-228.
644. W.J. Stewart (1994), *Introduction to the Numerical Solution of Markov Chains*, Princeton University Press, Princeton.
645. R. Stolletz (2003), *Performance Analysis and Optimization of Inbound Call Centers. Lecture Notes in Economics and Mathematical Systems Vol. 528*, Springer, Berlin.
646. U. Sumita and J.G. Shanthikumar (1985), APL software development for central control telephone switching systems via the row-continuous Markov chain procedure. In: *Teletraffic Issues in an Advanced Information Society, ITC-11* (Ed. M. Akiyama), pp. 439-445, Elsevier Science Publishers B.V., Amsterdam.
647. A.P. Suprun (1987), A stochastic model for estimating the influence of repeated calls in digital communication systems, *Moscow University Mathematics Bulletin* **42**, No. 3, 22-24.
648. N.B. Sutorikhin, K.I. Zaretsky and I.G. Fidelman (1984), Reattempt influence of grade of service after eliminating the total failure state of an exchange route. In: *Fundamentals of Teletraffic Theory. Proceedings of the Third International Seminar on Teletraffic Theory*, pp. 420-429, Moscow.
649. R. Syski (1986), *Introduction to Congestion Theory in Telephone Systems*, Elsevier Science Publishers, Amsterdam.
650. D.Y. Sze (1984), A queueing model for telephone operator staffing, *Operations Research* **32**, 229-249.
651. J. Sztrik (2005), Tool supported performance modelling of finite-source retrial queues with breakdowns, *Publicationes Mathematicae* **66**, 197-211.

652. J. Sztrik, B. Almási and J. Roszik (2006), Heterogeneous finite-source retrial queues with server subject to breakdowns and repairs, *Journal of Mathematical Sciences* **132**, 677-685.
653. H. Takagi (1991), *Queueing Analysis, Volume 1: Vacation and Priority Systems*, North-Holland, Amsterdam.
654. H. Takagi (1993), *Queueing Analysis, Volume 3: Discrete-Time Systems*, North-Holland, Amsterdam.
655. G.K. Takahara (1996), Fixed point approximations for retrial networks, *Probability in the Engineering and Informational Sciences* **10**, 243-259.
656. M. Takahashi, H. Osawa and T. Fujisawa (1999), $Geo^{[X]}/G/1$ retrial queue with non-preemptive priority, *Asia-Pacific Journal of Operational Research* **16**, 215-234.
657. E. Takemori, Y. Usui and J. Matsuda (1985), Field data analysis for traffic engineering. In: *Teletraffic Issues in an Advanced Information Society, ITC-11* (Ed. M. Akiyama), pp. 425-431, Elsevier Science Publishers B.V., Amsterdam.
658. J. Teghem (1986), Control of the service process in a queueing system, *European Journal of Operational Research* **23**, 141-158.
659. J.G.C. Templeton (Ed.) (1990), *Retrial Queues, Queueing Systems* **7**, No. 2, 125-227.
660. J.G.C. Templeton (1999), Retrial Queues, *Top* **7**, 351-353.
661. M.H. Ter Beek, M. Massink and D. Latella (2006), Towards model checking stochastic aspects of the thinkteam user interface. In: *Proceedings 12th International Workshop on Design, Specification and Verification of Interactive Systems. Lecture Notes in Computer Science Vol. 3941* (Eds. S.W. Gilroy and M.D. Harrison), pp. 39-50, Springer-Verlag, Berlin.
662. H.C. Tijms (2003), *A First Course in Stochastic Models*, Wiley, Chichester.
663. P. Todorov and S. Poryazov (1985), Basic dependences characterizing a model of an idealized (standard) subcriber automatic telephone exchange. In: *Teletraffic Issues in an Advanced Information Society, ITC-11* (Ed. M. Akiyama), pp. 752-759, Elsevier Science Publishers B.V., Amsterdam.
664. F. Toledano and J.R. De Los Mozos (1973), An analytical model to describe the influence of the repeated call attempts. In: *Proceedings of the 7th International Teletraffic Congress, ITC-7*, pp. 133/1-6, Stockholm.
665. P. Tran-Gia and M. Mandjes (1997), Modeling of customer retrial phenomenon in cellular mobile networks, *IEEE Journal on Selected Areas in Communications* **15**, 1406-1414.
666. B. Tsankov (1988), Blocking probability in digital link systems with step-by-step selection and restricted number of retrials. In: *International Seminar on Teletraffic Theory and Computer Modelling*, pp. 205-215, Sofia.
667. G.V. Tsarenkov (2001), A robust algorithm for distribution determination for a queueing system with batch Markov input, semi-Markov serving, retrials, and finite orbit, *Automatic Control and Computer Sciences* **35**, No. 5, 63-69.
668. G.V. Tsarenkov (2001), Evaluation of the embedded distribution of the $BMAP/SM/1$ queue Markov modulated linear strategy of retrials. In: *Queues, Flows, Systems, Networks. Proceedings of the International Conference "Modern Mathematical Methods of Investigating of the Information Networks"*, pp. 190-193, Minsk.
669. G.V. Tsarenkov (2002), Investigation of $BMAP/SM/1$ queueing system with modulated repeated request Markovian process, *Automatic Control and Computer Sciences* **36**, No. 4, 47-52.

670. G.V. Tsarenkov (2003), Evaluation of the performance characteristics for the $BMAP/SM/1/N$ queue with Markov modulated linear strategy of retrials. In: *Queues, Flows, Systems, Networks. Proceedings of the International Conference "Modern Mathematical Methods of Analysis and Optimization of Telecommunication Networks"*, pp. 244-249, Gomel.
671. K. Udagawa and E. Miwa (1965), A complete group of trunks and Poisson-type repeated calls which influence it, *Electronics and Communications in Japan* **48**, No. 10, 42-53.
672. P.V. Ushakumari (2006), On (s, S) inventory system with random lead time and repeated demands, *Journal of Applied Mathematics and Stochastic Analysis*, Article ID 81508, pp. 1-22.
673. N.M. Van Dijk (1993), *Queueing Networks and Product Forms: A System Approach*, John Wiley & Sons, Chichester.
674. M. Vázquez (1996), A retrial model in a nonstationary regime, *Top* **4**, 121-133.
675. N. Vlajic, C.D. Charalambous and D. Makrakis (2004), Performance aspects of data broadcast in wireless networks with user retrials, *IEEE/ACM Transactions on Networking* **12**, 620-633.
676. J. Wang, J. Cao and Q. Li (2001), Reliability analysis of the retrial queue with server breakdowns and repairs, *Queueing Systems* **38**, 363-380.
677. J. Wang (2006), Reliability analysis of $M/G/1$ queues with general retrial times and server breakdowns, *Progress in Natural Science* **16**, 464-473.
678. J. Wang and Q. Zhao (2007), A discrete-time $Geo/G/1$ retrial queue with starting failures and second optional service, *Computers and Mathematics with Applications* **53**, 115-127.
679. J. Wang and Q. Zhao (2007), Discrete-time $Geo/G/1$ retrial queue with general retrial times and starting failures, *Mathematical and Computer Modelling* **45**, 853-863.
680. J. Wang and J. Li (2008), A repairable $M/G/1$ retrial queue with Bernoulli vacation and two-phase service, *Quality Technology & Quantitative Management* **5**, 179-192.
681. J. Wang, B. Liu and J. Li (2008), Transient analysis of an $M/G/1$ retrial queue subject to disasters and server failures, *European Journal of Operational Research*, in press.
682. R. Warfield and G. Foers (1985), Application of Bayesian teletraffic measurement to systems with queueing or repeated attempts. In: *Teletraffic Issues in an Advanced Information Society, ITC-11* (Ed. M. Akiyama), pp. 1003-1009, Elsevier Science Publishers B.V., Amsterdam.
683. K. Wesolowski (2002), *Mobile Communication Systems*, John Wiley & Sons, New York.
684. W. Whitt (1999), Improving service by informing customers about anticipated delays, *Management Science* **45**, 192-207.
685. R.I. Wilkinson (1956), Theories for toll traffic engineering in the U.S.A., *The Bell System Technical Journal* **35**, 421-514.
686. R.I. Wilkinson and R.C. Radnik (1967), The character and effect of customer retrials in intertoll circuit operations. In: *Proceedings of the 5th International Teletraffic Congress, ITC-5*, New York.
687. R.I. Wilkinson and R.C. Radnik (1968), Customers' retrials in toll circuit operation. In: *IEEE International Conference on Communications*, pp. 9-14, Philadelphia.

688. R.W. Wolff (1982), Poisson arrivals see time averages, *Operations Research* **30**, 223-231.
689. R.W. Wolff (1989), *Stochastic Modeling and the Theory of Queues*, Prentice-Hall, New Jersey.
690. M.E. Woodward (1994), *Communications and Computer Networks: Modelling with Discrete-Time Queues*, IEEE Computer Society Press, Los Alamitos.
691. X. Wu, P. Brill, M. Hlynka and J. Wang (2005), An $M/G/1$ retrial queue with balking and retrials during service, *International Journal of Operational Research* **1**, 30-51.
692. X. Wu and X. Ke (2007), Analysis of an $M/\{D_n\}/1$ retrial queue, *Journal of Computational and Applied Mathematics* **200**, 528-536.
693. P. Wüchner, H. De Meer, G. Bolch, J. Roszik and J. Sztrik (2007), Modeling finite-source retrial queueing systems with unreliable heterogeneous servers and different service policies using $MOSEL$. In: *Proceedings of the 14th International Conference on Analytical and Stochastic Modelling Techniques* (Eds. K. Al-Begain, A. Heindl and M. Telek), pp. 75-80, Prague.
694. F. Xue, S.J. Ben Yoo, H. Yokoyama and Y. Horiuchi (2005), Performance analysis of wavelength-routed optical networks with connection request retrials. In: *Proceedings of the IEEE International Conference on Communication. ICC 2005*, pp. 1813-1818 (Vol. 3), Seoul.
695. T. Yamada, M. Kanedo and K. Katou (2005), A mobile communication simulation system for urban space with user behavior scenarios. In: *High Performance Computing and Communications, MPCC 2005. Lecture Notes in Computer Science Vol. 3726* (Eds. L.T. Yang, O.F. Rana, B. Di Martino and J. Dongarra), pp. 979-990, Springer-Verlag, Berlin.
696. T. Yang and J.G.C. Templeton (1987), A survey on retrial queues, *Queueing Systems* **2**, 201–233; Correction, *Queueing Systems* **4**, 94 (1989).
697. T. Yang, M.J.M. Posner and J.G.C. Templeton (1990), The $M/G/1$ retrial queue with nonpersistent customers, *Queueing Systems* **7**, 209-218.
698. T. Yang, M.J.M. Posner and J.G.C. Templeton (1992), The $C_a/M/s/m$ retrial queue: A computational approach, *ORSA Journal on Computing* **4**, 182-191.
699. T. Yang and H. Li (1994), The $M/G/1$ retrial queue with the server subject to starting failures, *Queueing Systems* **16**, 83-96.
700. T. Yang, M.J.M. Posner, J.G.C. Templeton and H. Li (1994), An approximation method for the $M/G/1$ retrial queue with general retrial times, *European Journal of Operational Research* **76**, 552-562.
701. T. Yang and H. Li (1995), On the steady-state queue size distribution of the discrete-time $Geo/G/1$ queue with repeated customers, *Queueing Systems* **21**, 199-215.
702. T. Yokoi (1988), End-to-end blocking probability in telecommunication networks and repeated call attempts, *Review of the Electrical Communications Laboratories* **36**, 23-28.
703. F. Zhang and G. Gupur (2005), The resolvent set of $M/M/1$ retrial queueing model, *International Journal of Pure and Applied Mathematics* **20**, 449-469.
704. W.H. Zhou (2005), Analysis of a single-server retrial queue with FCFS orbit and Bernoulli vacation, *Applied Mathematics and Computation* **161**, 353-364.
705. A. Zolfaghari (1991), Network simulation study as a base for survivability standards. In: *Teletraffic and Datatraffic in a Period of Change, ITC-13* (Eds. A. Jensen and V.B. Iversen), pp. 377-382, Elsevier Science Publishers B.V., Amsterdam.

706. M. Zukerman and C.M. Lee (2001), Performance bounds for cellular mobile communications networks with repeated attempts. In: *IEEE VTS 53rd Vehicular Technology Conference, VTC 2001*, pp. 996-1000 (Vol. 2).

Author Index

Abate, J. 28, 63, 101, 130, 151
Abramov, V.M. 93
Agmon, N. 47
Aguir, M.S. 6, 93
Ahn, W.C. 8, 32
Aissani, A. 34
Ajmone Marsan, M. 7, 93
Akar, N. 204
Akşin, O.Z. 6, 93
Alfa, A.S. 34, 92, 207, 239, 240
Alhassid, Y. 47
Amador, J. 173, 183
Anisimov, V.V. 34, 73, 92, 119, 266
Apaolaza, N.M. 183
Arivudainambi, D. 33, 92
Artalejo, J.R. 6, 15, 19, 20, 22, 24, 25, 27–29, 32–35, 47, 49, 50, 58, 66, 70, 73, 74, 79, 91–93, 119, 124, 125, 129, 130, 134, 147, 158, 173, 183, 207, 208, 210, 212, 230, 239, 240, 244, 266
Asmussen, S. 192, 193, 204
Atadzhanov, Kh.L. 34
Atencia, I. 33, 34, 58
Avram, F. 33, 239

Basharin, G.P. 7
Bini, D.A. 200, 204
Birukov, A. 267
Bocharov, P.P. 33, 34, 239
Bolch, G. 34, 56, 158, 204
Boyer, P. 33
Breuer, L. 32, 239, 267
Bright, L. 189

Brun, M.A. 33
Bruneel, H. 10, 56

Cao, J. 204
Chakravarthy, S.R. 32, 129, 192, 204, 207, 208, 210, 212, 219, 230, 239
Chang, Y. 33, 35, 239
Chaudhry, M.L. 56
Chauvet, F. 6, 93
Choi, B.D. 8, 32–35, 91, 92, 130, 158, 239
Choo, Q.H. 20, 129
Choudhury, G.L. 28, 63, 101, 130, 151
Chung, Y.H. 34
Çinlar, E. 194, 204, 262
Clos, C. 4
Cohen, J.W. 4, 14, 92
Conolly, B. 20, 129
Cooper, R.B. 20, 28, 35
Cozzolino, J. 91, 158

D'Apice, C. 33, 34, 239
Dallery, Y. 6
De Carolis, G. 7, 93
De Kok, A.G. 41
De Meer, M. 34, 56, 158, 204
Dennis, J.E. 82
Diamond, J.E. 92, 207, 239
Doshi, B.T. 35
Dudin, A.N. 32–34, 201, 202, 204, 239, 257–259, 262, 263, 267
Dupuis, A. 33

Economou, A. 6, 20, 34, 92, 124, 125, 129, 244, 266

Efrosinin, D. 239
Elldin, A. 4, 10
Evans, R.V. 203

Falin, G.I. 3, 4, 14, 17–19, 21, 23–25,
    27–29, 31–35, 40, 47, 49, 50, 65, 66,
    68, 69, 73, 84, 86, 88, 91–93, 105,
    129, 130, 154, 158, 182
Fapojuwo, A.O. 7
Farahmand, K. 32
Fayolle, G. 32, 33
Flannery, B.P. 35, 63, 158
Fredericks, A.A. 4, 80
Fricker, C. 28, 29, 158
Fuhrmann, S.W. 35
Fujisawa, T. 34, 63

Gail, H.R. 204
Gans, N. 5
Gaver, D.P. 93, 189, 194, 204
Gelenbe, E. 35
Giambene, G. 10
Gómez-Corral, A. 6, 15, 18, 20, 29,
    32–35, 47, 91–93, 129, 130, 134,
    158, 183, 219, 237, 239, 240, 244,
    266
Graham, A. 205
Grassmann, W.K. 203
Greenberg, B.S. 78, 79
Greiner, S. 34, 56, 158, 204
Grier, N. 93
Grishechkin, S.A. 28, 33

Hammond, J.L. 10
Han, D.H. 158
Hanschke, T. 18, 92
Hantler, S.L. 204
Harris, C.M. 4
He, Q.M. 239
Hoffman, K.L. 4
Houck, D.J. 32
Hunter, J.J. 56, 205, 246, 247, 258

Iasnogorodski, R. 35
Iversen, V.B. 33

Jacobs, P.A. 189, 194, 204
Janssens, G.K. 8, 32, 91
Johnson, R.W. 45

Jonin, G.L. 92
Joshua, V.C. 219, 267

Kapyrin, V.A. 92
Karaesmen, F. 6, 93
Karolik, A.V. 33
Keilson, J. 91, 158
Kelly, F.P. 5
Kharoufeh, J.P. 34
Khelladi, A. 33
Khomichkov, I.I. 8, 32
Kim, B.G. 10, 56
Kim, C.S. 32, 267
Kim, J.W. 34
Kim, Y.C. 18, 73, 92, 119, 154
Kleinrock, L. 16, 17, 27, 129
Klimenok, V.I. 32, 33, 201, 202, 204,
    257–259, 262, 263, 267
Koole, G. 5, 192
Kornyshev, Y.N. 88
Kosten, L. 3
Kouvatsos, D.D. 44, 45
Krishna Kumar, B. 33, 34, 92
Krishnamoorthy, A. 219, 267
Kulkarni, V.G. 8, 14, 16, 24, 33–35, 91,
    92, 204
Kumar, A. 10
Kuri, J. 10

Lai, W.S. 32
Langaris, C. 32–34
Latouche, G. 4, 70, 122, 189, 191, 193,
    194, 196, 198, 200, 204, 248, 252
Lebedev, E.A. 183
Lee, S.C. 32
Lehoczky, J.P. 93
Leonardi, E. 7, 93
Levine, R.D. 47
Li, H. 34, 51, 53, 55–58, 92, 239, 240
Li, Q.L. 32, 34, 204, 267
Li, W. 239
Liang, H.M. 34, 35, 92
Lind, G. 10
Liu, X. 7
Lo Cigno, R. 7, 93
Lopez-Herrero, M.J. 19, 20, 27–29, 32,
    91, 92, 106, 111, 124, 125, 129, 147,
    158, 230
Lucantoni, D.M. 204, 267

Malyshev, V.A.   13
Mandelbaum, A.   5, 6, 93
Mandjes, M.   7, 32, 93
Manjunath, D.   10
Martin, M.   24, 33, 47, 49, 50, 91
Massey, W.A.   6, 93
McKoy, T.   93
Mead, R.   47
Medhi, J.   34
Meini, B.   200, 204
Meo, M.   7, 93
Merkulov, V.E.   7
Moreno, P.   34, 58, 92
Moutzoukis, E.   32–34

Nelder, J.A.   47
Neuts, M.F.   4, 13, 19, 32, 66, 69, 70, 91, 111, 116, 123, 129, 196, 197, 199, 203, 204, 239, 266, 267
Nobel, R.D.   34, 141–143

O'Reilly, P.J.P.   10
Orlovsky, D.S.   34, 239, 267
Osawa, H.   34, 63

Park, K.K.   33, 92, 130, 158
Parthasarathy, P.R.   92
Pavai Madheswari, S.   34
Pearce, C.E.M.   14, 92, 158, 239
Pechinkin, A.V.   34, 239
Phong, N.H.   33
Pla, V.   6
Posner, M.J.M.   33, 51, 53, 55, 56, 92, 93
Pourbabai, B.   34, 92
Pozo, M.   19, 22, 66, 70, 74
Press, W.H.   35, 63, 158

Ramalhoto, M.F.   18, 93
Ramaswami, V.   4, 70, 122, 189, 191, 193, 196, 198, 200, 204, 248, 252
Rao, B.M.   19, 66, 69
Reeser, P.K.   92
Reiman, M.I.   6, 93
Reisner, G.A.   4, 80
Remiche, M.A.   183
Rhee, K.H.   130, 158
Rider, B.   6, 93
Riordan, J.   4, 34, 82, 84

Rodrigo, A.   182
Rom, R.   10

Salerno, S.   34, 239
Sapna Isotupa, K.P.   240
Saunders, P.B.   4
Schnabel, R.B.   82
Sedol, J.J.   92
Serfozo, R.F.   127, 129
Sherman, N.P.   34
Shin, Y.W.   8, 32, 33, 73, 92, 93, 119, 154, 239, 266
Shore, J.E.   45
Sidi, M.   10
Sohraby, K.   204
Stanford, D.A.   203
Stepanov, S.N.   19, 33, 66–68, 76, 92
Stewart, W.J.   158, 204
Stolletz, R.   10
Stolyar, A.   6, 93
Sudhesh, R.   92
Syski, R.   4, 10, 20, 34

Takagi, H.   28, 30, 35, 56
Takahara, G.K.   34
Takahashi, M.   34, 63
Taylor, B.A.   204
Taylor, P.G.   189
Teghem, J.   35
Templeton, J.G.C.   3, 4, 14, 17, 18, 23, 27, 29, 33–35, 40, 51, 53, 55, 56, 65, 69, 73, 84, 86, 88, 91–93, 105, 129, 154
Teukolsky, S.A.   35, 63, 158
Tijms, H.C.   41, 63, 91, 101, 141–143
Tran-Gia, P.   7, 32, 93
Trivedi, K.S.   34, 56, 158, 204
Tsarenkov, G.V.   267
Tsitovich, I.I.   92

Vázquez, M.   182
Vetterling, W.T.   35, 63, 158
Vijayakumar, A.   34, 92

Wesolowski, K.   10
Whitt, W.   6, 28, 63, 93, 101, 130, 151
Wilkinson, R.I.   4, 19, 65, 92
Wolff, R.W.   34, 40, 78, 79
Woodward, M.E.   10, 56

Yang, T.   14, 33–35, 51, 53, 55–58, 92, 93, 240
Ying, Y.   32, 34, 267
Young, H.   91, 158
Zhao, Y.Q.   32, 34, 204, 239, 267

Subject Index

Absorbing states 125, 158, 216
Accessible states 243
$AQTMC$ 200, 204, 257
Asymptotic behavior
   heavy traffic 29, 66
   high retrial rate 20, 29, 85
   low retrial rate 23, 29, 66, 84
Attempts since the last service completion 159

Balking 6, 33
Batch arrivals 32, 191
Bessel function 19
Birth and death
   standard $M/M/c$ queue 16
   structure 14, 18
   transient analysis 81
Bisection method 196, 248
Blocked primary arrivals 164
Blocked retrials 164, 167
Blocking probability
   finite population 90
   $M/M/1$ retrial queue 18
   $M/M/c$ retrial queue 21, 73
$BMAP$ process 191, 204
Branching process 28
Breakdowns 33, 34
Burke's theorem 35, 40, 52
Busy period, length of 95, 129
   $Geo/Geo/c$ retrial queue 249
   $M/G/1$ retrial queue 27, 95
   $M/M/c$ retrial queue 112
   $MAP/M/c$ retrial queue 220

   moments 19, 27, 101, 120, 222, 252
   numerical inversion 101, 118, 223
   standard $M/G/1$ queue 27
   standard $M/M/c$ queue 19

Call centers 5
Catastrophes, method of 99, 106, 129
Cellular networks 5, 6
Censored process 195, 196, 202, 259
Coefficient of correlation 192
Computer networks 7, 10
Continued fractions 92
Cyclic reduction 200

Departure process 30
Disasters 33
Discrete-time queues 34, 56, 241

Elapsed retrial time 52
Elapsed service time 134, 144, 159
Elsner's algorithm 196
Embedded Markov chain 35
   $BMAP/SM/1$ retrial queue 257
   $M/G/1$ retrial queue 24, 39
Erlang loss system 23, 84
Extreme values 129

Finite population 32, 44, 87, 233
Fluid and diffusion approximations 93
Folding method 204
Forward recurrence time 54, 135, 145
Foster's criterion 16, 244, 266
Fredericks and Reisner approximation 80
Fundamental period 116, 190

316   Subject Index

G-EAS policy   57, 242
Gamma distribution   29, 141
Gaussian distribution   29, 48
Geometric arrival process   57, 242
Geometric distribution   17
$GI/M/1$ structure   190, 198, 203
Guard channel   7

Hypergeometric function   18, 72

Impatience   3, 6, 33
Infinitesimal generator
    $BMAP$ process   191
    finite population, PH service and retrial times   234
    finite QBD process   189
    $M/M/c$ retrial queue   12
    $MAP/M/c$ retrial queue   219
    $MAP/PH/1$ retrial queue   208
    QBD processes   188
Integral approximation   63
Interpolation, approximation by   84, 92
Invariant subspace algorithm   204

Kolmogorov equations
    finite population   88
    $Geo/G/1$ retrial queue   58
    $M/M/c$ retrial queue   13
    $M/M/c$ retrial queue, finite truncation   65
Kronecker
    delta   12
    product   188
    sum   188

Lagrangian multipliers   46
Limiting distribution of the system state   39
    $BMAP/SM/1$ retrial queue   261
    finite population   88
    $Geo/G/1$ retrial queue   58, 60
    $M/G/1$ queue with general retrials   52, 54
    $M/G/1$ retrial queue   24, 39
    $M/M/1$ retrial queue   17
    $M/M/c$ retrial queue   13, 64
    moments   17, 18, 25, 47, 60, 90, 163
    numerical inversion   63

standard $M/G/1$ queue   24
standard $M/M/c$ queue   17
Little's formula   18, 28, 151, 156
Local area networks   8, 32
Logarithmic reduction algorithm   197, 204

$M/G/1$ structure   190, 198, 201, 260
MAP process   191
Markov renewal theory   14, 204
Matrix **G**   198, 199, 202, 260
Matrix exponential   188
Matrix-analytic formalism   187
Matrix-geometric distribution   196, 199, 203, 247
Maximal queue length
    $M/G/1$ retrial queue   108
    $M/M/c$ retrial queue   124
    $MAP/PH/1$ retrial queue   215, 239
Maximum entropy approach   44
    $Geo/G/1$ retrial queue, system state of   63
    $M/G/1$ retrial queue, busy period of   95
    $M/G/1$ retrial queue, system state of   47
    $M/G/1$ retrial queue, waiting time of   132
McLaurin series   20
Mean number of customers in orbit
    finite population   90
    $M/G/1$ retrial queue   25
    $M/M/c$ retrial queue   18
$MMPP$ process   192
Multiclass queue   33

Negative arrivals   33
Negative binomial distribution   18
Nonpersistent customers   33
Number of customers served
    $M/G/1$ retrial queue   105
    $M/M/c$ retrial queue   123
    $MAP/M/c$ retrial queue   225, 227
    moments   105, 225
Number of retrials made by a customer
    $M/G/1$ retrial queue   143, 158
    $M/M/c$ retrial queue   155
    $MAP/M/c$ retrial queue   230, 232
    moments   147, 156, 231

Subject Index   317

One-step transition probability matrix
  $AQTMC$ 200
  $BMAP/SM/1$ retrial queue   257
  $Geo/Geo/c$ retrial queue   242
  Markov chain of $GI/M/1$-type   190
  Markov chain of $M/G/1$-type   190
Orbit   4, 6
Overloading systems   34

$PASTA$ property   40, 41, 52, 78, 150
Permutation matrix   263
PH distribution   91, 170, 192, 204, 211, 233, 239, 252
Poisson process   11
Polling systems   34
Priorities   33

QBD process   69, 189, 207
  finite case   194
  level independent   195
Queueing networks   3, 34

Random access protocol   7
Rate matrix   197, 198, 210, 247
Reallocation of customers   169
Regenerative approach   41, 91
Regenerative process   19, 27, 41
Residual service time   135, 144
Retrial distribution
  general   51, 92
  PH 233
Retrial policy
  classical   11
  full access   239
  generalized   32
  random order   20, 131, 242, 254
RG-factorization   204, 267
RTA approximation   77, 84

Secant method   196, 248
Server assignments   174
Server idle and busy periods   15, 174
Shannon's entropy   45
SM service process   193
Small-rank adjustment   247
Spectral method   204
Spectral radius   196, 199, 212, 248, 249
Standard $M/G/1$ queue   24
Standard $M/M/c$ queue   16

Star topology   8
Stationary distribution of the system state
  $AQTMC$ 201
  $BMAP/SM/1$ retrial queue   258
  finite population, PH service and retrial times   235
  finite QBD 195
  $Geo/Geo/c$ retrial queue   244, 246, 247
  level independent QBD 195
  $MAP/M/c$ retrial queue   220
  $MAP/PH/1$ retrial queue   210, 212
  Markov chain of $GI/M/1$-type   199
  Markov chain of $M/G/1$-type   200
Stochastic decomposition   25, 35, 52
Stochastic monotonicity   35, 119
Successful primary arrivals   164, 166
Successful retrials   124, 164, 165
Successive substitution algorithm   197, 204
Supplementary variables   14
System of equations, solution of
  linear case   158, 204
  non-linear case   82

Taboo probabilities   129
Tauberian results   96, 102, 119, 133, 136, 152, 223
Telephone systems   4, 10
Time to reach a certain orbit level   179, 183
Transient behavior   93
Truncated models
  $BMAP/SM/1$ retrial queue, censoring   259
  $Geo/Geo/c$ retrial queue, finite truncation   244, 252, 254
  $Geo/Geo/c$ retrial queue, generalized truncation   247
  $M/G/1$ retrial queue   43, 99, 134, 145, 166, 167
  $M/M/c$ retrial queue, finite truncation   64, 92, 112, 123, 150, 155
  $M/M/c$ retrial queue, generalized truncation   68, 92, 116, 117, 123
  $MAP/M/c$ retrial queue, finite truncation   219, 220, 228

$MAP/PH/1$ retrial queue, generalized truncation 208
Truncation threshold
  $BMAP/SM/1$ retrial queue 260
  $Geo/Geo/c$ retrial queue 249
  $M/G/1$ retrial queue 102, 137, 147, 172
  $M/M/c$ retrial queue 73, 118, 152, 157
  $MAP/M/c$ retrial queue 222, 230
  $MAP/PH/1$ retrial queue 212

Vacation models 25, 34, 35

Waiting time 131
  $Geo/Geo/c$ retrial queue 254
  $M/G/1$ retrial queue 28, 131
  $M/M/c$ retrial queue 149
  $MAP/M/c$ retrial queue 228
  moments 28, 136, 151, 229, 255
  numerical inversion 136, 151, 230
  standard $M/G/1$ queue 28
  standard $M/M/c$ queue 20
Wald's identity 42, 105, 156

Printed in the United States
116318LV00002B/115-150/P